T0276941

Earthly Order

Earthly Order

How Natural Laws Define Human Life

Saleem H. Ali

OXFORD
UNIVERSITY PRESS

Oxford University Press is a department of the University of Oxford. It furthers the University's objective of excellence in research, scholarship, and education by publishing worldwide. Oxford is a registered trade mark of Oxford University Press in the UK and certain other countries.

Published in the United States of America by Oxford University Press
198 Madison Avenue, New York, NY 10016, United States of America.

Library of Congress Control Number: 2022000232
ISBN 978–0–19–764027–2

DOI: 10.1093/oso/9780197640272.001.0001

3 5 7 9 8 6 4 2

Printed by Sheridan, United States of America

To my students who have made me a life-long learner of fields beyond my comfort zone

Nature, that framed us of four elements
Warring within our breasts for regiment,
Doth teach us all to have aspiring minds.
Our souls, whose faculties can comprehend
The wondrous architecture of the world
And measure every wandering planet's course,
Still climbing after knowledge infinite,
And always moving as the restless spheres,
Will us to wear ourselves and never rest,
Until we reach the ripest fruit of all,
That perfect bliss and sole felicity,
The sweet fruition of an earthly crown.

—Christopher Marlowe (1564–1593),
Tamburlaine the Great

Contents

Preface

Natural systems are in a constant state of flux between randomness and order. As human beings, we seek to organize our lives, but we also shun being "ordered" to do so. Our decisions and our overall lack of understanding about the connections between various forms of order undermine our quest for sustainability. The initial inspiration for this book comes from reading two classic works and genres which have seldom communicated with each other: James Gleick's *Chaos* and Francis Fukuyama's *The Origins of Political Order*—both of which were *New York Times* bestsellers but likely read by very different audiences. I wanted to write a book which would cut across these genres but not feel contrived. Further inspiration for this work came from Tyler Volk's book *Quarks to Culture*, which admirably offered a bridging of genres by showing a nested set of patterns from fundamental physics to anthropology. On the structural literary aspects of learning, I drew insights from Clifford Siskin's book *System: The Shaping of Modern Knowledge*, which provides a broad intellectual history of how making connections between nodes of knowledge that constitute a system is so consequential. The connections create a synergy which is more functional than just the sum of the parts. Geoffrey West's fantastic synthesis on the importance of power laws across natural and social systems as presented in his book *Scale*, as well as Vaclav Smil's book *Growth*, gave me useful insights on structural macro-arguments.

Earthly Order builds on the foundations laid forth by these works by taking the conversation to the next step of tangible topics for the public to gain ecological literacy. Stories are told to hopefully draw the broadest audience, but substantive material is covered in the explanation of core concepts so that the book may also be suitable for teaching sustainability science and policy. The ultimate aim is contemporary global problem-solving by understanding basic tenets of *functional* order in natural, social, and political systems. Unpacking the term "order" requires us to consider its roots in the Greek term *Cosmos*—which roughly translates as "natural order" but also *Taxis*—which connotes "arrangement" or artificially created order. This book is aimed at helping the reader navigate the congruence of these two aspects of order with a long-term trajectory toward sustainability of human populations on the planet. I am focused on providing the reader an understanding of what may be termed "order by purpose" within different contexts. For example, a book in a library may be ordered by the author's alphabetical name or by topic. The former is less useful for research but more useful for a universal bibliographic listing. The latter is more useful for research but less so for finding a particular author in multiple genres. Functionality is linked to purpose and a beacon for prioritizing what material is presented in the book, though by no means presuming that we fully comprehend what is purposeful or not in the universe.

There are some seminal insights around the human quest to categorize natural systems through some objective physical concepts such as naming of particles, classes of compounds, taxa of species, or even languages for ease of understanding. When are such categories useful, and when do they hinder creativity? What should be the delineating features by which we cast "borders" in physical and cognitive space? How do physical laws consider order at the level of elementary particles, and how can the insights from such states be applied to human systems? Does considering these constraints help us in developing socioecological systems in human settlements to meet ecological, economic, and social sustainability targets? These are just some of the underlying questions which help to inform the conversation in this book.

The tension between order and chaos has been a fundamental feature of natural change and has animated human interactions with our environment. As an environmental systems scientist and geographer, I recognize that "order" is itself a fundamentally contested concept in both natural and social science. Writing such a panoramic book is an ambitious enterprise and has required me to consult with a wide range of scholars throughout my career, many of whom make appearances in the book's narrative. It has been a joy to write this narrative to inform the general public about some of the core structural concepts in science which may seem esoteric but are actually beautifully connected to our daily lives. The book strives to objectively present underlying natural constraints to how we order social, economic, and political life without being judgmental. Given the public's interest in "heroism" and personality ascendance, I use a cast of characters who have been instrumental in key discoveries and decisions and highlight their accomplishments and stories. These characters are also useful subjects for the accompanying multimedia elements of the project, such as podcasts and video content. It is worth mentioning that, throughout this text, awards such as the Nobel Prize are mentioned alongside their affiliations as a marker for the recognition they have received. However, such a mention does not imply any filtering of key scientific or literary personalities for the volume itself. Indeed, there are many scholars mentioned in these pages who were deprived of global recognition due to misfortune of circumstances or biases.

A central political argument of the book is that our current polarization of social perceptions stems from an inability to reconcile natural order with economic and social order. The key challenge is our unwillingness to engage with order as a concept that, by necessity, exists in flux with chaos or anarchy—both states have their own purpose and functionality in natural and social systems. Yet we are selective in when we want to justify our actions in using order as a structural necessity. This is most directly manifest in debates about climate change and environmental policy, but this disjunction is far wider. For example, we shun categorization of race as a form of exclusion but also want to assert some level of categorization in tribal order to recognize indigenous peoples. Which norms dictate such inclusion and exclusion need to be considered through a conversation between ecological and social systems.

Our current debates in politics are often about what should constitute a "world order," while scientists have wrestled with what truly are the fundamental conditions of "natural order." Yet there has been little effort by public intellectuals to wrestle with how natural and social systems have both resisted and embraced order under particular conditions. These discussions are now particularly timely as the world attempts to find consensus pathways between science and politics through entities such as the Earth Commission. Indeed, my own role on two United Nations science panels has motivated me to write this book to find that tough sweet spot between depth and breadth to galvanize informed decision-making. Learning from humanity's successes and failures in understanding natural order while constructing our own economic, social, and political order can mitigate the global culture wars and cognitive conflicts that haunt our times.

I am grateful to a range of scholars across numerous disciplines who reviewed parts of this book for accuracy, including Jed Buchwald, Roald Hoffmann, Delphis Levia, Asghar Qadir, Will Rifkin, , Monika Shafi, Qaisar Shafi, and four anonymous reviewers. My special gratitude goes to science communicators in print and on audio and video media who have helped me digest a breadth of ideas from disciplines and helped to make this narrative coherent. Online science communicators such as Sabine Hossenfelder, Brian Keating, Robert Lawrence Kuhn, Derek Muller, Matt O'Dowd, Anton Petrov, and Michael Shermer have also been immensely helpful. The panoply of presentations shared online by the Royal Institution, the Santa Fe Institute, the Perimeter Institute, World Science Festival, and numerous university seminars helped me engage in meaningful discussions through webinars during the pandemic.

The Rockefeller Foundation's Bellagio Writers' Fellowship was instrumental in securing sabbatical time from the University of Delaware to work on this book. In particular, architect Laura Lee and sociologist Reuben Miller who were in my cohort in Bellagio provided valuable critical feedback on the writing style and content of the manuscript. Although the Coronavirus pandemic derailed some aspects of my book-writing retreats, I was fortunate to have the support at home from my wife Maria and sons Shahmir and Shahroze to work on the manuscript. Conversations with my colleagues at the United Nations International Resource Panel and the Global Environment Facility (GEF) were extremely helpful in sharpening the educational and policy relevance of the narrative. In particular, Rosina Bierbaum and Thomas Lovejoy, both members of the National Academy of Sciences, whose current and past roles as chairs of the seven-member Scientific Panel for the GEF have been great mentors since I joined the panel in 2018. Sadly, Tom Lovejoy passed away on December 25, 2021 and it will be my permanent loss that he did not get to see the final book, though he did see earlier versions of the manuscript. My research assistant Zachary Shulman provided invaluable service from checking on references and diagrams to proof-reading and assisting with focus groups. I also benefited immensely from a peer writers' group of National Geographic Explorers comprising M. Jackson, Alex Carr Johnson, Kelly Koller, Danilo Thomas, and Emily Toner, all of whom provided me with candid feedback on writing style and substance.

Finding a publisher who is willing to take on such an ambitious project is always challenging, and I am deeply grateful to Jeremy Lewis, the senior acquisitions editor for science at Oxford University Press, for persevering with the manuscript which went through four peer reviews and subsequently had to be approved by three different boards of delegates at Oxford University Press (natural science, economics, and politics). To all those anonymous reviewers, my gratitude for holding my feet to the fire and making the manuscript better through your comments. Thanks to Yale University Press for allowing me to use a modified version of a small excerpt in Chapter 9 from a contribution I had authored in their anthology edited by Daniel Esty titled A Better Planet: 40 Big Ideas for a Sustainable Future (2019). Dan Esty was my graduate advisor at Yale for my Masters degree more than 25 years ago and it was great to reconnect with him for that book.

My family in the United States, Pakistan, and Canada have always been immensely supportive of my writing adventures. Earlier books have been deservedly dedicated to my wonderful wife, children, and parents and have acknowledged the support of my dear sisters and their families. For this book, I owe a special debt of gratitude to my nephew Ali Shahid, a mathematical finance professional in Canada, who gave me a twentieth anniversary edition of the 750-page masterpiece by Douglas Hofstadter *Gödel, Escher, Bach* for my birthday a decade ago. Browsing through those daunting pages, I felt encouraged to embark on such an ambitious writing venture across disciplinary genres.

Finally, this book would not be possible without my students worldwide, in the classroom, in the field, and through online learning who have questioned my lectures and pedagogy over the years. In particular, I feel fortunate to have taught largely in excellent public universities in the United States and Australia (University of Vermont, University of Queensland, and University of Delaware) that cater to a wide swath of demographics. I dedicate this book to my students for making me cross disciplinary boundaries in order to find cogent explanations and endlessly spurring curiosity in my life.

Introduction

The Limited Logic of Order?

> Deep in the human unconscious is a pervasive need for a logical universe that makes sense.
> But the real universe is always one step beyond logic.
>
> —**Frank Herbert, Author of the *Dune* saga**

Shattered pieces of glass fragments funneled into a metal tube with angled mirrors and a peep-holed cap give you a remarkable little invention. The Scottish inventor, Sir David Brewster, patented this little science toy and called it a *kaleidoscope*. The name stemmed from the ancient Greek words *kalos*, meaning "beautiful," and *eidos*, meaning "a shape which is seen." The random shapes of the glass, when confronted with the regularity of light reflecting infinitely between the angled mirrors, creates order to the beholder. *Order* can thus be defined as the manifestation of *patterns* that can be perceived as differentiators by the observer. *Structures*, such as rules or laws, are needed to maintain such order. Patterns emerge from every shaking movement of the object. Each one is different and yet predictably fascinating. What captivates the observer is this seamless transition from the seemingly chaotic act of shaking to the viewing joy of structure. Brewster's simple invention tapped into a human yearning for being surprised by order. Within 3 months of his commercial release of the kaleidoscope in 1817, more than 200,000 were sold in Paris and London.[1]

As a child growing up in Pakistan, the kaleidoscope was my favorite toy. You could buy a variety of paper-tubed ones which local street toymakers could make easily with cheap Chinese mirrors and some broken glass bangles or pebbles. Light beamed through the translucent end of the tube to illuminate a colorful pattern of the glass fragments. This happened simply because of the structure imposed on these random objects by three glass mirrors. Beams of light could functionally enchant an inanimate world. The velocity of light is considered the upper speed limit of the cosmos. I imagined the light passing through the glass, shifting and scattering as its speed was reduced ever so slightly. Still further, I studied the chemistry of water and saw the formation of snow crystals analogously to my beloved kaleidoscope. Each flake is unique insofar that its creation captured the moment a dust particle was encircled by cold water molecules in a crystal embrace. Yet all snowflakes also have a predictable

Earthly Order. Saleem H. Ali, Oxford University Press. © Oxford University Press 2022.
DOI: 10.1093/oso/9780197640272.003.0001

six-fold structure due to the fundamental nature of how water molecules arrange themselves in crystalline form. What one may call "ordered diversity" in natural phenomena has been a lifelong fascination for me.

Attending an all-male school in Pakistan, I experienced how strict social order was considered an essential prerequisite for success. There were clearly defined views of natural order that were linked to social order. Perceptions of how gender roles were to be defined with ostensible predication on the "natural" properties of each gender. There were notions of how lives should be ordered around particular milestones of college, marriage, children, caring for elderly parents, and then memorializing their passing through prayer. As I grew older and was exposed to a broader range of learning, my views of order changed, and I reveled in rebellious anarchy. Yet the need for some modicum of order to manage resources on our planet (and indeed our own personal lives) remained ever-present. Writing this book is a quest to find key insights from the coexistence and interplay between functional order and disorder and to link it to the quest for sustainability in human interactions with our planet.

My aim here is to test the limits of order as a concept of inherent value to us as custodians of nature as well as cultivators of sustainable human societies. How and why have human beings at once sought order in nature but are all too often tempted to indulge in chaos? What can we learn from natural order in systems while recognizing the limits of imposing or imagining structure where none might exist? Where and when is it appropriate for us to artificially impose order in social and political systems even when this may seemingly be at odds with natural order? These are questions I have considered from a panoramic perspective throughout my career. Defying disciplinary boundaries has been my own act of order defiance, one that has illuminated the path to this book and indeed is fundamental to contemporary narratives on sustainable development of human societies.

During my doctoral days at MIT in the late 1990s, I would frequently visit the Peabody Museum at Harvard. The brooding brick building full of enigmatic skeletons and precious mineral archives was my secret happy place in the cold New England winter months. On one such visit to the museum, I had the good fortune of encountering the late doyen of natural history, Stephen Jay Gould. This was a man whom the *New York Times* would later call, in his obituary, the best-known evolutionary biologist since Charles Darwin. Gould was all bundled up in a woolen coat and wearing a peculiar cap, ready to exit the building on a brisk Thursday night. He was running to catch a train to South Station and on to New York City, from where he commuted back and forth to Boston. I followed him persistently with a cavalcade of compliments which did not stir his attention. He was known to be prickly and dismissive. Yet despite an initial rebuff to my servility, we were able to strike up a conversation on the fabled Red Line "T" train into the city.

During the 20-minute train ride, I learned from Gould that the culminating book he was working on was to consider *The Structure in Evolutionary Theory*.[2] He reminisced on how his work as a public intellectual had forced him to look for structure and order to convey complex ideas to the general public. In return, this quest for structure had motivated him to reconsider his life-long study of Darwinian evolution to seek patterns. This magnum opus was to be Gould's last book—a 1,400-page beautifully illustrated tome–published only 2 months before his death. Gould's life's work had questioned structure on temporal terms. Along with his collaborator Niles Eldredge, his most notable contribution to evolutionary theory was the concept of "punctuated equilibrium."[3] This view suggested that evolution did not occur in smooth, orderly ways but in spurts of rapid change followed by a period of stasis or equilibrium that was dependent on environmental and genetic factors. Yet, as Gould himself acknowledged, this was itself imposing another form of structure—albeit a nonlinear one. Evolution was frequently described as a process emanating from random mutations which nonrandomly got selected to posit structure. Later research would suggest that even mutations in genes were not random and that fewer mutations are observed in highly expressed genes. Evolution would also become structurally embraced not just as a biological process but also as a more universally applicable algorithm for change, with the development of new academic fields such as "evolutionary economics."

That encounter with Gould more than 20 years ago, spurred me to study order at a fundamental natural and social level—the genesis of this book came about there. In addition to being a phenomenal scientist, Gould was a remarkable writer who could effortlessly convey the complexity of nature to a general reader with elegant prose and metaphor. His ability to tell stories of natural wonder from observing oddities of organisms (such as *The Panda's Thumb*) was legend. He was also very cautious about how observable patterns could tempt us into errant presumptions about the way the world works. Even a robust scientific concept like evolution by natural selection could lead us to darker paths of Social Darwinism and errant racial categorization, as eloquently argued in his book *The Mismeasure of Man* (published in 1981). One of the earlier inspirations for these pages was Gould's classic work *Ontogeny and Phylogeny* (published in 1977). Fancifully esoteric as the name may sound, the book was simply a critical history of how early biologists like Ernst Haeckel (who brought the term "ecology" into the scientific disciplinary lexicon from the Greek root *oikos*—or "home") tried to make a misguided connection between how individual animals develop (ontogeny) and how entire species evolve (phylogeny). Thanks to the popular exposure to such concepts by Gould, I became fascinated by how minds like Haeckel were seduced by structure and made leaps of conjecture.[4]

I learned how "The World Riddle" became a popularizing concept at the end of the nineteenth century, whereby Haeckel and his more infamous contemporary, Friedrich Nietzsche, posited ways of making sense of the universe through natural and social inquiry. Haeckel stated the riddle quite simply as a question: "What is the nature of the physical universe, and what is the nature of human thinking?" Implicit in this quest was the search for order and meaning across the wide span of human inquiry and practice. Composer Richard Strauss embraced this in his music, most notably in the unresolved harmonic progression at the conclusion of his tonal poem *Also sprach Zarathusta* (Thus Spoke Zoroaster). Contemporary readers would recognize the drumbeat crescendo of this musical quest for eternal meaning from the theme of the Stanley Kubrick film *2001: A Space Odyssey*, which had similar grand goals through the art of film.[5]

As I further traveled down the path of questing order, I discovered a frequently cited journal called *The Monist*, whose name intrigued me as a non-philosopher. Further prying into this curious name for a journal led me to discover that the roots of *monism* could be traced back to a seminal eighteenth-century book titled *Logic*, written by the German philosopher Christian von Wolff. At its core, Wolff's coinage of "monism" with its roots in "mono"—or one—was the search for a unified structure for human inquiry. This was a radical departure from the mind-body "dualism" that had defined human views of intrinsic order for the preceding periods of inquiry.[6]

Two's Company

Much of philosophical tradition had sought to find order through the binary: on and off, male and female, yin and yang, science and religion, good and evil. Dualism was perhaps captivating due to the allure of sexual complementarity that was quite literally built into our DNA. Tribal societies have also considered duality as core to many numbering systems. The Torres Strait Islanders who inhabit the archipelago between Papua New Guinea and the northern tip of the Cape York Peninsula in Australia used a system of two digits (*urapun* for one and *okosa* for two). All other numbers were a combination of these two digits. The simple elegance of the binary system has been essential to the development of computer science, with "on" and "off" switches, "true" and "false" circuity, and 0 and 1 numerology. The English mathematician George Boole gave us the insights to develop this form of simple "Boolean" algebra that has been foundational in developing the binary system on which modern computation infrastructure is based. Ultimately, this would lead to the everyday dualism of software and hardware or, in human terms, of the more controversial dualism of *genes* and *memes*. Genes were physical carriers of biological information, while memes were cultural carriers of information that infected our brain's biology and had only material presence in terms of neural circuitry and memory.[7]

It was dualism that had so entranced Nietzsche and attracted him to the centuries-old scriptures of the Persian sage Zarathustra. The founder of the Zoroastrian faith, Zarathustra, had proposed balance in the universe through the existence of good and evil divine forces and posited the concepts of *Asha* (Truth and Order) and *Druj* (which is often translated as Deception). Just like the two balancing arms of a scale, the universe's harmony was considered through such countervailing, coexisting, and inexorable forces.[8] The physical earth also seemed to ripple with dualism, where magnets had two poles and particles could have two charges. There were further insights on the power of the couplet from the galactic orbits of planets, wherein any more than two celestial objects cannot have a closed form (simple equation) solution of orbit trajectories. For example, since the gravity of the sun, the moon, and the earth are acting on each other, the orbital rotation of each cannot be expressed as a simple mathematical equation and requires more complex analytics.

Yet such structure and notions of dualist order seemed illusory to many other religious traditions as they considered the multifaceted behavior of individuals. *Monotheism* emerged as an effort to reconcile multiple patterns of existence within the individual. There were spectrums and complexity between the binaries. Contradictions and inconsistencies abounded through such theist trajectories, too, as humans still tried to find meaning and purpose in the universe. Order was still eluding us, and yet we yearned for it. The paradox of monism was that while it rejected dualism and suggested a single structural coherence for mind and body, it also gave us more nuance and plurality of explanations for the "World Riddle."

As a chemist by training, I began to see how such a unified explanation of patterns would be appealing to the experimenter. The chemical elements were the functional building blocks of our existence, and the periodic table could be thought of as the unifying matrix of elemental order. To validate my impulses, I was further intrigued to learn then that the founder of *The Monist* journal was not a literary philosopher but rather a chemist and metallurgist! Edward Carl Hegeler was a German-American zinc merchant who passionately felt that dividing human inquiry into dualist structures was an affront to rational learning. As a mining engineer, he had burnished his view of the world through literally digging deep into the earth and finding a central reality of what he viewed as elementally determined existence.[9]

The detractors of monism, and indeed the "World Riddle," saw the search for unified meaning and purposeful order as a futile quest for simplicity in a chaotic realm. Among the foremost critics was William James, who taught the first ever course in "psychology" at the turn of the nineteenth century. In his lectures on *pragmatism*, James, who had been trained as a medical doctor, made the following rebuke of such quests for order:

> All the great single-word answers to the world's riddle, such as God, the One, Reason, Law, Spirit, Matter, Nature, Polarity, the Dialectic Process, the Idea, the Self, the Oversoul, draw the admiration that men have lavished on them from this oracular role. By amateurs in philosophy and professionals alike, the universe is represented as a queer sort of petrified sphinx whose appeal to man consists in a monotonous challenge to his divining powers. THE Truth: what a perfect idol of the rationalistic mind![10]

The views put forward by scholars like James would hold sway for much of the twentieth century as natural and social sciences began a further move toward smaller units of study—a process termed "reductionism." The Nobel Prizes, first awarded in 1901, were delineated in terms of specific fields of natural science: chemistry, physics, and medicine, and separately in literature and peace. Academic departments developed accordingly, with growing exclusion between even chemists and physicists. In the social sciences, French anthropologist Claude Levi-Strauss founded the "structuralists" school, which proposed that human societies could be studied through key structural attributes such as language, food, matrimony, and so on. There was some push-back to such moves as well—but mainly in the humanities. Most notably, French-Algerian philosopher Jacques Derrida gave a notable lecture at Johns Hopkins University in 1966, titled "Structure, Signs and Play in the Discourse of the Human Sciences."[11] Derrida questioned the value of structure when so many variations and creative impulses were afoot and no single "event" could be linked to giving them common connectivity. Such debates also raged in political philosophy and art between "modernists" and "post-modernists." I shall spare you the intricacies of such academic lore since it is of less consequence to the core conversations about disciplinary order that defined how we learned and managed knowledge.

An antidote to this imposition of disciplinary order eventually came from a new brand of dualism that was at once embraced and then rejected by British author C. P. Snow in his classic work, *The Two Cultures of Scientific Revolution.* Snow recognized that there was an established dichotomy between two academic cultures—the sciences and the humanities—but he also lamented the lack of conversation between these two cultures. This reticence was a great hindrance to solving global problems. Snow's book was a turning point in human realization that reductionism was stifling academic creativity. Disciplinary order had to be once more shaken. "Interdisciplinarity" (conversation between disciplines) and "transdiciplinarity" (purposing beyond the original discipline) began to gain currency. The "two cultures" mode of thinking has been further challenged and refined by considering the intersectionality among science, technology, engineering, arts, and mathematics (the STEAM paradigm).[12]

It was as if we were returning to the Old World views of learning, wherein philosophers like Aristotle were at once scientists, too, or, later still, Da Vinci and other Renaissance polymaths who could concurrently be artists and aeronautic inventors. The American "liberal arts" curriculum had traced its origins to the Greek and Roman tradition of *liberalia studia*, wherein comprehensive learning necessitated a study of the humanities and science. Yet even though students were studying many fields to get a degree, they were not necessarily connecting the courses, which is what Snow was exhorting us to do. He was himself a physical chemist and a novelist and tackled fiction and nonfiction with equal zeal. In British academia, where he was trained and spent most of his career, there was a tendency to be "intensive" about one field in learning. His views found a more welcome reception in America where the tradition had been to instead be "extensive."

Snow's time at Wesleyan University in Connecticut during the early 1960s further motivated American academia to build across fields and disrupt departmental order. It would be such epistemic anarchy which would ultimately allow for the field of complex systems science to emerge and flourish and for the ideas that have shaped this book to form. Innovative team-taught courses began to be listed. As an undergraduate student at Tufts University in the 1990s, I was a beneficiary of such courses taught through the university's pioneering "Experimental College," and I benefited as well from a proliferation of new eclectic majors. Environmental Studies was my major—though the trepidation of the university was evident in their requirement that all such interdisciplinary majors must also choose a primary anchoring double major. In my case that primary major was Chemistry, which has remained my first learning-love and remains for me the most *functionally* fundamental science.

Chemical Chaos

As a chemistry major, one of my greatest fascinations with natural order was the "octet rule": we were told as students that there was a mysterious charm about the order of eight electrons in the atomic shell of all elements larger than boron which rendered them stable. It was the octet rule which made me understand why the "noble gases" like krypton and xenon did not engage in ignoble chemical reactions. These happy atoms were content to exist on their own without any particle exchanges and bonding with their periodic peers. This rule of thumb also explained why carbon (the element just after boron on the periodic table), which has four electrons in its outer shell, was such a prolific bond-builder. It was in that ambivalent spot of being able to share its electrons in a give-or-take assemblage of eight that made it possible to form the vastest array of compounds. Later, I would discover that there was far more complexity to this rule, and that the ease of bonding had a lot to do with nuclear assemblages of protons and neutrons in the atom as well. Physicists Eugene Wigner and Maria Goeppert Mayer would call this stability of certain configurations "magic numbers" as they worked on the Manhattan Project to find the best assortment of elements for nuclear fission.

Carbon, the emperor of elements, had its own orderly branch of chemistry—organic chemistry—which was far more predictable in its structural components.[13] As the name suggests, carbon chemistry is the literal backbone of life-giving molecules. Applying the octet rule shows us how carbon can form long, orderly chains that are the fulcrum of petroleum, plastics, and indeed DNA. Such is the ubiquity of carbon atoms in natural molecules that when we draw them, they are usually unlabeled, constituting a corner in a geometric diagram of the molecule. Chemists simply know the corner has to be carbon unless otherwise stated. Organic chemistry was one of my favorite subjects because it also helped me to bridge my two majors. I was in my sophomore year when the Earth Summit in Rio De Janeiro was held in 1992—the largest assembly of world leaders ever gathered together for planetary concerns. Carbon chemistry figured prominently in the conversations at the summit, where the United Nations Framework Convention on Climate Change was opened for signature. From my elementary studies until then in chemistry, I could understand how carbon dioxide was so easy to release into the atmosphere. Any time you burn an organic molecule you end up producing this potent and abundant greenhouse gas.

The octet rule became a structural, albeit flawed, fascination for me. I began to even make far-fetched leaps between science and religion using it as a metaphor. In a conversation with one of my suite mates who was majoring in Asian Studies, I noted how the Buddhist "Eightfold Path" had perhaps gained some metaphysical insights about the centrality of the octet rule. Further entrenching my fascination with octet structures was a reading of Nobel Laureate Murray Gell-Mann's 1961 paper that led to the development of quark physics in which he, too, recognized the power of eight in elemental order and alluded to Buddhist scripture in its title: "The Eightfold Way: A Theory of Strong Interacting Symmetry." Gell-Man's paper, despite its deceptive title, was later to shatter many of the enshrined views about atomic order that were previously held. The Eightfold Path was not necessarily deterministic. Soon I learned that the octet rule also had many exceptions. Through Gell-Mann's work, we moved beyond the simple view of the atom as comprising protons, neutrons, and electrons to a much more complicated structure of understanding particle physics, which is discussed in Chapter 1.

Physicists still yearn for a unified theory that could explain the structure of the universe but that has until now eluded us, even though intense mathematical inquiry and distant astronomical observations continue in its quest. On this journey we are discovering new forms of matter and energy that remain elusive to cogent description but are mathematically real. This quest started perhaps with the seminal work of the first astrophysicist, Johannes Kepler, in his seventeenth-century *Harmonices Mundi* (The Harmony of the Worlds). In this treatise, he posited the laws of planetary motion that helped us to separate the fictional allure of astrology from the empirical excitement of astronomy. Constellations in the cosmos could thus provide two divergent epistemologies of structure. The attraction of patterns in the cosmos is akin to broader pattern-seeking behavior in humans. The founder of the Skeptic Society and historian of science Michael Shermer calls this tendency to find meaningful patterns in meaningless noise "patternicity." Studying the basic sciences, particularly chemistry and physics, is a "coming of age" for most high school students in differentiating real and illusory patterns in natural systems.

The atomic world forms that border where conventional chemistry and particle physics begin to interact with trepidation beyond the binds of disciplinary reduction. As I studied further, the octet rule became an important but incomplete heuristic device—functional for teaching—so long as you knew its limits of applicability. Order within the atom was elusive as more particles emerged on the scene, but there was still one constant thread that seemed to permeate the fabric of the universe: light. Reconciling the traveling ability of light as a wave with the persistence of particles required a new kind of order to be conceived. The brilliant physicist Paul Dirac was to provide us a way to reconcile the order of particles and waves through his discovery and mathematical description of *anti-matter*. Dirac's "wave-particle duality" was noticeably different from the aforementioned dualisms of the ancients. Instead of delineating difference, Dirac's dualism suggested differentiated coexistence and complementarity.

As an ardent *Star Trek* fan, I had become used to hearing the term "anti-matter" as a teenager but had considered it fictional until I read about Dirac's work. The skeptic within me had thought that anti-matter was just a conceptual way for us to consider symmetry in the universe. But, lo and behold, anti-matter did in fact exist. Within 3 years of the publication of Dirac's conceptual argument for anti-matter's existence in 1928, the Caltech physicist Carl Anderson was able to empirically observe a positive electron, what he called the *positron*. He simply titled his paper "The Positive Electron"—a wry provocation to question the dominant order of charged assumptions about particles at the time. Later, when teaching a course on technical writing, I would assign Anderson's original paper to my students as an example of pithy scientific prose. What entranced me most about the discovery of this anti-matter particle was how it showed that mathematical insights in conceiving order could lead to tangible discoveries in even the most esoteric ways. It gave me confidence in the use of intangible concepts like "imaginary," "irrational," and even "surreal" numbers in creating an edifice upon which we could build our understanding of the universe. Although such numbers defy a palpable grasp, their mathematical existence is essential for charting tangible outcomes such as the path of flights through complex computations in radar systems.

Further mathematical intricacies would help us to grasp the world of quantum mechanics, in which particles can be conceived as spinning in opposite directions simultaneously. Relations between space and time could be described with greater precision but also lead to further exasperation. Our mind's eye is unable to conceive such contradictory realities and highly differentiated order at the diminutive scale of the quantum realm. Indeed, Newtonian order, which we observe at human scale (laws of motion and energy conservation), exists and is functional for our daily lives, even if it fails at explaining subatomic phenomena. Ever so rarely we can now glimpse the impact of the intangible quantum world. Medical diagnostic equipment has made the quantum realm tangible for us. The "positron" is manifest in the positron emission tomography (PET) scanner, and the magnetic resonance imaging (MRI) machine is able to see through our flesh because of a specific spurring of quantum behavior in the hydrogen atoms in our bodies. Two parallel and seemingly contradictory orders thus serve our needs, and that is how we must be willing to consider the functionality of our world and indeed the universe.

Functional Order

Our simplest mechanism for understanding order functionality in mathematics is the equation. Even the most confounding attributes of existence, like the conversion of matter to energy, can often be expressed through an equation such as Einstein's immutable $E = mc^2$. Many equations create order by relating independent variables to dependent variables if the processes are irreversible. A constant term can convert a proportionality relationship into an equation as well. In the case of Einstein's equation, that constant is the speed of light (c), although this equation pertains to reversible processes and this dependence and independence are not relevant. Most of our physical "laws," like Newton's laws of motion or the laws of thermodynamics, are described through simple equations. In discrete mathematics, there are also terms such as "partial order" and "total order" linked to notions of reflexive, symmetrical, or translational properties of functions. Some of the major paradoxes in the field of mathematics and logic also come from a quest for defining order through "self-reference." Can an omnipotent God make a stone which can't be broken? The ancient circular symbol of the serpent or dragon consuming its own tail—the *Ouroboros* (Ancient Greek for "tail-eating")—is a metaphor for what can be perplexing and enigmatic quests for order.

The second law of thermodynamics is perhaps most consequential in our understanding of order for it posits that a system will increase its level of disorder or reach an equilibrium with its surroundings. To move from disorder to order requires an investment of energy. The law was proposed in different forms, first by the French physicist Sadi Carnot and then by German physicist Rudolf Clausius, with reference to heat transfer between objects. They observed that heat always flowed from hot to cold objects and never in reverse. A one-directional transfer of energy, similar to the arrow of time, was itself a form of order and one with profound effect. Instead of a regular equation, an inequality was used to appropriately describe the irreversible directionality of the law. Out of this law came the fundamental physical concept of "entropy," which defines the distribution of energy in a system in terms of its availability for useful work. The less available the energy is due to the extent and quality of order in a system, the higher the entropy. Crucially, however, ordered systems do not necessarily have low entropy. As we will see throughout the course of this book, the entropy of the system can be operationalized in a variety of ways from particles at the micro-state to information transfer at the macro-state. Indeed, the second law may well not be functionally adhered to at micro-scales because it is essentially a simplification derived from statistical probabilities of order. At very small scales of material interactions, it is possible for ordered states to arise more plausibly by statistical chance. Such observations have interesting implications for the development of fields such as nanotechnology.

Indigenous societies that lived through more natural temporal rhythms viewed time and the cosmic order quite differently from us. For example, Australian Aborigines do not view time as a horizontal linear process but as a series of feedback loops in which the present is connected to the past. The notion of the "Dreamtime" in Aboriginal tradition comes from such a worldview. There is a notion of primordial order to which the present is connected and with which we should align to maintain balance. Although such traditional worldviews may seem abstruse, they stem from a keen awareness of the natural world and explore explanations where empirical observations are not possible.

The American physicist J. Willard Gibbs provided a three-dimensional approach to understanding entropy, one linking energy, entropy, and volume, and introduced the concept of "free energy." His insights led us to the fascinating observation that the complexity of a system initially increased with entropy, but then reached a point after which "mixing" leads to a simplicity in overall form emerging. A common example

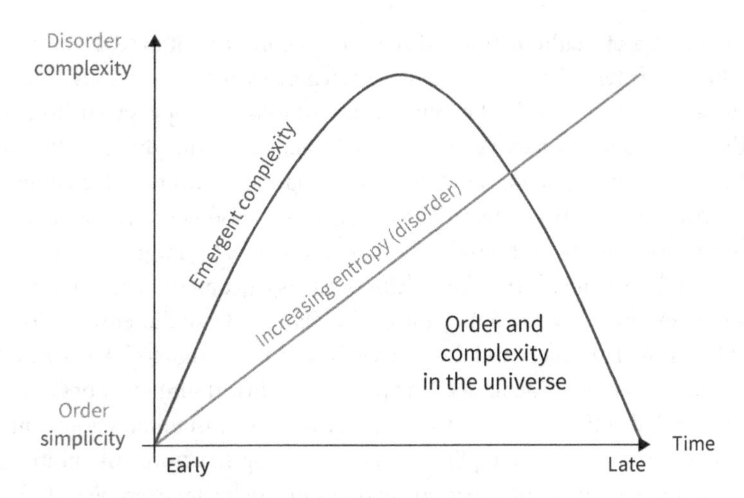

Figure I.1 Relationship between entropy and complexity over time.

to exemplify this is given by science educator Sean Carroll, who showed how coffee and milk, when initially mixed, have increasing entropy and complexity as plumes of coffee and milk form complex eddies and currents as they mix. However, once the latte is fully mixed and made, the system is entropic but not complex to the consumer (Figure I.1).

The flexibility of mathematical representation helps us in creating new forms of order where we may initially find a dead end. The Dirac equation had a special attribute in its mathematical representation that allowed for a repetitive transference and mirroring of variables. This is how anti-matter was theorized to exist and eventually proved to do so. As for quantum mechanics, if all else failed with equation structures, we could bring forth probabilistic mathematics to portray and measure concepts of this elusive realm, which we will explore in Chapter 1. But doing so would limit the extent of what we could know with certainty. This is enshrined in Werner Heisenberg's famous "Uncertainty Principle" that shows that, in order for the wave-particle duality to be mathematically coherent, we cannot know with certainty the spatial location and momentum of a particle. Around the same time as quantum mechanics was developing, the great Austrian mathematician Kurt Gödel provided us with his landmark "incompleteness theorem," which puts forward certain limitations of proof in mathematics more broadly. Intriguingly enough, the theorem also notes that you may have a true phenomenon which still cannot be proved because of certain inherent constraints of measurements.[14]

Gödel's insights are now also being applied to quantum physics itself in recognizing that certain key observations in physics, such as why certain particles have mass and others do not, can never be resolved.[15] Thus uncertainty and incompleteness became ensconced to some degree in limiting our knowledge about the universe and perhaps even in the explicability of nature more broadly. Such a call to humility coming from within the realm of mathematics and science is an important base from which this book is being written. However, even in cases where we have completely "rough" and unpredictable outcomes—like the infinitely multitudinous shapes of a pebble—a field of mathematics called *fractal geometry* was developed by the gifted mathematician Benoit Mandelbrot to provide a structural explanation. In his honor, a couple of his protégés, Adrien Douady and John Hubbard, used yet another branch of mathematics called *set theory* to study the bounds of a stable set of possible numerical inputs on the complex numbers plane.[16] This "Mandelbrot set" is a geometric construct to understand order and patterns when they escape conventional two dimensions. The x axis in this plane is real numbers and the y axis is "imaginary numbers" that are multiples of the square root of –1. The underlying premise of this "thumbprint of God," as it is sometimes called, is the principle of self-similarity which instantiates how complexity often emerges from a simple replication of an existing structure. The branches of a tree, veins in our bodies, and riparian channels in a delta are examples of this principle (Figure I.2). Several decades earlier, the great codebreaker and mathematical genius Alan Turing, whose work and tragic life story were recently popularized by the film *The Imitation Game*, also applied his insights to show how patterns like the spots on a leopard can emerge in nature. His pioneering 1952 paper on trying to understand the emergence of patterns in nature (morphogenesis) was a perspicacious precursor to complex systems science.[17] Turing's "machines" and their "completeness" are still the touchstone for the data manipulation ability of a computational system.

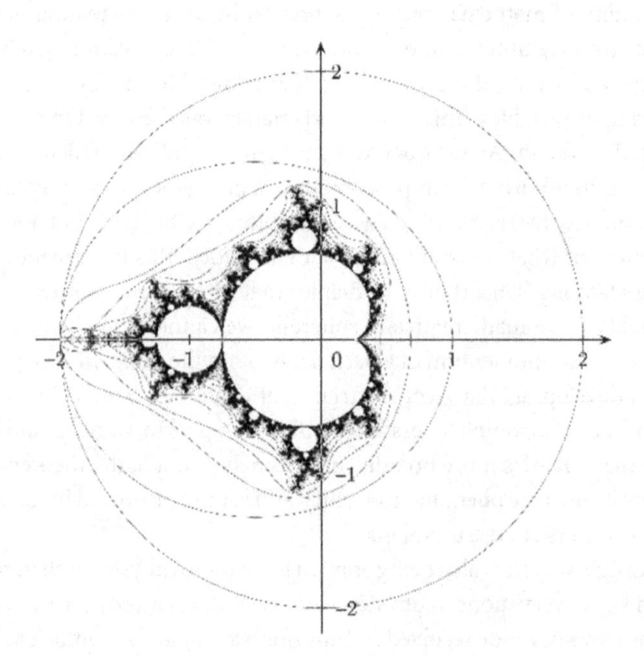

Figure I.2 The Mandelbrot set of emergent order from a self-similarity image of what is colloquially called by some science journalists as "the thumbprint of God."

In popular online videos one can find colorful representations of the Mandelbrot set that look like psychedelic symmetry warps in motion. My first exposure to this image was in a chance encounter with Benoit Mandelbrot himself in 2005, a few years before his death, at a meeting at the Museum of Science in Boston. In the introduction to the session, our host, the science communicator and entrepreneur Adam Bly, noted the Mandelbrot set as an example of how we were grasping the "mathematics of order." This encounter was yet another spurring notion which led to this book. Notably, the Mandelbrot set is bounded by the number 2 in both the real and imaginary number planes. For our purposes, the mathematics behind this is not of consequence. Rather, what is interesting is how a complex but ordered pattern can emerge from simple mathematical replication, particularly when we bring in additional dimensions. What is also intriguing is how this mathematical pattern can then be used as a vindication of some abstract, transcendent concepts in our mind's eye. Some see the mathematical image of a reclining Buddha while others see heart-shaped or "cardioid" shapes as a symbolism of universal love. Mathematics gives the image objective currency for many who might not otherwise show much respect for science and its exacting methods of inquiry.

This insight about how complexity and seemingly disordered outcomes still have latent structure led to the development of a spectrum of inquiry that is termed "Between Order and Randomness," (BOAR) by the complex systems theorist Scott Page. Natural phenomena, and consequently social phenomena, exist on this spectrum, wherein we must differentiate between what is complicated, complex, and approaching randomness. As a baseline, Table I.1 is provided to ensure that we understand this spectrum and the definitional presumptions that underpin the conceptual structure, with a simple example in each case. Note that I have not defined "chaos" as a category as some typologies tend to do because there are confounding colloquial definitions of chaos that are not particularly helpful for our purposes of understanding order. Suffice it to say that most definitions of chaos have substantive commonality with complexity as delineated in this table.

The last column of Table I.1 sheds light on the subtitle of this book. "Randomness" is an elusive concept and is defined primarily through the limitations of the human ability to process all variables in a given circumstance. What matters is whether a particular outcome is knowable within the powers of human measurement. Hence

Table I.1 Baseline for system differences

Ordered		Disordered	
Predictable		Unpredictable	
Simple	Complicated	Complex	Functionally random
Known knowns	Known unknowns	Unknown unknowns	Unknowables
Car ignition	Car engine	Traffic in city	Car hit by lightning

"functionally random" is the operative phrase, and the functionality of a concept will be our guiding principle in the chapters ahead. This does not define me as a "functionalist" by any arcane academic definition, but rather lays down the pragmatic goals for this book. My journey in this quest is guided by an understanding of the four key forms of order on which human beings operationalize much of their worldviews: natural, economic, social, and political. The book is thus simply divided into these four parts. This typology also echoes the "cynefin" framework for decision-making (after the Welsh word for "habitat"), developed by David Snowden at IBM, which substitutes "functionally random" with "chaotic" and "simple" with "obvious" or "clear." He further adds "disorder" at the intersection of all these four categories wherein categorization may be elusive. A change of scale is needed then to allow for categorization to be functionally useful for decision-making.[18]

Central to efforts at simplifying a bewildering array of causal variables within these orders, human ingenuity has focused on creating "models" with which we can make predictions. If there are some patterns to be found, the greater likelihood is that a formulaic model could be developed. Such models constituted the core of scientific endeavor, particularly in the earth sciences. For the general public, it seems perplexing at times that while we can develop fairly good models of long-term climate change that can predict temperature changes several decades hence, we are unable to predict short-term weather phenomena beyond a week to 10 days. This confounding paradox led to the development of *chaos theory* and complex systems science that has been a fascination of public intellectuals ever since James Gleick published the first edition of his bestseller on the topic more than three decades ago.[19] This book does not aim to revisit the specific episodes in the history of the development of complex systems science. That has admirably been tackled by Gleick as well as by numerous other writers affiliated with the Santa Fe Institute in New Mexico, which has become the Mecca for such insightful inquiry over the past three decades as well.

Instead, this book traces how *order* has been functionally useful and hindering in our understanding of the natural world and how human societies have developed as a result. It is thus a *functional* quest of how and when order has served a meaningful purpose and when it has failed us in both deciphering nature as well as in furthering our own well-being and sustainable existence as a species. To accomplish this task, contemporary views of economic and political order are essential to explore as well. Loosely, we throw about terms such as "new world order" without questioning whether order is even applicable in the context we are examining. Bridging the divide between understandings of natural order and social order is essential in this regard, but the "science wars" have often prevented a meaningful exchange. The term "science wars" is used to describe a period in the 1990s when many positivist scientists (who believe in the exclusive objective reality of natural phenomena) challenged their humanities counterparts to dispense with what they considered vacuous theories without clear evidence. Natural scientists, and even some empirical social scientists, questioned the abstract and nonempirical views of politics that have been presented in the humanities.[20] In one notable episode of this period, mathematical physicist Alan Sokal wrote a fancy-sounding paper for a humanities journal titled "Transgressing the Boundaries: Toward a Transformative Hermeneutics of Quantum Gravity." He sent it to test if the journal would publish it even though it was logically nonsensical and simply used abstract sentences and references that flattered the particular worldviews of the editors. When the paper was accepted, he went public to show that this was a prank and an indictment of obfuscation through language in the humanities.[21]

No doubt, rigor is essential across all human inquiry, but inspiration for new and often highly insightful worldviews gets dismissed by natural sciences as well. Sir David Brewster, the inventor of the kaleidoscope, whose story started this introduction, also tried to grapple with the challenge of abstract inspiration versus objective reality through correspondence with his contemporary, the novelist Sir Walter Scott. In a series of *Letters on Natural Magic*, in 1832, he explored ways of making sense between our views of the structure of reality versus illusion and the inspiration that lies in between. There are fascinating ways in which natural order and disorder intertwine. Observable reality provides a window on order. Yet we have also devised "imaginary" numbers and planes to make sense of nature. There are limits to what comparisons we can draw between order in nature and the order which humans create. Yet natural order has, for much of human history, been used as a pretext or foil for economic, social, and political order. Predicated on views of natural order, whether through divine or undetermined universal laws, we have also tried to develop order in our social norms and in our lives. But here, too, the spirit to rebel against order has persisted alongside the urge for uniformity, or at least an impulse for structure. In this book, we will question the ways order *could* orchestrate our lives by considering what function it serves in society and planetary sustainability. What patterns can be purposeful, and which can be a mirage that tempts us toward totalitarianism? Understanding the role of order in nature and society can, in essence, help us to engage constructively with many of our most fundamental misunderstandings and resulting conflicts.

Approaching such a magisterial topic necessitates humility. Our understanding of order in all its natural and socially developed forms is confined by our scientific methods and perceptions. Each chapter is an attempt to illuminate key assumptions which we make about order in our understanding of the world and indeed the universe. In contemporary environmental studies textbooks and a plethora of international documents about "sustainability" of planetary processes, Venn diagrams such as Figure I.3 are presented. Such diagrams do not consider the underlying structures of order that may be necessary to truly achieve sustainability. They simply present a stylized view of how humanity conceives its spheres of influence. This book is aimed at the next level in considering how we need to functionalize sustainability through engagement with the four cardinal dimensions of order which have necessary hierarchies of operation. However, these hierarchies should be considered in *functional* terms harkening back to the classic 1972 article in the journal *Science* by Nobel Laureate P. W. Anderson, "More Is Different: Broken Symmetry and Nature of Hierarchical Structure in Science." In this short inspirational commentary, Anderson set forth how

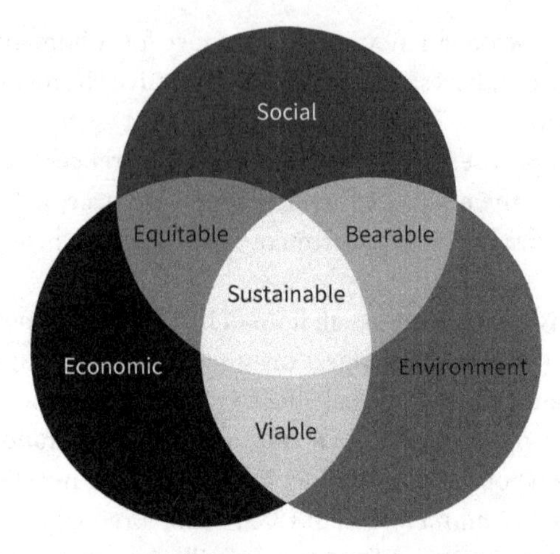

Figure I.3 Conventional view of the determining spheres of "sustainable development."

"intensive" and "extensive" fields of natural science, predicated on pure versus applied research, can also be linked to social sciences and indeed humanities. He was also quite prescient about how absolutist views can emerge in both science and the humanities about structure being imposed and transferred where it might not be appropriate. Thus, attempts at providing "unification" paradigms, such as the notion of "consilience" popularized by the great entomologist and naturalist E. O. Wilson in his book[22] of the same title, need to be evaluated with greater attention toward their accuracy but also their functionality. Unification narratives in some forms may be presented as a triumph of reductionism and could lead to a negation of analytical nuance at the macroscopic level. One of Anderson's intellectual proteges and fellow Nobelist Robert Laughlin wrote a book playing on Anderson's original paper, entitled *A Different Kind of Universe*, which suggests that reductionism has limited explanatory power in key systems phenomena. A philosophical differentiation in this regard occurs between ontology (the nature of being comprised of constituent parts) and epistemology (knowledge of functionality). The reductionist view has ontological validity but is functionally less useful for learning or understanding coherent knowledge about the natural world.

Anderson acknowledged that the symmetry we often see at the fundamental levels of natural phenomenon in terms of structure can be "broken" without "violating" the orderly elegance of the physical laws.[23] Indeed, the very existence of our universe was likely possible because the symmetry between matter and anti-matter in equal measure was broken with the triumph of physical matter. Had both existed in equal measure after the Big Bang, they would have annihilated each other into photons, as proposed in the plot of Dan Brown's book *Angels and Demons*—though with less theological fanfare. However, asymmetry, most likely caused by the elusive non-charged neutrino, gave matter an edge to constitute the universe.[24] Hypothetically, there could also be "negative mass" matter, which would have very peculiar properties and interactions with regular mass in which, instead of showing inertia, the material would hasten acceleration when force was applied. Though negative mass has not been found so far in the universe, Nobel Laureate Adam Reiss, in discovering the acceleration of the universe, jotted down in his lab notebook that negative mass could be an explanation for the observation that we now ascribe to "dark energy."

Symmetry at a core physical level remains central to our views of concepts such as the "conservation laws" of energy and momentum. The oft-neglected pioneering work of Emmy Noether (1882–1935) laid out the fundamental ways in which symmetry in space and time persists. She recognized astutely that in science symmetry is not just about mirror images (discrete symmetry) but about maintaining the order of transformation (continuous symmetry). Moving objects and particles through space and time without making an existential change to their fundamental properties is at the heart of the physical symmetry that Noether posited in her theorems that came from her studies of abstract algebra. She also showed that lack of symmetry at larger scales does not necessarily translate into local symmetry at a smaller scale. Yet symmetry is not immutable, and indeed, and as we shall see further in our inquiry, "breaking" of symmetry can be essential for functional forms to emerge. As we strive to find technical and social means of making the human presence on the planet more sustainable, such a multilayered understanding of order is essential. The conventional view of sustainability presented through diagrams such as Figure I.3 still begs the question of how we might functionally reach that coveted space at the center of overlapping spheres of influence.

The diagram also misses the inherent hierarchies which must be considered if we are to reach such an aspirational goal. This book provides a synthesis linking different manifestations of order and has been structured deliberately in a sequence from natural to economic to social to political order as a nested hierarchy. Figure I.4 shows the four functional orders around which this book is structured, nested in terms of their hierarchies of dependence. Through such a view of sustainability, we are more likely to have functional impact but also be a more informed society that is aware of its opportunities and limits for controlling our future. Such a depiction suggests a tiered hierarchy and may lead some to see a bias toward natural science. However, if we analyze such a system in terms of *proximate* and *ultimate* means of intervention, then each order has its particular value. While natural order will remain an ultimate confining feature in human systems, to have impact, proximate interventions may well be more appropriate within social, economic, or political order. The "humanities" provide the philosophical and experiential thread between these four orders and are also woven through the narrative. This book aims to consider how such a deep understanding of the fundamental building blocks of natural order can then lead us also to either literal or figurative insights that can make other forms of order more meaningful. It is important to note that this nested view of order does *not* imply the classical form of "environmental determinism" proposed by geographers such as Ellen Semple who, in 1911, infamously said, "Man is the product of the earth's surface."[25]

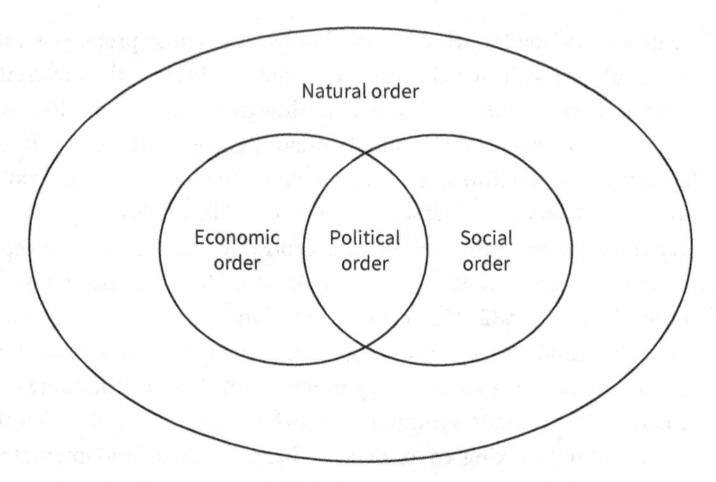

Figure I.4 Hierarchies of order: Transitioning from the conventional view of "sustainability" (Figure I.3) to a functional view of order that is considered in this book.

Rather, the hierarchy assumes influences and agency on outcomes and presents this diagram from the perspective of operational constraints. The Venn diagrams allow for influence across each domain of order and, indeed, a transformation of the natural order through the functioning of a complex adaptive system. Just as the organelles of a biotic cell have permeability, these heuristic cells delineating domains of order can have dynamic flows. However, a recognition of these nested hierarchies and the underlying structures of order on which they are built helps us to consider ways of achieving optimal functional outcomes for society. Thus, in a sense, I am moving from the diagnostic determinism of panoramic scholars like Jared Diamond, Daron Acemoglu, and James Robinson to consider deeper structural attributes whose knowledge can move us prescriptively toward a more sustainable future.

Such an approach recognizes that there will be priorities in terms of action when we consider large global efforts such as the United Nations Sustainable Development Goals. Out of the 17 goals set forth, with additional 126 targets for action between 2015 and 2030 for all countries of the world, we will need to consider the interactions between each goal, which range from poverty reduction to climate action to global partnerships.

Natural systems are the template of existence on which all other orders can be functionalized and hence hold primacy in presentation. Social and economic order are more closely coupled and overlapping in some areas of functionality but not exclusively so. It is possible to have social norms which are not predicated on economic variables of resource scarcity (e.g., our norms around family and tribal structures or ethical structures of human interaction). Less appreciated is that part of the Venn diagram which shows that you can also have economic order within natural systems that is not predicated on human social systems. Stocks and flows, input and output, and equilibrium, which we consider in economic models, can exist extant of the preferences of human social systems as well within the economy of nature, as we will explore in Part II of this book. Social order among humans and other organisms is operationalized through conflict and cooperation among individuals, often mediated by economic incentives but also by more intangible values. As we advance as a civilization and develop more complex hierarchies of organization, we must also consider how we are transcending primal biological derivations of order in terms of sexual roles and norms. Furthermore, technology is merging with human social order through the advent of revolutionary technologies, such as the internet, which can facilitate order or anarchy depending on their management. The advent of artificial intelligence and the conundrum of consciousness challenge our view of human centrality in a sustainable future for our species.

At the confluence of social and economic order is the emergence of political order whereby governance mechanisms can be enabled. The impact of this political order may well radiate outward to disrupt social, economic, and even natural order but that does not change the underlying determinants of long-term sustainability of the system. Territory and the physicality of political order has been a core challenge to effective planetary governance mechanisms. Nesting political order within the natural, social, and economic context in Part III provides more realistic mechanisms for dealing with issues such as nationalism. These challenges to planetary visions of natural order are manifest in notions of human "security" as a means of functionalizing political order. Human choice and liberty are themselves direct challenges to order emergence, and the calibration of democratic systems to balance functionality and inclusion will be discussed in the final chapter of the book.

How might the disjunctures between natural, economic, social, and political orders be reconciled? Fundamental to such a process is highlighting the prime dependence of all systems on core natural processes, many of which are beyond human agency despite our contention of living in "the Anthropocene" (the age of humans). A rather astounding set of analyses in this vein was conducted by researchers at the Weitzman Institute in Israel to highlight our impact. By their calculations, in 2020, the total mass of all human-created material now exceeds all natural biomass. The same team also calculated that the 95% of the total biomass of mammals on earth was now either human or animals cultivated by humans for their consumption![26] Second is the recognition that order is not normatively good or bad and must be considered in terms of functionality toward desired goals of the sustainability of natural systems and the development of progressively harmonious social systems. For example, borders, as a form of order, have an important ecological functionality in preventing the spread of disease, but they can be a hindrance to economically and ecologically efficient trade. Finally, we must always be sensitive to the temporal aspects of the emergence of order at geological, biological, and human time scales. Keeping perspective on the divergence in order and equilibrium processes in natural versus human social systems can avoid cognitive disjunctures and lead to more effective planetary planning.

By casting a panoramic view of the applicability of order in our existence, we can better reconcile the tensions that prevent us from reaching consensus on essential planetary problems, from climate change to social inequality. The narrative will hopefully provide the reader with the cognitive composure needed to recognize the opportunities and limits of human control. To have functional sustainability within our own planetary microcosm, we must understand underlying structures where they exist in more cosmic or universal terms. We must also grapple with areas of uncertainty and risk that are fundamental to the trajectory of various forms of order.

The scales of order presented in this book and their potential vertical integration engages, but also contests, the notion of "coarse graining," which has gained currency in complex systems science. Finer granularity of research provides fundamental insights about phenomena, but one must zoom out to see the coarser units that make a complex system functional—this is the logic of coarse graining. Returning to the kaleidoscope metaphor that started this narrative, zooming in to understand the chemical composition of the shards of glass may well be inconsequential to the purpose of the object and the patterns it conveys. Thus, a coarse-grained focus on its functional attributes of orderly patterns, such as the arrangement of mirrors, the colors of the fragments, or the shape of the shards, may be what most determines the outcome. At the same time, if we were to make the product more durable or consider other uses for it in a broader planetary context, finer granularity of analysis on its material composition might well be useful. When discussing environmental sustainability, we similarly use the metaphor of "missing the trees for the forest" or vice versa—yet why stop at just the tree or just the forest? Why not explore further backward and forward linkages to the existence of trees and forests? At the same time, we also must be careful that the vertical integration of knowledge and patterns does not lead to tempting errant paths of pseudo-science. Such forays may be good for fictional recreation and perhaps even for literary inspiration but need to be differentiated accordingly from empirically, or at least theoretically, plausible inference. The limits of transferability of paradigms and patterns in real and heuristic terms is part of the essential journey this book aims to traverse. The goal here is to stretch that specter of inquiry across the full spectrum of human learning about ordered systems to make the quest for sustainability more meaningful in both literal and figurative ways.

Notes

1. Brewster, David. 1858. *The Kaleidoscope: Its History, Theory, and Construction* (contemporary edition 2018). Kessinger Publishing, LLC.
2. Gould, Stephen Jay. 2002. *The Structure of Evolutionary Theory*. Harvard University Press.
3. Gould, Stephen Jay, and Niles Eldredge. 1986. "Punctuated Equilibrium at the Third Stage." *Systematic Zoology* 35 (1): 143–148. https://doi.org/10.2307/2413300.
4. An excellent contemporary translation of Haekel's work with commentary is Haeckel, Ernst. 2018. *The Riddle of the Universe: At the Close of the Nineteenth Century*. CreateSpace Independent Publishing Platform.

5. For a learned reflection on the philosophical underpinnings of the film, see Benson, Michael. 2018. *Space Odyssey: Stanley Kubrick, Arthur C. Clarke, and the Making of a Masterpiece* (reprint edition). Simon & Schuster.
6. A good review of the debates around monism can be found in Cornell, David M. 2016. "Taking Monism Seriously." *Philosophical Studies: An International Journal for Philosophy in the Analytic Tradition* 173 (9): 2397–2415.
7. The term "meme" was coined by Darwinian zoologist and public intellectual Richard Dawkins and later given attention by the inventor of Microsoft Word Richard Brodie in his 2009 book *Virus of the Mind: The New Science of Memes*. Henry Hay Publishing.
8. For an engaging discussion of dualism, particularly with reference to mind and body, see Golden, Kristen Brown. 2012. *Nietzsche and Embodiment: Discerning Bodies and Non-Dualism*. SUNY Press.
9. For the development of monism by Hegeler, see Guardiano, Nicholas L. 2018. "Monism and Meliorism." *European Journal of Pragmatism and American Philosophy*. https://doi.org/10.4000/ejpap.1072.
10. James, William. 2008. *Pragmatism: A Series of Lectures by William James, 1906–1907*. Arc Manor.
11. The lecture along with related critical commentary can be found in Newton, K. M. 1997. "Jacques Derrida: 'Structure, Sign, and Play in the Discourse of the Human Sciences.'" In *Twentieth-Century Literary Theory: A Reader*, edited by K. M. Newton, 115–120. Macmillan Education UK. https://doi.org/10.1007/978-1-349-25934-2_24.
12. Press, T. M. 2020. *Making Art Work*. MIT Press.
13. For a learned scientific history of carbon, see Hazen, Robert M. 2019. *Symphony in C: Carbon and the Evolution of (Almost) Everything* (1st edition). W. W. Norton & Company.
14. Goldstein, R. 2006. *Incompleteness: The Proof and Paradox of Kurt Gödel (Great Discoveries)* (reprint edition). W. W. Norton & Company.
15. This problem in theoretical physics is known as *Yang-Mills and mass gap*, which is considered one of the "Millennium Problems" by the Clay Mathematics Institute (https://www.claymath.org/millennium-problems), which has a $1 million prize ready for anyone who can prove this mathematically, even though it has been observed experimentally and through computer simulations. Applying the Gödel theorem to this problem suggests that it cannot be solved mathematically, even though it is true. Castelvecchi, D. 2015, December 9. "Paradox at the Heart of Mathematics Makes Physics Problem Unanswerable." *Nature News*. https://doi.org/10.1038/nature.2015.18983.
16. Mandelbrot, Benoit B. 1983. *The Fractal Geometry of Nature* (1st edition). W. H. Freeman and Company.
17. Turing, Alan Mathison. 1952. "The Chemical Basis of Morphogenesis." *Philosophical Transactions of the Royal Society of London. Series B, Biological Sciences* 237 (641): 37–72. https://doi.org/10.1098/rstb.1952.0012.
18. Snowden, D. J., and Boone, M. E. 2007, November 1. "A Leader's Framework for Decision Making." *Harvard Business Review*. https://hbr.org/2007/11/a-leaders-framework-for-decision-making..
19. Gleick, James. 2008. *Chaos: Making a New Science* (anniversary reprint edition). Penguin Books.

20. A good primer on linking human behavior to underlying natural patterns is Barabasi, Albert-Laszlo. 2010. *Bursts: The Hidden Patterns Behind Everything We Do, from Your E-Mail to Bloody Crusades*. Plume.
21. Berube, Michael. 2011. "The Science Wars Redux: Fifteen Years After the Sokal Hoax, Attacks on 'Objective Knowledge' That Were Once the Province of the Left Have Been Taken up by the Right." *Democracy: A Journal of Ideas* 19: 64–74.
22. Wilson, E. O. 1998. *Consilience: The Unity of Knowledge*. W. W. Norton.
23. Anderson, P. W. 1972. "More Is Different." *Science* 177: 393–396. https://doi.org/10.1126/science.177.4047.393.
24. Major experiments are under way in Japan and South Dakota to show the role of neutrinos as the agent of breaking this symmetry. Results in April 2020 from the Super-Kamiokande neutrino detector are providing compelling evidence that neutrinos played a role in breaking symmetry and leading to the mass-based order emerging in the universe. Abe, K. et al. 2020. "Constraint on the Matter–Antimatter Symmetry-Violating Phase in Neutrino Oscillations." *Nature* 580: 339–344.
25. Semple, Ellen Churchill. 1911. *Influences of the Geographic Environment: On the Basis of Ratzel's System of Anthropo-Geography*. Henry Holt. Semple was the first woman to be president of the American Association of Geographers. Some of her ideas of environmental determinists were misinterpreted to justify colonialism, just as Darwinism was used similarly by social Darwinists.
26. Elhacham, E., L. Ben-Uri, J. Grozovski, Y. M. Bar-On, and R. Milo. 2020. "Global Human-Made Mass Exceeds All Living Biomass." *Nature* 588:442–444. https://doi.org/10.1038/s41586-020-3010-5; Bar-On, Y. M., R. Phillips, and R. Milo. 2018. "The Biomass Distribution on Earth." *Proceedings of the National Academy of Sciences* 115 (25): 6506–6511. https://doi.org/10.1073/pnas.1711842115.

PART I

NATURAL ORDER

The best way to defeat nature is to obey its laws.

—Francis Bacon

My second law, your second law, ordains
That local order, structures in space
And time, be crafted in ever-so-losing
Contention with proximal disorder in
This neat but getting messier universe.

—Roald Hoffman (excerpt from the poem "The Devil Teaches Thermodynamics")

1
Seduction of Structure in Nature

> What I see in Nature is a magnificent structure that we can comprehend only imperfectly, and that must fill a thinking person with a feeling of humility.
>
> —**Albert Einstein**

As humans retreated to their residences during the great coronavirus pandemic, a strange sense of consolation was found in narratives about natural systems resetting to a new normal. A structural shift to a pre-human balance of earthly forces was noted worldwide. The Hong Kong zoo was enthralled to report that their panda couple, Ying Ying and Le Le, who had remained celibate for a decade, found the composure in the absence of human visitors to engage in rapturous sex. Enchanting vistas of the Himalayas were supposedly now visible due to a clearing of smoke from South Asian cities. The silver lining to the chaos in our lives appeared to be that nature was finding a lost order and equilibrating. While some of the social media memes were clearly doctored to accentuate impact, there was also a less than admirable side to the "natural healing" mantra. Shopping carts which had fallen into murky urban rivers were also now visible due to the renewed clarity of the waterways and were a fun counterpoint of rebuke to the natural healers. Others reminded us of how nature itself could violently disrupt ostensible order through volcanic eruptions or asteroid impacts. The pantheists argued back that such disruptions were merely a self-correction mechanism for a celestial order beyond our comprehension. Frantic and frenetic as the conversations were, they provided a learning moment for us all about understanding the very nature of disturbances to planetary order.

The public went through a crash course in virology, public health, and ecology in this crisis and at an unprecedented pace. The neat image of the pathogen as a prickly orb of spike proteins, lipids, and nucleic acids was emblazoned everywhere. Media-savvy doctors showed us diagrams of how a lowly soap molecule could so easily bind and pry open the lipid shell of the virus, leading to its demise. The average citizen was widely exposed to debates in science about whether or not viruses can be considered alive or dead—and were such dichotomous categories even meaningful? Some scientists pivoted their research to quell conspiracy theories about the virus being engineered in a lab. The method they chose for this proof was that the virus had followed a natural order of mutation rather than a human engineered one.[1] They argued that the

Earthly Order. Saleem H. Ali, Oxford University Press. © Oxford University Press 2022.
DOI: 10.1093/oso/9780197640272.003.0002

precision with which the COVID-19 virus was able to target and enter human cells could only be a result of a natural order of mutations rather than anything humans would have engineered.

Furthermore, the viral evolution could be tracked to mutations from benign viruses in bats, and, so the argument went, if humans had wanted to engineer a pathogenic virus, they would have focused on existing disease-causing agents first. These two insights about the virus are an intriguing starting point for us to reflect on how natural order might be considered in the absence of deliberate human intervention. The power of *incremental change*, which is the hallmark of mutative evolution, delivers a potently effective outcome of the virus entering the human cell and perpetuating its replication. Yet does such incremental change operate within specific natural structures that we might consider to be more immutable? Would an understanding of such structures lead to more of the predictive power that we so sought in the pandemic? These are the questions we consider in this chapter to conceptualize how structure in natural systems can be ascertained.

Molecular "Magic"

The Coronavirus has the ability to infect a host because of certain fundamental abilities of its constituent molecules. In the case of this virus, it is a large ribonucleic acid (RNA) molecule that has two vital properties: (a) *self-replication*, meaning that it can construct an identical copy of itself through the physical rebonding of its constituent molecular parts and (b) *catalysis*, the ability of a substance to create a more energy-efficient pathway for a chemical reaction to occur due its structural form. Clearly both these properties are a manifestation of a molecular order which seems almost magical in its spontaneity. In essence there is a natural elegance to nucleic acid molecules that allow for these properties to be manifest. These acid molecules seem to wrap around in certain ways and form specific shapes because of some underlying natural physical laws. Core to this understanding is the realization that molecules that comprise the chemistry of life do not just operate by *chance* but through some latent structural mechanisms.

Chance and random mutative processes are the hallmarks of natural selection, which is established science. However, a study of evolution without considering the underlying chemical and physical laws on which it operates has led to a widespread misunderstanding about the story of living organisms. Popular views of the scientific ascendance of natural selection in contrast with creationism tend to eschew conversations of order within the natural world. Yet evolution inevitably operates on an ordered template of existence. This template is what the geneticist Seward Wright referred to as a "fitness landscape" as far back as 1932, a term given contemporary usage by the complex systems researcher Stuart Kauffman.[2] The landscapes themselves have key structural attributes (such as the spatial ordering of proteins, as suggested by the theoretical geneticist John Maynard Smith) which have allowed the self-organization of molecular forms to emerge rapidly. Indeed, if we solely operated by chance at the molecular level, the probability of life emerging could not be possible within the known age of our universe. In his notable book *Chaos and Order*, Friedrich Cramer from the Max Planck Institute calculated that, considering the genome of a simple *Escherichia coli* bacterium, the probability that this arrangement could have emerged by pure chance is 1 in $10^{2,400,000}$ (1 followed by 2.4 million zeroes). Cramer further calculates that even an "intelligent" machine capable of checking the appropriate molecular permutations occurring by chance and organizing them at a rate of 1 per second would require far more time than has elapsed since the Big Bang, which is approximately 10^{17} seconds. He thus concludes that the "age of the universe is infinitesimal compared with the time needed for the emergence of order by chance."[3] There is now compelling evidence from botanical research that mutations themselves are also not random across a genome and occur with different probabilities in various locations.[4]

Cramer was a protégé of James Watson and Francis Crick, the discoverers of the structure of DNA, and a colleague of the Nobel Prize-winning chemist Manfred Eigen whose work on fast chemical reactions was instrumental to understanding *autocatalysis*, the self-organizing chemistry of life. Eigen built his efforts on the earlier work of Harold Urey and Stanley Miller in their famous 1953 experiment whereby they had successfully synthesized amino acids from a mixture of methane, ammonia, and water in a closed circulating system. Amino acids are the building blocks of proteins and gave validation to the prospect that organic biotic molecules could emerge from simple inorganic chemistry with opportune energy influx under certain environmental conditions. The next puzzle for Eigen was how such biotic molecules could then organize into replicating systems. A key challenge in this regard was that random collisions of these molecules could give rise to as many useless organic molecules as it would to useful ones (an error rate). Furthermore, there was a chemical "mutation" rate which would allow for useless molecules with no functional purpose to cannibalize emergent useful molecules by bonding with them. To produce a cache of useful organic molecules that could replicate and form orderly, large-strand molecules that were capable of genetic coding, one had to mitigate this negative mutation rate. The larger molecules (whether primordial RNA or rudimentary proteins is still debated) would in turn be able to facilitate the formation of enzymes, the catalysts whose unique forms had the capacity to physically reorganize their peer molecules in a more coherent, orderly way to deliver biotic outcomes. Eigen calculated that if merely random mutations were allowed to continue, the maximum useful molecular size that could be produced was far less than would be needed to produce the larger molecules. He was thus confronted with a paradox: in order to get a genome for higher order life, you needed enzymes, and in order to have enzymes, you needed a genome. His pathway out of this paradox was the hypothesis that a series of nested chemical reactions that were replicating small sets of molecules could in turn be feedstock for additional sets of reactions that would operate in a similar cycle. The molecules in a hypercycle would thus be "cooperating" with each other in a virtuous circle, allowing

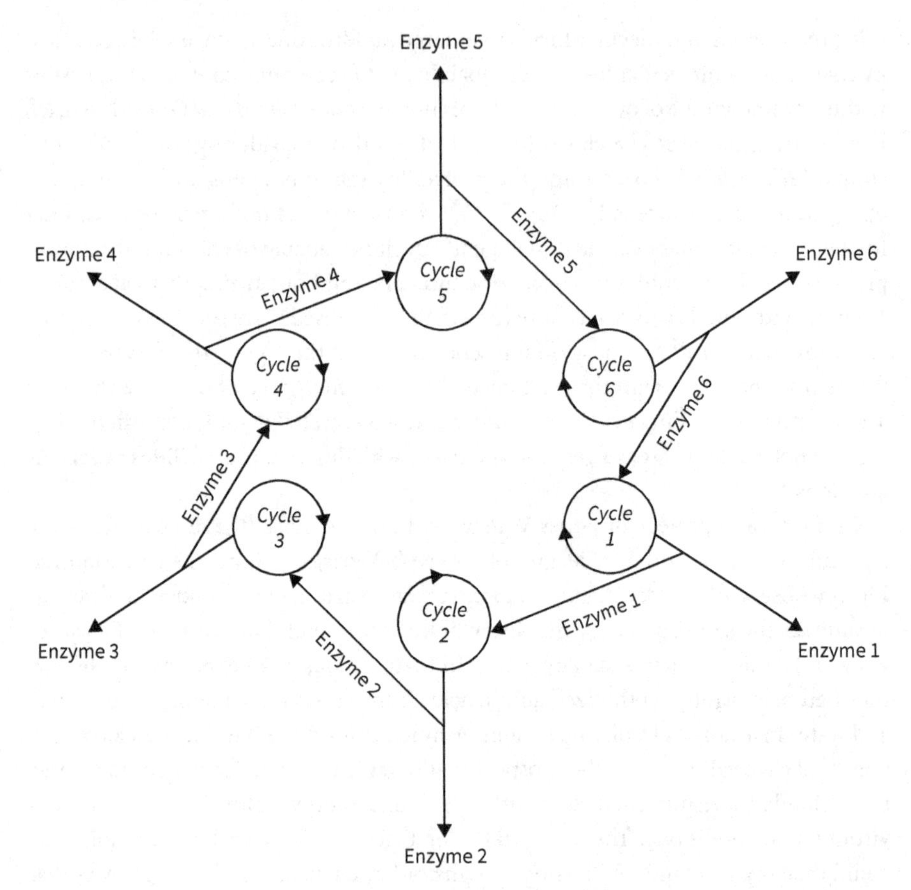

Figure 1.1 The Eigen Hypercycle. Within a protective membrane, such a cycle can lead to the emergence of living cells.
Adapted from an original figure by Gerald Ulrich.

for the emergence of the coveted larger strand molecules (Figure 1.1). However, such molecular cooperation would still need to be insulated from external "competition" from other molecules trying to engage in errant reactions.

Evolutionary order is often conceived through the lens of stress and survival of particular useful mutations. While this is certainly an important mechanism for driving and accelerating evolutionary change in particular directions, there is growing evidence that change would proceed in living systems in particular directions regardless of such pressures. The longest running experiment in microbial evolution has been carried out at Michigan State University by microbial ecologist Richard Lenski, since 1988. The Long-Term Evolution Experiment (LTEE) has been tracking genetic changes in 12 initially identical populations of asexual *E. coli* bacteria which go through six or seven generations every day. Thus, after 30-plus years, the bacteria in the flasks will have achieved more than 74,000 generations. This represents an equivalence of around 1.5 million years of human evolution. As predicted by evolutionary theory, the fitness of each subsequent generation increased dramatically for initial generations. However, this fitness increase slowed down as a balance between positive and negative mutations and the speed of mutation itself was modulated by this trajectory. In a natural environment, there are many stressors and disruptions which can further test and spur evolution. In a controlled environment, without such shifts, the slowing down of the fitness dividends of evolution is thus predictable. However, rather than flatlining toward an asymptote, the fitness followed a "power law model" whereby fitness still continued to increase at a steady albeit slower rate than in earlier generations. The most profound insight from this experiment in terms of evolutionary order according to Lenski was that "even in the absence of environmental change, there are so many opportunities of smaller and smaller magnitude to make progress that progress will never stop even in a constant environment."[5]

Emergent natural order of molecular transition from chemistry to biology rests on the key success of this chemical upscaling, and herein lies the major contestation between proponents of nature's "bottom-up order" proposed by Darwinists and the "top-down order" proposed by the proponents of "intelligent design." The latter group, championed by controversial think tanks such as the "Discovery Institute," often end their argument with the molecular improbability of chemical upscaling. The "intelligent design" proponents try to differentiate themselves from "creationists" by noting that they are not presuming theological origin but rather considering a process-of-elimination approach to scientifically plausible alternatives.[6] Indeed, one of the great challenges of presenting detailed and interdisciplinary perspectives on order is that they can also lead to attractive but specious arguments that can neither be refuted nor empirically tested. This is precisely why a more nuanced and critically evaluative view of evolution's role in order formation from a complex systems perspective is needed. In the words of renowned theoretical biologist and complex systems scientist Stuart Kauffman,

> [o]ur legacy from Darwin, powerful as it is, has fractures at its foundations. We do not understand the sources of order on which natural selection was privileged to work. As long as our deepest theory of living entities is the genealogy of contraptions, and as long as biology is the laying bare of the ad hoc, the intellectually honorable motivation to understand partially lying behind the creationist impulse will persist.[7]

However, there are numerous other enabling factors which can facilitate orderly emergence related to the template on which the reactions of life may take place. The aforementioned hypercycles might well be insulated from negative competition by protection within an oily lipid barrier that could have been produced in the extraordinary chemistry of hydrothermal vents, geysers, or tidal pools, where early life is believed to have originated. There is still underlying structural order which could have further enabled the kind of hydrogen bonding needed to create hydrophilic and hydrophobic molecular tails that ultimately lead to the lipids forming an insulating membrane. Such membranes would be the ultimate protectors and incubators allowing life-giving molecules to emerge. Enzymes and organisms that use them have shown remarkable ability to also adapt to and metabolize disruptions in natural order. For example, the pesticide atrazine is now decomposable by a specialized enzyme in *Pseudomonas* bacteria that is able to use the pesticide as a source of nitrogen for its metabolism.

Just as living forms can show evolutionary properties, there is the prospect that non-living materials can also go through a process of natural selection. After all, the dichotomy between living and non-living at a purely physical scale is less meaningful. Materials of all kinds can be "selected" for in earthly order by the most efficient process of absorbing and dissipating energy. Such a perspective has been presented by the physicist Jeremy England as "dissipative adaptation" in his book *Every Life Is on Fire* and considered an important insight for nanotechnology development.[8] However, this effect is a permutation of an earlier observation of enduring natural designs by mechanical engineer Adrian Bejan who suggested a *constructal law*, which states that "for a finite-size flow system to persist in time (to live) it must evolve such that it provides greater and greater access to the currents that flow through it." This law determines vascular forms from trees to human capillaries in living systems, but also riparian deltas. This is in essence a functional manifestation of the power of fractals that were briefly noted in the Introduction of this book. Engineers have been able to mimic the power of such structures in human construction as well and consider its resilience.[9] The geometry of molecules in particular triangular forms that may give strength to a material may be mimicked in bridge structures to accord them strength as well.[10] The complex molecular chain structure of a spider's web as well as its patterns in spinning a macro-structure can also be a "biomimicry" opportunity for engineering. The biologist Francois Jacob who made essential discoveries in our understanding of enzyme synthesis, suggested that the best metaphor for natural selection and evolutionary adaptation leading to useful "design" was not of "engineering" but of "tinkerking."[11]

Human ingenuity can further harness both evolution and natural structural processes for further accelerating functional order, as demonstrated by the remarkable work of chemical engineer Frances Arnold. In her revolutionary approach to enzyme synthesis, Arnold and her team have captured the power of evolution to synthesize targeted enzymes that can serve a range of industrially relevant purposes from cleaning up pollution to making laundry detergents more effective. They even were able to develop enzymes with the help of Icelandic microbes that could incorporate silicon into biogenic molecules and make bonds between carbon and boron atoms. In her Nobel Prize lecture in 2018, Arnold referred to a short story by Argentine essayist Jorge Luis Borges called "The Library of Babel," which conceives the universe as a library of all possible books, most of which are gibberish. The same is true of the vast assemblage of possible compounds we could synthesize. The goal of the librarian, or of the applied scientist or engineer, is to develop tools whereby we can strategically use the structural attributes of nature alongside the power of evolution to find functional material forms.

Such a search for structural attributes in genomes has been a fascinating quest that has opened new doors for fields such as gene therapy. The discovery of simple recurring patterns of DNA sequences in prokaryotic organisms in the 1980s and 1990s was revolutionary. Such repeating patterns were termed clustered regularly interspaced short palindromic repeats (CRISPR), and they alerted researchers to the ability of specific protein molecules to edit out fragments of RNA that would potentially be reflective of an external agent like a virus. Jennifer Doudna and Emmanuelle Charpentier (who won the Nobel Prize in Chemistry in 2020) were able to further refine this process, which has led to a proverbial revolution analogous to the invention of the internet. The ability to edit naturally occurring genomes to correct "errors" or enhance particular traits has enormous material and ethical implications which are still being debated. But can a vertical transmission of order from the subatomic to the molecular level occur spontaneously? To answer this question, we first need to understand the elusive world at the very core of material existence: quantum order.

Quantum Order

The discrete world of the "quantum realm" popularized in the 2018 Marvel film *Antman and the Wasp* reveals some enigmatic features that are partially predicated in science. Marvel filmmakers were careful to consult with notable physicists and science educators, such as Caltech's Sean Carroll, to fictionalize within the bounds of scientific plausibility—albeit with some poetic license. The phantom anti-villain in the film, Ava Starr, is a tormented young woman who phases in and out of material existence like a spooky ghost because her body is functioning as if she were in the quantum realm. Although such ghostly qualities of passing through physical matter is not plausible at human scale, it is within the orderly limits of quantum mechanics at the subatomic level. Key to understanding quantum order is an appreciation for the concept of *probabilistic reality*, whereby existence in space and time is impossible to predict with certainty. Hence the prospect of the seemingly unreal possibilities of "quantum tunneling," whereby particles have the probability of passing through material existence because of their wave functionality. A key area of contention in the field of quantum biology is whether this subatomic process can also be impacting molecular-level processes beyond the mechanics at the subatomic level. In other words, do the subatomic processes *functionally* impact the reactions between much larger atoms and molecules beyond the quantum realm?

One of the doyens of quantum mechanics, Edwin Schrodinger, whose famous cat in the box metaphor made famous the paradoxical realities of quantum mechanics, conjectured of such a prospect in a notable little book from 1944 titled *What Is Life?* The complexity and self-organization of living systems, and even the palpability of consciousness, were explored in this deep treatise through the filter of quantum mechanics. The emergence of catalysis or other key functional processes in natural order being facilitated by some of the "weird" properties of quantum systems continue to be explored from Schrodinger's initial inspiration. Four examples of such impacts have been proposed in general to showcase this transference:[12]

1. The mechanism by which light energy is harnessed in photosynthesis
2. The way birds are able to detect magnetic fields during long-range flights
3. Our sense of smell and the molecular process of transport to sensory systems
4. Enzymatic action in facilitating the speed of reactions

Empirically testing these effects remains a vibrant area of research inquiry and may ultimately help us understand the resilience of natural systems as well as new forms of adaptive order in natural systems. Although recent research and empirical observations suggest that the impact of quantum effects may be overstated in earlier studies,[13] there are still other fascinating avenues of natural order that emerge from even questioning this line of inquiry. Another intriguing area of quantum biological research pertains to understanding the intrinsic nature of consciousness itself. The mathematician Sir Roger Penrose had made the provocative conjecture that quantum phenomena at the level of neurons in an animal brain could have a role in giving the literal spark of the conscious mind. His 1989 book, *The Emperor's New Mind*, made this case but was dismissed by most scientists at the time as purely speculative. However, in recent years, nano-neurobiology has developed techniques to study the subatomic impacts of anesthesia, which could test some of the hypotheses that were presented by Penrose. Contentious as such approaches may be, the quest to understand this most consequential form of order—the emergence of conscious intelligence—is essential for charting our path toward sustainability and will be further explored in Chapter 6.

One of the fundamental principles in quantum order is that of *coherence*, which suggests that harmonizing the discrete nature of quantum theory in terms that preserve the continuity of thermodynamic systems requires us to consider particle behavior in wave functions. Particles can thus exist as waves with probabilities of physical occurrence in a particular spot once the system is exposed to the environment or an observation event. That exposure can cause a "de-coherence" and a shift from the quantum realm of probability to the thermodynamic realm of our day-to-day world. There are plenty of detailed explanations in Royal Institution lectures and other online tutorials on these matters as well as numerous popular books on the quantum world (coherence is also functionally equivalent to the concept of *entanglement*) that can further elucidate these concepts. For our purposes, what matters most is to understand the implications of this coherence on natural order that stems from its unique properties. One of the key aspects of entanglement is that it suggests the existence of particles that may be physically distant from each other but could still instantly be linked in terms of their behavior through the fabric of space-time. Such a phenomenon can occur instantaneously between distant objects, where information is transmitted seemingly faster than the speed of light. This has also been empirically shown to be possible in the Nobel Prize-winning work of David Wineland (b. 1944) and opening up new frontiers for possible natural order to be realized. Note that this process does not involve physical movement and hence does not negate the upper speed limit of light posited by Einstein. What Einstein said was not that we cannot travel faster than the speed of light but rather that we cannot accelerate from below the speed to above the speed limit. Also, light does not always travel at its maximum speed (in a vacuum) for it can be slowed by passage through matter; thus, other particles such as neutrinos in such contexts could travel faster than light in certain situations. In the context of space-time and quantum dynamics, it is possible to conceive of existential speeds that are inherently above the speed of light and do not need to accelerate past the discontinuity of light speed itself. Furthermore, the bending of space-time through investments of vast amounts of energy also suggest that a "warp drive" may become physically plausible. Two papers which consider the prospect of wave packets or bubbles called "solitons" have recently confirmed the physical possibility of such a warp drive, much to the joy of science fiction enthusiasts.[14]

Quantum coherence is essential for the operation of the emergent field of *quantum computing*. Unlike conventional computing, which relies on the storage of information in binary "bits" that form a binary code of on and off switches in an integrated circuit, quantum computers harness the power of probability to store information in "qubits" that can process information much faster because they can capitalize on the ability of quantum-mechanical systems to possess multiple "superpositions" (due to a probability-based framing of their existence) rather than the binary limitations of a transistor. The main challenge of harnessing this coherence is that functional use of quantum properties requires specific materials called *superconductors*—materials which are able to pass electric current (streaming electrons) without any resistance and subsequent loss of heat. Superconductors exhibit these properties at very low temperature, as often shown in contemporary science shows by cooling a superconducting disc in liquid nitrogen and levitating it over a magnet. Quantum computing is now a reality as well thanks to these materials but so far is also only possible at very low temperatures. This may change in coming years if we are able to use photons of light instead of ions or atoms within crystals to store information at room temperature.[15] Regardless of the current low-temperature limitation, in October 2019, Google declared "quantum supremacy" (or quantum advantage), indicating that it had successfully performed a complex mathematical calculation in 200 seconds using a quantum computer. The same calculation would otherwise have taken a conventional computer at least 2.5 days (according to IBM) and as much as 10,000 years (according to Google) to perform.[16]

The quantum order being exhibited by specific substances, such as floating discs on a magnet or quantum computers, has implications for many of the yearnings we have for energy transmission, storage, and broader sustainability concerns as well. For example, the World Economic Forum noted, in 2018, that quantum computing could assist in combatting climate change through the simulation of large complex molecules in terms of their viability. These simulations could in turn uncover new catalysts for carbon capture or superconductive materials that could store battery charge without degrading. In May 2020, Microsoft announced that it was launching a new "planetary computer for a sustainable future" initiative employing artificial intelligence technologies that could likely benefit from advances in quantum computing as well.[17] The manifestation of quantum order from the subatomic level to large-scale phenomena is an intriguing field of further inquiry. The Center for Quantum Coherent Science at the University of California, Berkeley was established, in 2016, to explore such transferability of quantum phenomena to macro-level applications. Numerous other centers are emerging worldwide to further such clearer applications of quantum order in contemporary sustainability challenges. Maglev trains and their evacuated tubular corollary—such as the Hyperloop being championed by Elon Musk—are important prospects in this vein.

Much of quantum order resonates in terms like "super" and "hyper"—a parable of extreme existence. The most potent measure within human reality of such extreme states is temperature. Our bodies are ordered around a very specific temperature, and we literally "feel" temperature most acutely as a metric of change. Yet there is a visceral quantum feature of temperature which can lead to extraordinary material properties and the absorption or release of energy. Heat is the most palpable form of energy not just for humans but also figuratively for the universe at large. What is the relationship then between the quantum effects dominating at very low temperatures versus very high temperatures, and how does this inform natural order? Temperature is the most palpable mechanism by which we measure particle energy levels, and, as the temperature decreases, we reduce particle movement. There is a point at which the particles all revert back to the lowest energy levels of their quantum "ground state," and particle motion essentially ceases. Absolute zero (–273.15° Celsius) is that temperature at which these kinetics cease: one can't go any lower because there is no reduction of motion beyond the stationary state. On a side note, it is worth noting that while one can't dial down the temperature to absolute zero or beyond, there is such a phenomenon as "negative temperature" on the other side of absolute zero. This phenomenon is based on a definition of temperature as the tradeoff between energy and entropy in a system. Under normal thermodynamic conditions, as you inject energy into a system, its entropy increases. However, under specific circumstances, where particles may be in elevated states of energy in a system, injecting more energy may decrease the entropy of a system. This would be an example of "negative temperature"; however, this does not mean that the body at negative temperature would be physically colder to human perception. Rather the body would be in a quantum state similar to a laser, wherein stimulated excitation of particles when injected with more energy leads to more ordered forms of light being generated.

In the next section of this chapter, we will further explore the full range of the temperature spectrum on the various states of matter. Extreme low temperature thus confers a fundamental state of order which allows for the specific properties of superconductivity and superfluidity to be realized in which resistance to electric charge movement and viscosity tend toward zero. The quantum dominance that comes forth at very low temperatures also has potential implications for some of the continuing mysteries that persist for science in understanding the order of the universe. For example, *dark matter* has been theorized to potentially exist as a superfluid by some physicists, as a way of explaining its elusive existence.[18] Quantum physicists intriguingly use the term "cooperative" to describe the behavior of particles in this state. As we reduce temperature, there is lessened agitation and a greater degree of "cooperation" between particles and hence a "topological order" that seems to take root. The temperature dependence of quantum effects has to do with a very fundamental aspect of particle motion and state that relates to energy levels.

Even though particles have wave-like properties, quantum mechanics is predicated on the principle that energy has discrete levels, or "quanta," that can be occupied by elementary particles. These particles fall into two broad categories of order: (1) mass-bearing particles (fermions, which include quarks and leptons) and (2) force-bearing particles (bosons, which include W-bosons, Z-bosons, Higgs-bosons, gluons, and photons). The term "hadron" refers to the functional particles that arise from the interaction of mass and force particles (essentially protons and neutrons), which is what we end up colliding in particle physics experiments at the Large Hadron Collider outside Geneva that has gained much popularity in the public's imagination.

Three of the four fundamental forces of nature (or more precisely "interactions" in physics) can be described by the mathematical intricacies of what is called the "standard model" of particle physics that currently comprises 25 particles.

1. *Electromagnetic interaction*, which acts between all charged particles and is responsible for a vast range of phenomena eponymously including electricity, light, and magnetism. Originally it was believed that light, electricity, and magnetism were separate forces, but the experimental work of James Maxwell in the mid-nineteenth century proved that they were all essentially manifestations of the same force. This was later linked to statistical mechanics and a geometric approach to thermodynamics by J. Willard Gibbs.
2. *Weak nuclear interaction* is related in particle lineage to the electromagnetism in so far that it is the result of a bifurcation of the "electroweak" force caused by the now empirically observable Higgs boson. Due to this process of its emergence, the weak nuclear force is the only one of the four forces which is asymmetric in physical terms. The force is responsible for radioactive decay and plays the pivotal role in nuclear fission energy release as well as in starting nuclear fusion reactions that power stars including our sun. Yet it is thankfully "weak," otherwise the sun would have burned up its nuclear fuel before our evolution!
3. *Strong nuclear interaction*, which holds particles in the nucleus of an atom together, is the most potent of the fundamental forces but operates only at small subatomic levels. It is able to overcome the repulsion of the electromagnetic force to keep close together like-charged particles like protons. The strong interaction is mediated through a set of three particle states called "colors" (of

no relevance to visual color) and hence gives rise to an area of research called *quantum chromodynamics*.

4. *Gravity* is the fourth fundamental "force" or interaction; it is one that we feel directly, yet it continues to elude this model. Instead, gravity is best explained through Einstein's theories of relativity as the result of a curvature in space-time. A key prediction of Einstein's theory on gravity has recently been validated by the 2017 empirical observation of "gravity waves." Interestingly enough, Einstein's formulation of a cogent theory of gravity helped dispel an errant empirical observation in astronomy. The wobbly rotation of the planet Mercury was believed to be caused by a hidden planet, Vulcan, nearer the sun whose gravity was believed to be causing the phenomenon. Einstein's theory of relativity explained this wobble through the curving of space-time due to the sun's massive gravity.[19] However, neither Einstein's elegant and empirically sound exposition of gravity nor the equally celebrated Higgs boson help to bring gravity mathematically within the ambit of the standard model. Perhaps gravity, being an inherent property of space-time, is intrinsically divorced from the standard model? MIT physicist Frank Wilczek has proposed that we dispense with the use of the term "standard model" and simply present the current disjointed setup of physics as a "core theory."[20]

One of the great popularizers of particle physics, Richard Feynman (1918–1988), gave us a series of elegant diagrams (which bear his name) that show how to connect forces to particles; these are among the most useful ways of functionally describing the standard model. If we were to go with a sense of visceral order through these diagrams, a hypothetical particle called a "graviton" might be a suitable proposition to explain "quantum gravity." Yet no such particle has been detected so far. Figure 1.2 is an accessible and accurate summary of order in physical and chemical systems adapted from a diagram prepared by science educator and nuclear physicist Jim Al-Khalili.[21] Of particular interest in this diagram is the simple "Current Paradigm" of cosmology that presents the term ΛCDM. The Greek letter lambda (Λ) refers to the cosmological constant that is most often explained as "dark energy" and comprises 68.3% of the universe. Cold dark matter (CDM) makes up around 26.8%; physical matter, which comprises all the observable multitudes of galactic conventional mass objects, including ourselves, accounts for only 4.9%. The awarding of the 2019 Nobel Prize in Physics to Jeffrey Peebles suggests a clear measure of confidence in the ΛCDM model which he has helped to harmonize despite the lingering loose end of quantum gravity.[22] The two major contenders for unification are *string theory* and *quantum loop gravity theory*, both of which require us to make fundamental assumptions about the structure of particles or the universal scaffolding in which they operate and remain unproved.

Whether or not unification of physical theories is a wishful monist quest for order remains to be seen. The mysteries of the universe in terms of order emergence have yet to be resolved. There are many more massive experimental apparatuses still

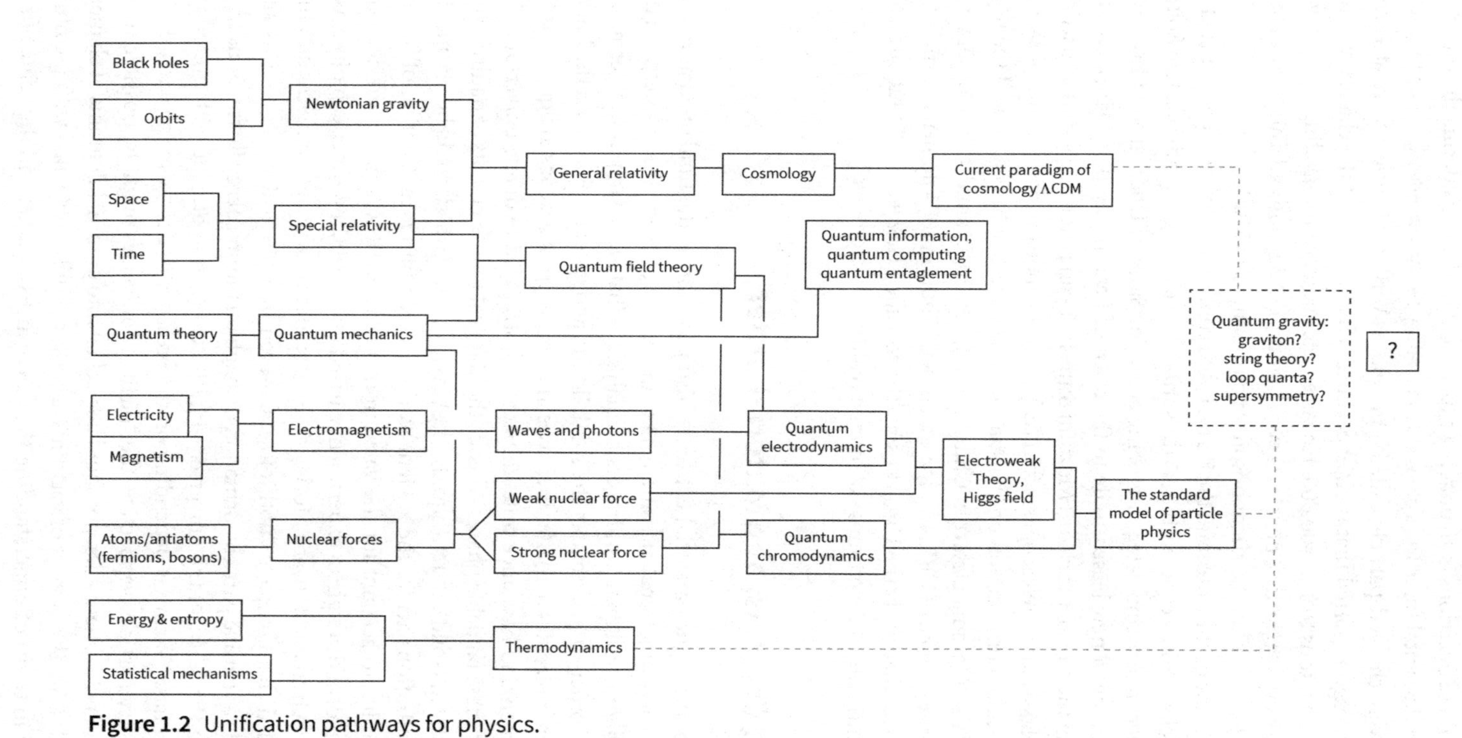

Figure 1.2 Unification pathways for physics.
Diagram adapted from J. Al-Khalili, 2020. *The World According to Physics*, with guidance from A. Qadir and Q. Shafi.

being built to help us grasp this order. Among the most promising signs of some new answers to these questions is an ongoing experiment that is considering the prospect of a fifth fundamental force to explain a discrepancy in the observed dipole moment of a negatively charged particle called a *muon*. In April 2021, researchers at Fermi lab near Chicago reported that their experiments had narrowed the chance of error in their observations to 1 in 400,000. However, the standard for labeling an observation a "discovery" in particle physics is even higher: 1 in 3.5 million! This bar was met when the "discovery" of the Higgs boson was announced in 2013.

From the cow pastures around the Fermi Lab, a conduit is also being constructed to beam particles 4,800 feet below the earth's surface to a closed gold mine shaft in South Dakota where a neutrino detector is being built. The Sanford Underground Research Facility is giving a new lease on life to the town of Lead in the Black Hills. On the other side of the world, the closed Mozumi mine in Japan is housing a similar particle detector. Massive new telescopes, often defying indigenous peoples' views of sacred order, are under construction on the summit of Mauna Kea in Hawaii and the high plateaus of the Atacama Desert in Chile. On June 19, 2020, amidst the Coronavirus pandemic, the council of CERN endorsed a new $25 billion 100 kilometer particle accelerator to further study the Higgs boson (the so-called *God particle*). Our quest for understanding the underpinnings of human reality is far from over.

Phases, Crystals, and Material Order

Such explorations into material existence also posit for us the fundamental role of temperature as a functional measurement of material phases and the link between the thermodynamic and quantum frames of analysis. What we see in physical terms are "phases" of order—most notably the solid, liquid, and gaseous phases. As the bonding between molecules of matter increases, they move from gaseous to liquid to solid states. Humankind has, since time immemorial, struggled to define universal order in terms of these palpable phases—hence the classical four "elements of nature" were defined as earth (solid), water (liquid), sky (gas), and fire (which could be plasma) by human civilizations across the ancient world. Later, a fifth errant notion of "ether" was also added to account for the vacuum of space, but it was proved to not exist in material form through an experiment carried out in 1887 by American physicists Albert Michelson and Edward Morley. A few decades later, quantum mechanics further revealed that complete "nothingness"—even in a physical vacuum—was also not possible as elusive particle formation through quantum energy fluctuations were inherent to the fabric of the universe.[23] Physicists began to grapple with notions of "true" and "false" vacuums depending on the presence of energy in the latter. From an elementary particle perspective, the "birth" of universes might be possible theoretically if massive energy density could be packed into even 10 kilograms of elementary particles. Yet the physics of vacuums are still not fully reconciled with quantum mechanics in terms of understanding the emergence of matter and the value of the cosmological constant (which is measured in units of energy per unit volume or energy density as well).

For practical purposes, solid matter is an ostensible mark of stability for humans, but its strength can be deceptive if this matter is brittle or fragile, as is the case with glass or certain forms of metal. Glass, in physical terms, also connotes a particularly disordered state of molecular arrangements in a solid which we will also revisit when discussing magnetic disorder in the next chapter. It is worth noting that bulletproof "glass" is a composite which often combines layers of plastic organic compounds alongside glass layers and is in some cases entirely polycarbonate plastics (such as Lexan). Solid order in brittle objects is deceptive because when energy is applied to them, their structure is unable to absorb and dissipate it effectively. Introducing molecules within the structural bonds of solids, which can more effectively absorb energy and dissipate it throughout the structure, is a key attribute which can lead to orderly adaptation through change in physical form—like baking dough. Gases and liquids have this property, and that is why they can remain ordered at a molecular level while taking the shape of their vessels; we often refer to this as *fluidity*. Solids can also have fluid properties through appropriate molecular interventions.

We might form alloys or use other mechanical techniques to allow for the material to become malleable or ductile like bronze (an alloy of tin and copper) or even aluminum foil, owing to the way we might extrude the metal. Among natural materials, rubber has had particularly special properties in this regard, and here, too, with simple molecular changes in its structure we can get products which are highly adaptive and resilient to their physical environment. The most revolutionary of synthetic material interventions in recent history has of course occurred not with simple elemental structures but with complex organic molecules, with the advent of plastics. The highly adaptive order which plastics have given us range from materials that can be used for sandwich wraps to bulletproof vests. Yet it is the relative durability of that order which also makes them such a scourge in terms of pollution concerns on land and in the oceans. It is also interesting to note that the advent of plastics recycling coopted a partial illusion of order for the consumer when the plastics industry created categories and labeled different classes (1–7) of plastic compounds to supposedly ease the recycling process. Through a carefully crafted marketing campaign, the plastic industry actually alerted the public to plastics pollution in the late 1970s and 1980s and effectively shifted the onus of conservation to consumers, urging them to follow this order of recycling and pushing local waste collectors to accept all forms of plastics. The consumer believed that the problem of plastics waste was taken care of, but the actual recycling being undertaken was largely only for labels 1, 2, and 5, and the rest was mostly being landfilled. By one estimate, only 7% of plastics ever made has actually been recycled due to continuing technical challenges.

Apart from the well-understood three phases from solids to liquids to gases, there are also more extreme phases of matter. Under very high temperatures gas molecules liberate electrons and become dispersed with the surrounding molecules to form the *plasma phase*. This is the state of the sun's interior and is also observable in neon signs, bursts of lightning in the sky, or in the ornamental plasma globes that adorn many night clubs or museum shops. On the other end of the material spectrum, at very low temperatures, we have unusual states of matter that emerge as particles begin to revert to more wave-like forms, and this constitutes yet another form of matter known as "Bose-Einstein condensates." These were named in honor of a historic correspondence between the Indian scientist Satyendra Bose and Albert Einstein in the 1930s, which led to their discovery. Bose's contributions to particle physics were also recognized by Paul Dirac, who named the fundamental class of force particles "bosons" in Bose's honor. Indeed, at very low temperatures, it is the dominance of bosons in a cooperative state which gives us insights into underlying quantum-mechanical properties. The physical laws of existence are, however, transferable between very low and very high temperatures. Many of the experiments carried out on these extremely cold condensates can also give us insights into what happens in the extremely hot interior of neutron stars, which could have a temperature of as much as 3 billion degrees Celsius/Kelvin.

There are quantum-mechanical limits to how hot we might get as well based on the inextricable linkage between any temperature of a body and electromagnetic radiation that is always emitted therein. Longer wavelengths of radiation, such as radio waves, come from colder objects; humans emit infrared radiation at our stable temperatures: this is what allows those nifty thermal scanners to check our temperatures now at airports in the post-COVID world. The sun's core emits gamma radiation at much shorter wavelengths. As we get hotter and hotter, there is a limit at which the wavelength of radiation emitted gets so small that it can no longer be explained by our current equations of thermodynamic or quantum order. This lesser-known upper limit is the *Planck temperature*—a mind-blowing 1.4×10^{32} degrees Kelvin. Such a temperature has also been achieved for a fleeting moment at the Large Hadron Collider. Beyond this temperature, there could be a sudden material collapse into a special kind of black hole called a *Kugelblitz* (the German word for "ball lightning"), which is theoretically plausible within the general theory of relativity but still purely speculative. A related limit attributed to Planck and linked to a similar black hole eventuality is the minimal length which could be possible in physics—this could be the minimal wavelength in the aforementioned example, but it is a value that has not been physically measured. Through extrapolation with measured values, this is estimated to be around 10^{-35} meters and is known as the *Planck length*.

Black holes do provide us with ways of exploring material order in various dimensions of human inquiry. These mysterious entities are the ultimate concentrations of mass and gravity in the universe. They constitute an example of a physical "singularity," which in essence means that the usual "laws" of physics do not apply in the way Einstein conceived in his general theory of relativity. Mathematical singularities are points at which mathematical functions are undefined. This could be the tip of a water drop or the discontinuity in longitude measurements at the pole, or, as Stephen Hawking asked, "What lies north of the north pole?" The Big Bang is also considered a singularity although observably different from a black hole. The discovery by astronomers of the physical existence of these massive gravitational forms, coupled with the mathematical elegance of their explanatory power, has led to a vast genre of orderly, but speculative, research about energy, time, and the pliability of the universal constants of nature. The Canadian physicist Lee Smolin has, for example, used black holes to suggest the possibility of cosmological evolution that is analogous to biological evolution. Black holes could thus be portals from which other universes could evolve with a different range of universal constants. The aforementioned ubiquity of "cold" dark matter once elucidated might well further explain whether any such notions are plausible. Dark matter is usually termed "cold" (as in the aforementioned ΛCDM model of the universe) because of the linkage of particle motion to our conception of temperature. For structures to evolve in the universe, slower moving dark matter particles—hence, cold—are likely to be more conducive to order formation than "hot" dark matter.

Heat energy creates a kind of data noise in evaluating subatomic systems, and the ability to cool matter to low temperatures helps to create a sort of analytical order through which insights about fundamental laws of nature can be gleaned. Much as cold temperatures are revelatory and we have seen their extraordinary value in allowing for superconductivity and superfluidity, absolute zero temperature cannot be attained, although we continue to get closer and closer to that goal. This is analogous to having a zero in the denominator of a fraction (e.g., the simple equation $y = 1/x$), which leads to asymptotic functions from basic middle school math, where the line drawn on a graph from this equation never touches either axis. The laws of quantum mechanics have at their core the notion that we are unable to predict the location and energy state of a particle at the same time with certainty (the *Heisenberg Uncertainty Principle*). It is this principle which allows for the paradoxical properties of the quantum realm, such as wave-particle duality, superposition, and entanglement, to be mathematically reconciled. Absolute zero would suggest no movement at all of a particle, which would make its location and energy inherently predictable and would thus violate the Uncertainty Principle. As we will see later in our discussion of economic order, the concept of "quantum cooperation" can also have implications in *game theory*, which is fundamental to many aspects of understanding economic systems and their implications for social systems. Games have value in entertaining us because uncertainty about the behavior of other players creates opportunities for testing our ability and our "luck." These uncertainties can be different in the quantum realm compared with the classical physical realm that functionally defines our daily lives. In social order, uncertainty also plays an essential role in how game theoretic models are developed and understood.

As we reduce the temperature, there is also a greater propensity for another macro-level order to manifest in the form of crystals. These beautiful structures have captivated humanity precisely because they reflect anorderly form of elemental expression. Most gemstones are crystals, and they are the favored form of supernatural expression and healing materials for those with such proclivities. Yet, regardless of any such unproved powers, crystals charm us because of their intrinsic symmetric form. They are also a window into molecular structure, wherein we can see an upscaling of atomic arrangements to the visual realm. Key to understanding this upscaling in the role of phase transitions is to consider two concepts which allow for such order to arise: *criticality* and *nucleation*. Order emergence in physical matter follows certain key "tipping points" in terms of thermodynamic variables of pressure and temperature. In the quantum realm, these variables may also be expressed in similar terms, but the dominant variable is electromagnetic wave frequency rather than temperature (although, as we noted earlier, any substance above absolute zero emits some level of electromagnetic radiation). Figure 1.3A B shows the way these phase transitions can be represented in terms of both quantum and thermodynamic (classical) phases.

The *quantum critical point* (QCP) is at absolute zero. The ordered state in the quantum phase diagram gives us entities like superconductors and superfluids. The shaded area in Figure 1.3A shows where the classical phase transitions (unpacked in Figure 1.3B) would lie. Criticality at the phase transition is linked to positive feedback loops, which essentially means that a small change or perturbation in a system signals

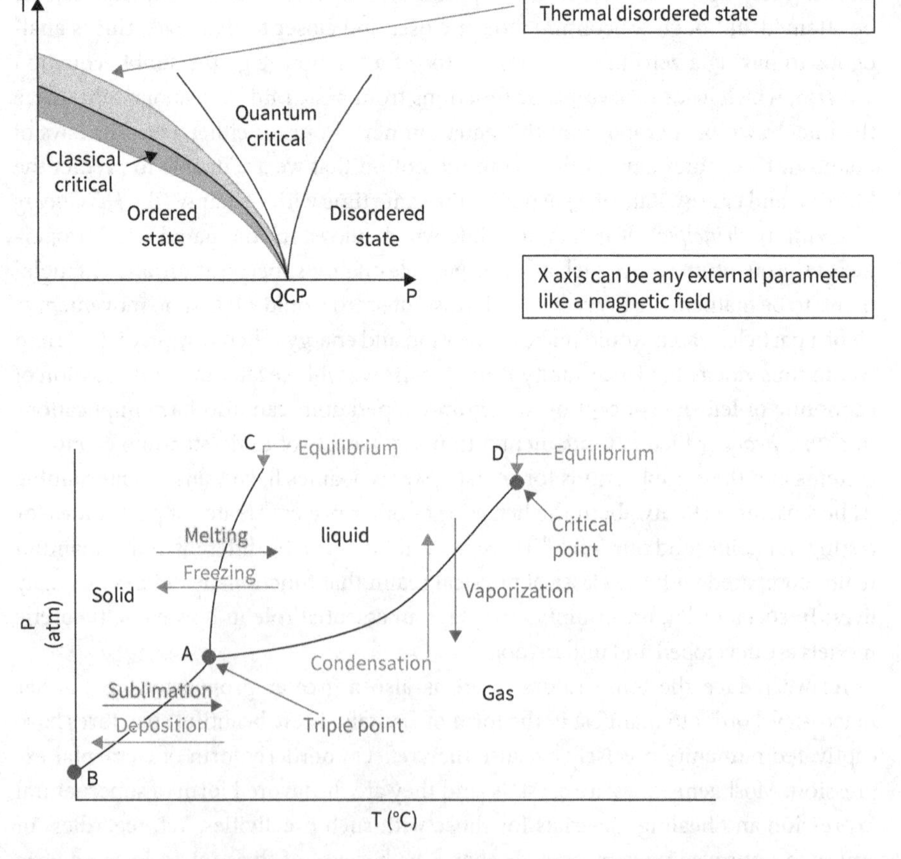

Figure 1.3 Phases of order in quantum systems (a) and thermodynamics (b).

a cascade of reinforcing signals or stimuli that lead to qualitative physical change in that system. In a linear (straight-line) process each change in one unit of a stimulus leads to a concomitant unit change in the output. However, in a nonlinear system there is either greater or less per unit change incrementally (leading to curved lines). Because of the positive feedback process this change is nonlinear (often exponential) and by its very nature unstable. On the other hand, negative feedback loops in a linearly growing system can lead to a "steady state" or dynamic equilibrium that is also represented in phase diagrams. *Chemical equilibrium*, which is different from dynamic equilibrium, refers to a quiescence between reactants whereby there is no net change occurring without any external intervention. Energy transfer is key to how we understand equilibria processes and how that energy is then translated into the ordered structuring of the output, thus determining the shape and condition of the end product.

Ultimately, materials will attempt to adapt their physical shape to reach equilibrium with the environment in the most energy-efficient way possible. This observation is linked to a fundamental principle of the universe which permeates forms of order discussed in this book: the *principle of least action*. This principle essentially suggests that the path of least resistance is taken by natural phenomena and leads to the structures we observe, or as Jennifer Coopersmith has stated in her book on the topic, we have a "lazy universe."[24] This purported laziness may lead to the creation of feedback loops as systems equilibrate with their surroundings. The same sort of feedback loops and properties of rapid phase change phenomena can also be observed in economic, social, and political systems, as we will explore in subsequent parts of this book. Shapes arise through interactions between various forces acting on the material. With large-mass bodies, gravity dominates and stability tends toward spherical shapes. Hence planets and stars tend to be spherical. Protrusions or anomalies on celestial spheres may appear due to other forces acting on the material (such as plate tectonics, which may form mountains). In another context, hexagonal structures are able to optimize space in mosaic patterns; this explains why beehive cells form in such shapes as bees try to minimize wax usage (and hence reduce the energy investment) in forming their homes.

Natural systems have a mechanism by which the process of critical phase change can be instigated by a "nucleating" agent. A single particle can facilitate nucleation, which then sets into motion a chain reaction when the requisite energy is available. In the context of crystal formation, nucleating agents can be physically used to stimulate crystal growth at particular speeds and sizes. Snowflakes and gemstones emerge from liquid forms through such a process, and they exhibit a signature order that reflects their core molecular structures. Even though there is a range of weather-related variables that give individual snowflake crystals the unique shapes in which they solidify, they all have an underlying six-fold hexagonal symmetry. This fundamental order of a snowflake comes from the angles at which hydrogen atoms are connected to an oxygen atom in water. The largest crystals form under very specific conditions of temperature change and perturbations. It is also possible for no crystals to form during cooling if the nucleation process cannot be facilitated at all due to physical properties of the cooling substance under certain environmental conditions, as is the case with cooling lava on the earth's surface. However, the same lava may be able to crystallize under different conditions of pressure and cooling—indeed, this is what gives rise to diamond formation in the mineral kimberlite.

There are three key principles that mineralogists have set forth with regard to crystalline order.

1. All pure substances can form crystals as long as there is enough time for the atoms and molecules to move into an orderly arrangement.
2. All crystals are periodic arrangements of atoms composed of a repeating cluster of atoms with equal spacing.
3. Every periodic atomic arrangement can be categorized according to its symmetries, and there is a finite number of symmetries.

Symmetry again figures prominently in natural order and in the context of crystals. The most important form of symmetry is *discrete rotational symmetry*, meaning that rotating the object in discrete units gives you the same shape. The packing of molecules into a structure that can be easily repeated is required for the formation of a crystal lattice. As shown in Figure 1.4, there are essentially four kinds of rotational symmetries that allow for complete "packing" of a lattice and thus less energy wasted in formation. These are two-fold (parallelograms), three-fold (triangles), four-fold (squares), and six-fold (hexagons). These are the most observed forms of natural order in crystals for this very reason. Most minerals tend to occur in some permutation of these four forms of rotational symmetry. Many natural engineering feats, like honeycombs, have hexagonal symmetry, as do the massive basalt columns at Giant's Causeway in Northern Ireland.

The pentagon configuration in Figure 1.4 poses a challenge to fit into a lattice because there is empty space left between patches if we attempt to tile a surface with identical shapes. However, if we fold the pentagon faces into a three-dimensional object, it is possible to get a completely closed orb with identical faces. The ancient Greeks were fascinated by the resulting dodecahedron that emerges from such a folding. The five "Platonic solids" which were set forth by the Greeks included this unusual structure because it still held two key properties in common with the other solids shown in Figure 1.4: (1) these are the only existing solids in which all the faces of a given solid are identical and equilateral, and (2) each of the solid's vertices exactly touches a surrounding sphere.

The Greeks associated each of these solids with the essential elements: earth (cube), fire (tetrahedron), air (octahedron), and water (icosahedron). As for the enigmatic dodecahedron, this was associated with the elusive "ether" or with the universe as a whole. The perceptive surrealist painter Salvador Dali noticed this geometric history and emblazoned his painting, *The Sacrament of the Last Summer*, with a floating dodecahedron above Jesus and the apostles. Though the ether itself was a figment of

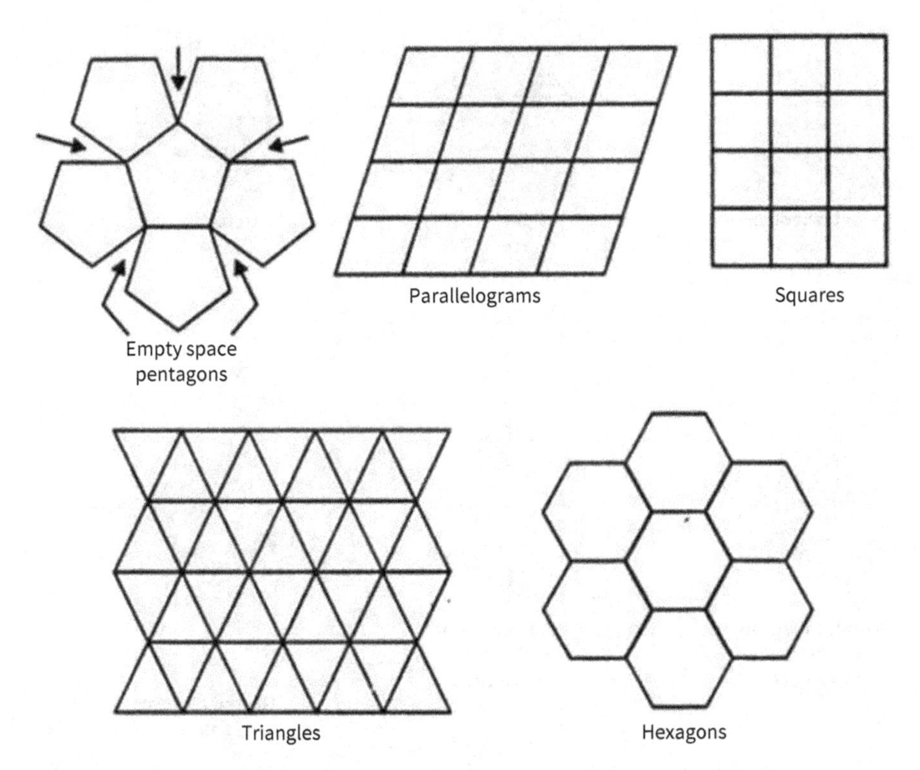

Figure 1.4 Four kinds of rotational symmetries and the pentagon challenge.

fanciful imagination rather than observed reality, the unusual geometric properties of this structure led to some other notable realizations about natural order. The two-dimensional replication of pentagons could not give us a closed-form tiling, and yet it yielded a three-dimensional structure with partial symmetric properties. This insight heralded the way for scientists to consider underlying variations and constants in natural order. Shattering a crystal or breaking the form of any solid can also reveal its underlying propensity for order. The notion of *isotropy* in material science connotes a uniformity of physical properties that allows uniform forces. Crystals are often isotropic when struck in this regard, but wood is *anisotropic*, with a divergence of force impacts depending on whether the blow is along or against the grain. Deep molecular arrangements manifest such differences, and understanding that both propensities can be found in nature is essential as we consider the vulnerability of materials.

Constancy and Hybrid Natural Order

Could a crystal be formed in an ordered lattice with five-fold symmetry but also partial translational symmetry, and if we moved the structure in physical space without rotation, would it not be identical to its starting point? This would have been an

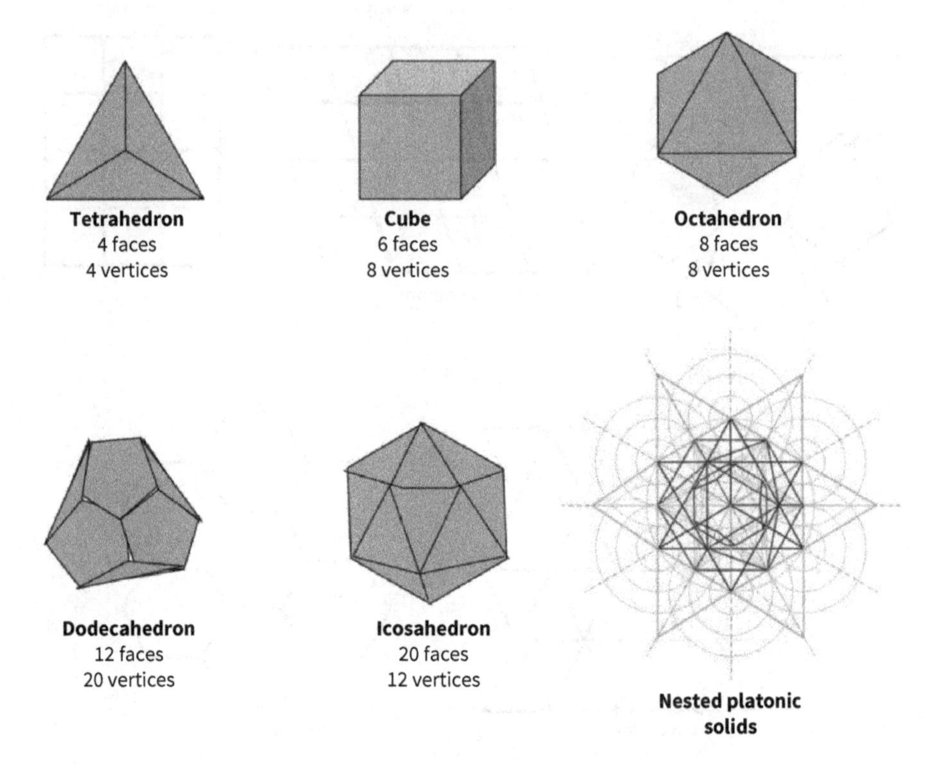

Figure 1.5 The five platonic solids and their nested form presented by Kepler in 1596.

intriguing question to answer but it never directly occurred to scientists because there was a fundamental fixation with the "law" of crystal formation that had "forbidden" five-fold symmetry. In 1981, Israeli physicist Dan Schectman was on sabbatical at Johns Hopkins University and the National Bureau of Standards studying rapidly transitioning aluminum alloys. Per chance, he found that one of his samples upon analysis was showing the "forbidden" five-fold symmetry. He was astounded and hesitant to believe this observation, especially after vile rebuke from his colleagues and friends. Eventually he published his findings, buoyed partially by a mathematical tiling sequence that had been proposed by mathematician Roger Penrose in the 1970s, in his exploration of periodicity of pentagonal forms. Artistic manifestations of such patterns can also be found in Islamic art on the ceilings of buildings such as the Alhambra Palace in Spain and the Darb-e-Imam Shrine in Isfahan, Iran, and these have been studied in detail by physicists Peter Lu and Paul Steinhardt.[25] The allure of this new form of matter would later take Steinhardt, his geoscientist son, and a minerals museum curator from Italy on a treasure hunt to find natural occurrences of these quasi-crystals as the ultimate proof. From a storage room in a Florentine museum, the team traced a specimen to the wilds of Chukotka in far eastern Russia and showed definitely that quasicrystals could be found in nature—perhaps in the form of meteor fragments hitting the earth.[26] The team also considered how the Japanese paper-folding art of Origami often reflects the human quest for unlikely patterns. The process of folding itself is a key feature of natural order whereby the most efficient usage of space and energy can be set forth. The crinkles in leaves or the complex folding of protein molecules are essential ways by which natural systems negotiate a response to forces acting on materials.

Despite having received validation of his revolutionary discovery of essentially a new form of matter, Dan Schectman had a formidable intellectual foe: two-time Nobel Laureate Linus Pauling who, until his death in 1994, was not convinced of their existence. The correspondence between the two scientists shows respectful persistence on the part of Schectman and polite intransigence on the part of Pauling. The discovery of quasi-crystals is an instructive and cautionary tale about how we may errantly perceive certain attributes of natural order as immutable. Even in current "advanced" times of atheistic scientific inquiry, there are plenty of notable figures within the establishment of science who can prevent progress due to a stubborn adherence to established views of natural order. Hybridity and texture constantly challenge dogmatic absolutism within natural order, as the discovery of quasi-crystals shows us. Pauling, despite his brilliance, was unwilling to consider a disruption of what he considered established natural order even when presented with convincing evidence.

Until the discovery of quasi-crystals, the definition of a crystal was simple and beautiful but clearly inaccurate and could be phrased in various forms as a solid composed of atoms arranged in a pattern, periodic in three dimensions. In 1992, a decade after their discovery, the International Union of Crystallography changed the definition to the following: "by *crystal* we mean any solid having an essentially discrete diffraction diagram and *aperiodic crystal* we mean any crystal in which three-dimensional lattice periodicity can be considered to be absent."[27] Note how the definition became longer, more nuanced, and, as Dan Schectman noted in his Nobel Prize lecture in 2011, more humble. He went on to say: "The definition is open; it does not say a crystal 'is' it says by crystal 'we mean.... [A] humble scientist is a good scientist." Simple but profound words for all to heed as we are tempted by the hubris of discovering hidden order. The question of dimensional order beyond our palpable three remains an enchanting aspect of physical mathematics. Vector analysis allows you to do calculations in any number of dimensions, not just in the three dimensions where we can observe coordinates. Einstein believed that, due to the flexibility of vector analysis, geometry was most likely to deliver us a unified theory of the universe. Extra geometric dimensions of order have been suggested to accommodate various fundamental forces, time, or even a curled-up version of physical space and higher harmonic forms (like music) as suggested by the early twentieth-century physicist Oskar Klein. Yet, with unusual humility, he stated and we should concur: "whether these indications of possibilities are built on reality has of course to be decided by the future."[28] An important cautionary reminder not to be beguiled once again by what we hope would be order unless it is appropriately observed.

Underlying the remarkable geometric anomaly of quasi-crystals was another important insight on natural order that had its root in the wisdom of Euclid of Alexandria (circa 300 BC). The pentagonal structure of quasi-crystals enshrines a mathematical ratio which is manifest numerous times in natural systems and is referred to as the "Golden Ratio." The ratio is derived from a simple sequence built by adding preceding numbers, a process laid down by the twelfth-century mathematician Leonardo Bonacci of Pisa (later known as Fibonacci). Quite simply, the ratio of the size of two entities (which may even be two fragments of a line segment) is said to be "golden" if it is the same as the ratio of the size of the longer entity to the size of the sum of both entities together. This ratio is what we call an "irrational" number, which means it cannot be expressed as a simple fraction and goes on without a recurring pattern. Another example of such an irrational number would be the square-root of 2 ($\sqrt{2}$). The Golden Ratio is commonly designated by the Greek letter Φ (phi) or in technical mathematics as τ (tau); it equals approximately 1.618033.

In the case of pentagonal forms, the Golden Ratio emerges as the length of the diagonals if we embed a pentagram within the structure shown in Figure 1.5A. The ratio continues to be repeated as we go down into smaller and smaller embedded pentagrams. The same ratio emerges in various aspects of the geometric growth of nautilus shells, spiraling plants, the shape of galaxies, and even in musical fugues in particular harmonic forms. No doubt this order has also led to superstitious attributions of the pentagram and to a range of other geometric and mathematical sequences. The astrophysicist Mario Livio has written a beautiful book on the topic called *The Golden Ratio* with a front cover endorsement from Dan Brown, author of the *Da Vinci Code*. Yet there may be cosmic limits to the scale of such ordering of universal systems. Most astronomers are convinced that beyond 300 million light years distance scale the universe loses its lumpiness of structures (such as galactic clusters) and becomes "isotropic" and smooth. Such a limit is termed "The End of Greatness" by cosmologists. The observable universe in terms of light reaching us from the cosmic microwave background remnants of the Big Bang—while accounting for its expansion since its origin—is believed to be around 91 billion light years. However, recent discoveries have suggested that even at distances as large as 3 billion light years, there are massive galactic structures. In June 2021, the American Astronomical Society was graced with a presentation about such a massive structure in the constellation Boote. "The Giant Arc"—if real—may well reflect the same structures and embodiment of structural features which we are considering at the micro-scale.

Figure 1.6B shows another ratio which emerges from octagonal space, one that is less widely known. The realization that there are such ratios that show up in nature is at once awe-inspiring and puzzling. We also have a wide range of "constants" that adorn the mathematical fabric of the universe and that also provide insights about order. Some constants that are expressed in units are often misleading in our analysis, such as the speed of light. This is because any constant with units is linked to our own measurement mechanisms. For example, the International Union of Pure and Applied Chemistry (IUPAC) or the International Standards Organization

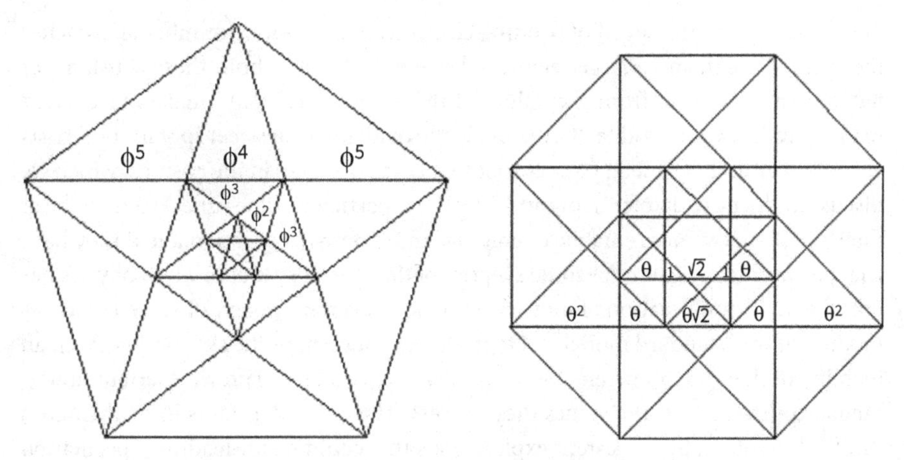

Figure 1.6 (A) Diagonals of a star pentagon intersect in the golden ratio (Φ:1); (B) diagonals of a star octagon intersect in the sacred cut ratio (θ:1).

uses a range of physical attributes to define exact measurements. Functionally, we say that the speed of light is a constant in a vacuum, but its numerical units make it a somewhat illusory constant for mathematical purposes. Einstein was prompted to consider the question of constants of nature[29] in correspondence he had with a German-Australian physicist named Ilse Rosenthal-Schneider. In this correspondence, Einstein relayed that there were essentially four types of "constants" in physics that we can consider.

1. *Geometric constants*, such as π (ratio between circumference and diameter), the aforementioned Φ (Golden Ratio), or e (the base for the natural logarithm related to $1/x$ functional outputs)
2. *Conversion factor constants*, such as the Boltzmann constant between energy and temperature or Avogadro's number related to the number of particles in a mole of a sample
3. *Dimensionless constants*, such as the "fine-structure constant" (~1/137) which characterizes the strength of electromagnetic field between elementary charged particles
4. *Cosmological parameters* (still to be determined, such as dark energy)

The notion of "fundamental constants" is often invoked across science, and one can find lists of around 20 or so numbers which could fit the bill for being "fundamental" because they reflect a sort of fine-tuning of the universe in terms of their values. They also help us define key limits within their accepted context. For example, if we have accepted the inviolability of the speed of light in a vacuum as a cosmic limit as well as the fine structure constant (α) as the tightness of electron binding to the nucleus, we find that the maximum size of an element on the periodic table would be 137 protons.

This is so because the speed of orbiting electrons at this atomic number approaches the speed of light, and a larger elemental size would violate both these constants of nature. Any deviation from the values of these constants would make our current reality physically impossible; they provide the mathematical sweet spot for our existence, similar to the planetary sweet spot that Earth occupies in the cosmos, which we discuss in the next chapter. How and why these particular "fine tunes" came to be at their sweet spot values remains an enigma, and it remains questionable if they have changed over time. There are some skeptics of the trajectory that modern physics has taken, with its proliferation of more theoretical or even speciously observed particles to force-fit the "standard model."[30] Of particular concern for the skeptics has been an inability of theories or current empirical observations to explain why certain fundamental constants have the values they possess. The use of constants in mathematics can lead to internally consistent explanations that could be misleading. Speculation in this regard is instead explained through abstruse concepts such as "multiverses," whereby a range of universes could have been instantiated with "fine-tuned" values, but the one we live in is the way it is because it had to have evolved intelligent life to observe the order that it presents. This fundamental insight about the centrality of human perception in defining our view of universal order is humbling and referred to as the *anthropic principle*.[31]

For our purposes of understanding functional natural order, it is sufficient to recognize that at least the geometric constants also resonate some level of abstract symmetry in the universe. There may also be a "supersymmetry" whereby force and mass would have equivalence, but this would require us to identify a series of "partner particles," which have thus far not been observed. Symmetry is seductive in its appeal to humans just as much as equilibrium states are in terms of their temporal stability. Many of the insights about quasi-crystals also have transference to a separate crystalline form that is even more perplexing. A year after Schactman won his Nobel Prize in chemistry, another laureate in physics, Frank Wilczek, proposed the existence of "time crystals." The logic behind this proposal was that just as regular crystals are periodic in space, there could be a form of matter that is periodic in time. In a subsequent lecture at Uppsala University in Sweden, Wilczek noted that the human heart could in essence be considered a time crystal of sorts because it was a composite of matter that pulsed at regular intervals. Indeed, this notion had been introduced years earlier by theoretical biologist Arthur Winfree in his mysteriously titled book, *The Geometry of Biological Time*.[32] However, Wilczek provided the physical context to this new form of matter and purer time crystals, which led to a whole field of research into how oscillations in time could be the lowest ground state of energy for a material to exist. Lasers, which are themselves specially well-ordered electromagnetic waves with a known period or frequency, were subsequently also used to synthesize such crystals by two different methods in 2016.[33] Although these time crystals are still highly unstable for industrial use at present there could be potential applications in quantum computing further down the road. They also give us an example of how established physical laws, such as the law of conservation of energy, might not be universally applicable. Energy conservation only applies when we have time translation symmetry (according to Noethers's theorem, presented in the Introduction).

The advent of time crystals and the daring of physicists to explore the hybridity of material forms also opens opportunities for theories of unification to be considered alongside the elusive cosmological constant or particles such as "axions" (also proposed by Wilczek). If matter can settle into discrete lattices in time, could there be intriguing opportunities for reconsidering the possible unified order that Einstein aspired to? In his book *The Order of Time*, theoretical physicist Carlo Rovelli considers how the trajectory of time is analogous to entropy or functional disorder. Our view of both lies in the eye of the beholder as focusing on a particular scale or moment can shift our understanding of order. A disorderly room appears so because we are viewing it from the macroscopic level of the observer at the door, but it would not seem so at the micro-scale of the fibers of the bed sheets or pillows lying in a crumpled pile on the floor. Similarly, Rovelli argues that the conception of time itself is a function of human perception.[34] Entropy and time are not just linked by analogy, but also could have more direct connections as well in terms of physical reality. As Julian Barbour explains in his provocative book *The Janus Point* (2020): "for many scientists, the growth of entropy and with it disorder, is the ineluctable arrow that puts the direction in time." Barbour, questions this arrow of time and suggests we revisit the opposite of entropy: a tendency toward order, termed "negentropy" (a term coined by Edwin Schrodinger) or "syntropy" (coined by Buckminster Fuller), could potentially even lead to a better understanding of gravity's linkage to quantum mechanics. In its more "New Age" form, such a concept has been promoted as "extropy," in terms of the unique human ability to create such long-term functional order. Fuller also posited the structural notion of "tensegrity" as a portmanteau of "tensional integrity": this is an engineering principle which shows how isolated components under compression can support a system under tension. The Kulripa pedestrian bridge in my second home city of Brisbane, Australia, is an example of such a system. Biological systems have also followed such structures, including the muscles in the heart, which are in turn tested by tension over time.

Time crystals may provide further insights into how we may yet discover new aberrations in what is conceived to be natural order. Functionally speaking, time is of course very real for humanity regardless of its "existence" in some abstract reality. The importance of "free will" and the agency of our actions in determining the fates of human societies is linked to that reality of time in which you are reading this book. Physicists prefer the term "statistical independence" over "free will" because it gets at the root of the issue from a quantum mechanics perspective that is predicated on probabilities of existence in various states. Some of the mathematical quirks of the quantum world also suggest a way out of these multiple realities by proposing a particular trajectory for the universe or "super-determinism." Such an approach might make all scientific inquiry seem futile at first. After all, if the course of the universe is already predetermined, what is the point of studying it? It seems exasperatingly similar to the theological notion of destiny. However, if we zoom in and consider nested choice at smaller scales within our functional existence, but recognize that this is bounded by determinism at the cosmic scale, we might still have some use for this concept.[35] Such propositions are useful mental gymnastics, spurred by what is being termed the "second quantum revolution" as more experimental work is done at the subatomic scale and more practical applications for quantum phenomena are developed.

This is an exciting age of exploration in natural science, similar to the great age of planetary navigation, wherein we knew the parameters of the planet's geography and developed the tools to undertake some of those voyages. Now we know the broad parameters of the universe and have the tools to even experimentally test many of our theories or observe the cosmos to validate our presumptions. For example, the Dark Energy Survey (DES) project (a collaboration of more than 400 scientists) has been taking a fascinating array of measurements worldwide and recently reported a deviation from Max Planck's prediction of the current "lumpiness" of matter.[36] Although conventional and dark matter were evenly distributed in the early universe, that is not the case within currently observable galaxies. Part of the challenge lies with how studying structure itself requires us to develop certain standard mechanisms for comparisons that can themselves be tested. Thus, for cosmic measurements, the "standard candle" of luminosity measurements developed by the oft-neglected astronomer Henrietta Storm Leavitt (1868–1921) gave us the "cosmic distance ladder." This has served us well until recent refinements in astronomical measurements that are leading us to consider other prospects and explanations which might explain the paths taken by galactic formation. Such structural paths may require calibration to ancient acoustic wave patterns after the Big Bang or that we consider the way we are observing the bending of light around massive gravitational objects such as quasars. With ever more refined astronomical observations, the "standard model/ core theory" continues to be tested. Natural order may defy our views of elegance or balance or beauty altogether. In this spirit, complex systems theorist and founder of Mathematica Steve Wolfram stunned the physics world in April 2020, in the midst of the Coronavirus pandemic, by launching such a unification quest in a series of public tutorials on his website alongside a 400-page treatise on his approach.[37] Wolfram was perhaps inspired by Newton, who is believed to have formulated the structure of calculus while in quarantine during the great plague of 1665. Unlike Newton, he is following a crowdsourced approach to getting feedback and ideas for developing his model. He has come under criticism for not following a peer-reviewed pathway for his research presentation, but his track record of accomplishments in computer science applications of complex systems research is quite admirable.

Perhaps we need iconoclasts to move us beyond the intellectual inertia around accepted explanations of natural order. Yet a cautionary note about structural elegance and "originality" in the context of physics from Werner Heisenberg should hold our attention. In one of his many public lectures he warned against drawing too many comparative inferences between science and art: "the difference between an abstract artist and a physicist is that the artist aspires for being original in his or her product;

the physicists aspires to be conservative."[38] A related matter that we should be careful of is what has been termed in philosophy as "the naturalistic fallacy," which states that just because something is natural does not mean it is good (also referred to as "the is/ought" fallacy). However, it can be surmised that what is natural and has existed through a range of resilience-testing processes possibly has some "staying power" and hence has qualities that should be evaluated. The ultimate staying power of our existence is no doubt subject to certain natural parameters which cannot be ignored in their mathematical convergence. Sir Martin Rees, in his landmark book simplified these parameters to *Just Six Numbers*—key proportionalities of fundamental forces or dimensions—that allow the universe to exist in its observed anthropic form, which he stated as follows:

1. N in nature, equal to 1,000,000, 000,000,000,000,000,000,000,000,000,000. This number measures the strength of the electrical forces that hold atoms together, divided by the force of gravity between them. If N had a few less zeros, only a short-lived miniature universe could exist: no creatures could grow larger than insects, and there would be no time for biological evolution.
2. The number Ɛ, whose value is 0.007, defines how firmly atomic nuclei bind together and how all the atoms on Earth were made. Its value controls the power from the Sun and, more sensitively, how stars transmute hydrogen into all the atoms of the periodic table. Carbon and oxygen are common, whereas gold and uranium are rare because of what happens in the stars. If Ɛ were 0.006 or 0.008, we could not exist.
3. The cosmic number Ω (omega) measures the amount of material in our universe–galaxies, diffuse gas, and "dark matter." Ω tells us the relative importance of gravity and expansion energy in the universe. If this ratio were too high relative to a particular "critical" value, the universe would have collapsed long ago; had it been too low, no galaxies or stars would have formed. The initial expansion speed seems to have been finely tuned.
4. Measuring the fourth number, λ (lambda), was the biggest scientific news of 1998. An unsuspected new force—a cosmic "anti-gravity" (now referred to as dark energy)—controls the expansion of our universe, even though it has no discernible effect on scales of less than a billion light years. It is destined to become ever more dominant over gravity and other forces as our universe becomes ever darker and emptier. Fortunately for us (and very surprisingly to theorists), λ is very small. Otherwise, its effect would have stopped galaxies and stars from forming, and cosmic evolution would have been stifled before it could even begin.
5. The seeds for all cosmic structures—stars, galaxies, and clusters of galaxies—were all imprinted in the Big Bang. The fabric of our universe depends on one number, Q, which represents the ratio of two fundamental energies and is about 1/100,000 in value. If Q were even smaller, the universe would be inert and

structureless; if Q were much larger, it would be a violent place, in which no stars or solar systems could survive, dominated by vast black holes.

6. The sixth crucial number has been known for centuries, although it's now viewed in a new perspective. It is the number of spatial dimensions in our world, D, and equals 3. Life couldn't exist if D were 2 or 4. Time is a fourth dimension, but distinctively different in terms of our agency to control its directionality.[39]

Another permutation of insights on patterns in natural systems pertains to what ornithologist Richard Prum has termed *The Evolution of Beauty.* Many of the features which are observed as patterns and attractive features might not be instrumentally useful in physical terms. The mate choice behavior of birds can lead to highly impractical plumage which can even reduce the survival potential of a male bird, but the artistic beauty of the plumage itself has some intrinsic worth. The perception of beauty and patterns in advanced organisms is a fascinating area of further inquiry. There is a part of the human brain called the "default mode network" that is stimulated when we experience art and reflect on its connections with broader cognition. The same may well happen with deep physics reflections. Ultimately, the conservative impulse of science that Heisenberg alerted us to can lead to a transference of knowledge. This impulse can be coupled with the unique set of inspirations that art can galvanize to action to serve complementary purposes. Yet before we can realize the appropriate agency of art and science in driving sustainability for human systems (as noted toward the end of this book), we must understand several more scales of order.

Notes

1. Andersen, Kristian G., Andrew Rambaut, W. Ian Lipkin, Edward C. Holmes, and Robert F. Garry. 2020. "The Proximal Origin of SARS-CoV-2." *Nature Medicine* 26 (4): 450–452. https://doi.org/10.1038/s41591-020-0820-9.
2. Kauffman, S. A., and E. D. Weinberger. 1989. "The NK Model of Rugged Fitness Landscapes and Its Application to Maturation of the Immune Response." *Journal of Theoretical Biology* 141: 211–245. https://doi.org/10.1016/S0022-5193(89)80019-0.
3. Cramer, Freidrich. 1993. *Chaos and Order: The Complex Structure of Living Systems* (translated from German by D. I. Loewus). VCH Publications.
4. Monroe, J. G. et al. 2022. "Mutation Bias Reflects Natural Selection in Arabidopsis Thaliana." *Nature* 602 (7895): 101–5. https://doi.org/10.1038/s41586-021-04269-6.
5. Richard Lenski, Interview with Derek Mueller, Veritasium YouTube channel, https://www.youtube.com/watch?v=w4sLAQvEH-M, video posted June 17, 2021.
6. A readable book review that lays out non-creationist perspective on intelligent design was presented by computer scientist David Gelertner in Stephen Meyer's book *Darwin's Doubt.* See Gelertner, David. 2019. "Giving Up Darwin." *Claremont Review of Books*, Spring.
7. Kauffman, Stuart A. 1993. *The Origins of Order: Self-Organization and Selection in Evolution* (1st edition). Oxford University Press, 643, Epilogue).

8. England, J. 2020. *Every Life Is on Fire: How Thermodynamics Explains the Origins of Living Things* (illustrated edition). Basic Books. See also England, J. L. 2015. "Dissipative Adaptation in Driven Self-Assembly." *Nature Nanotechnology* 10 (11): 919–923. https://doi.org/10.1038/nnano.2015.250.
9. The engineer J. E. Gordon has beautifully presented the power of such structures and how engineering developed accordingly in his classic work *Structures: Why Things Don't Fall* (Da Capo Press, 2003).
10. An excellent book on the importance of geometry in this regard is Ellenberg, J. 2021. *Shape: The Hidden Geometry of Information, Biology, Strategy, Democracy, and Everything Else*. Penguin Press.
11. Turner, J. Scott. 2010. *The Tinkerer's Accomplice: How Design Emerges from Life Itself*. Harvard University Press.
12. Brookes, J. C. 2017. "Quantum Effects in Biology: Golden Rule in Enzymes, Olfaction, Photosynthesis and Magnetodetection." *Proceedings in Math, Physics and Engineering Science* 473. https://doi.org/10.1098/rspa.2016.0822.
13. Cao, J., R. J. Cogdell, D. F. Coker, et al. 2020. "Quantum Biology Revisited." *Science Advances* 6. https://doi.org/10.1126/sciadv.aaz4888.
14. Bobrick, A., and G. Martire. 2021. "Introducing Physical Warp Drives." *Classical and Quantum Gravity*. https://doi.org/10.1088/1361-6382/abdf6e and also Lentz, E. W. 2021. "Breaking the Warp Barrier: Hyper-Fast Solitons in Einstein-Maxwell-Plasma Theory." *Classical and Quantum Gravity* 38 (7): 075015. https://doi.org/10.1088/1361-6382/abe692.
15. "Path to Quantum Computing at Room Temperature." 2020, May 1. https://www.sciencedaily.com/releases/2020/05/200501184307.htm.
16. Arute, F., et al. 2019. "Quantum Supremacy Using a Programmable Superconducting Processor." *Nature* 574: 505–510. https://doi.org/10.1038/s41586-019-1666-5.
17. Refer to the Microsoft Innovation site at https://innovation.microsoft.com/en-us/planetary-computer (accessed July 14, 2020).
18. Berezhiani, L., B. Famaey, and J. Khoury. 2018. "Phenomenological Consequences of Superfluid Dark Matter with Baryon-Phonon Coupling." *Journal of Cosmology and Astroparticle Physics* 021–021.
19. An excellent book that lays out the rise and fall of Vulcan in science is Levenson, Thomas. 2015. *The Hunt for Vulcan: ... And How Albert Einstein Destroyed a Planet, Discovered Relativity, and Deciphered the Universe*. Random House.
20. Wilczek, F. 2016. *A Beautiful Question: Finding Nature's Deep Design* (reprint edition). Penguin Books.
21. Al-Khalili, J. 2020. *The World According to Physics*. Princeton University Press.
22. Peebles, P. J. E., and B. Ratra. 2003. "The Cosmological Constant and Dark Energy." *Reviews of Modern Physics* 75: 559–606. https://doi.org/10.1103/RevModPhys.75.559.
23. A fantastic book for a general audience on the enigmas of voids and vacuums with reference to quantum physics is Barrow, J. D. 2001. *The Book of Nothing: Vacuums, Voids, and the Latest Ideas About the Origins of the Universe* (1st edition). Pantheon.
24. Coopersmith, J. 2017. *The Lazy Universe: An Introduction to the Principle of Least Action*. Oxford University Press.
25. Lu, P. J., and P. J. Steinhardt. 2007. "Decagonal and Quasi-Crystalline Tilings in Medieval Islamic Architecture." *Science* 315: 1106–1110. https://doi.org/10.1126/science.1135491.

26. Bindi, L., N. Yao, C. Lin, et al. 2015. "Natural Quasicrystal with Decagonal Symmetry." *Scientific Reports* 5: 9111. The extraordinary journey of the discovery and validation of quasicrystals is narrated for a general audience in Steinhardt, P. 2019. *The Second Kind of Impossible: The Extraordinary Quest for a New Form of Matter* (reprint edition). Simon & Schuster.
27. Definition of aperiodic crystals (accessed June 19, 2020) at https://dictionary.iucr.org/Aperiodic_crystal.
28. Klein, O., J. L. Heilbron, T. S. Kuhn, and L. Rosenfeld. n.d. Oral history interview with Oskar Benjamin Klein. American Institute of Physics. https://www.aip.org/history-programs/niels-bohr-library/oral-histories/4709-3.
29. An excellent book on this topic is Barrow, J. 2009. *The Constants of Nature: The Numbers That Encode the Deepest Secrets of the Universe*. Vintage. Also noteworthy in this genre is Atkins, P. 2018. *Conjuring the Universe: The Origins of the Laws of Nature*. Oxford University Press.
30. Recent books that have raised cautions about physics losing its way through obfuscation and internally consistent but empirically distant theorizing include Baggott, Jim. 2014. *Farewell to Reality*. Pegasus Books; Hossenfelder, Sabine. 2020. *Lost in Math: How Beauty Leads Physics Astray*. Basic Books; and Lindley, David. 2020. *The Dream Universe: How Fundamental Physics Lost Its Way*. Doubleday.
31. Barrow, J. D., and F. Tipler 1988. *The Anthropic Cosmological Principle*. Oxford University Press.
32. Winfree, A. T. 2010. *The Geometry of Biological Time* (2nd edition; softcover reprint of the original 2001 edition). Springer..
33. Sacha, K., and J. Zakrzewski. 2017. "Time Crystals: A Review." *Reports on Progress in Physics* 81: 016401. https://doi.org/10.1088/1361-6633/aa8b38.
34. Rovelli, C. 2018. *The Order of Time*. Riverhead Books.
35. Hossenfelder, S., and T. Palmer. 2020. "Rethinking Superdeterminism." *Frontiers in Physics* 8: 139. https://doi.org/10.3389/fphy.2020.00139.
36. Details at https://www.darkenergysurvey.org/the-des-project/overview/ (accessed June 10, 2020).
37. Refer to https://writings.stephenwolfram.com/2020/04/finally-we-may-have-a-path-to-the-fundamental-theory-of-physics-and-its-beautiful/ (accessed June 16, 2020).
38. Heisenberg, W. 1958. *Physics and Philosophy (Gifford Lectures 1955)*. Harper & Row.
39. Rees, Martin. 2000. *Just Six Numbers*. Basic Books.

2
The Elements of Earthly Order

> It is the function of science to discover the existence of a general reign of order in nature and to find the causes governing this order. And this refers in equal measure to the relations of man—social and political—and to the entire universe as a whole.
>
> **—Dmitri Mendeleev**

Our fascination with the cosmos and the ambitions we have for understanding the universe are still largely anchored to our planet. Even attempts at exploring the potential for life on "exo-planets" is often predicated on assumptions about the primacy of earthly order. All our fundamental understandings of particles and quantum theory is only of value for the foreseeable future if we can understand the mechanisms by which life has been sustained on Earth. The functional building blocks of planetary processes that are most consequential for understanding earthly order are what we label as *the* chemical elements. The most celebrated ordering of these elements graces the walls of millions of classrooms worldwide: the *periodic table*. We all seem to revere that asymmetric crown of abbreviated letterings with hydrogen and helium at the extreme peaks of the tiara and two disembodied radioactive rows emanating from the third column. Many entertaining histories of the periodic table and the discoveries of its elements have been written,[1] alongside celebrated melodies such as Tom Lehrer's playful classic, *The Element Song*. Yet the elemental order we have enshrined in the shape and structure of this table continues to be challenged.

In 2019, the world celebrated with much fanfare the 150th anniversary of the first organization of the elements of the periodic table by the Russian chemist Dmitri Mendeleev. Science museums held element-themed galas and explosive experiments reenacting memorable discoveries. Amid all this revelry, science educator Martyn Poliakoff literally turned the table on its head and published a short paper in the journal *Nature Chemistry* which suggested that the table is better organized upside down (Figure 2.1)! This reminded me of my fellow Australians reveling in tourist maps which reverse the polarities of usual geographies to show the southern hemisphere at the top of a chart! Unlike the fairly arbitrary north–south reversal on a map, the rationale for the periodic table reversal is that such a structure is better-suited to explaining how electrons fill orbitals in elements. The reordering of the table has been a constant point of debate, and numerous other versions of the table have been

Earthly Order. Saleem H. Ali, Oxford University Press. © Oxford University Press 2022.
DOI: 10.1093/oso/9780197640272.003.0003

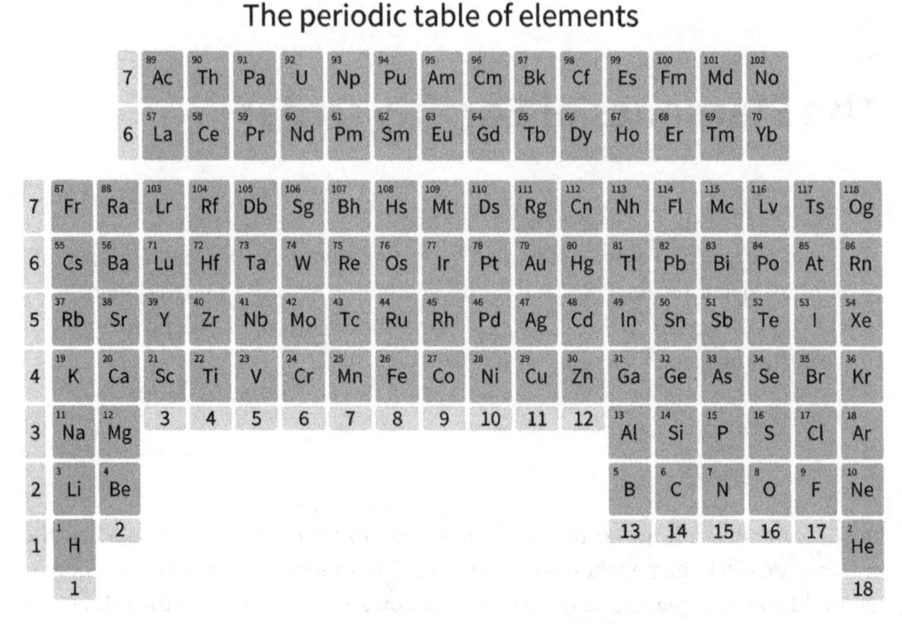

Figure 2.1 Reconfiguring of the Periodic Table offered by Poliokoff et al., in *Nature Chemistry*, considering electron functionality.

published with shapes ranging from Mayan wheels to infinity curves. What was special about this article was that Poliokoff collaborated with a psychologist (who happened to be his daughter) to consider the human perception of ordered shapes. The paper revealed that when students view the table, their sense of order focuses on the center of the diagram regardless of the shape and then moves to the extreme ends of helium and hydrogen as areas of priority.

What this playful but publishable exercise showed was that chemical order as manifest in two versions of the table can be prioritized differently but can still elicit substantive patterns of meaning. Regardless of how the table is constructed, the primacy of the elements in defining Earth sciences is unquestionable. In one of my earlier books, *Treasures of the Earth*, I had analyzed how particular metals and their alloys defined human civilizations and temporal order. A humble alloy of tin and copper—bronze—defined our earliest technological leap from the lithic lethargy of an elementally nondescript "stone age." From there we moved to the age of iron—an element which is literally at the core of planetary order on Earth, which we will explore later in this chapter in considering magnetism. Carbon has defined much of our Industrial Age for the past 200 years, while its neighboring element, silicon, has helped bring us the information revolution from microchips to fiber optics.

Much to the chagrin of many hard-core chemists, broad categorizations within the periodic table are often presented in terms of "metals" and "nonmetals," with the interstitial category of "metalloids" lying in between. Metals dominate the elemental taxonomy and have been roughly described in terms of their macro-physical properties of being "malleable" and "ductile," or carrying electric charge. The "transition" metals constitute the middle corpus of the table and are chemically defined in terms of electron orbital states that give them more varied bonding properties with other elements. This includes the two separate rows of large atomic elements, some of which are often classified as "rare earths" even though they are found in small amounts quite widely distributed in the earth's crust. Non-metals are generally defined in terms of lacking the properties of metals. The twilight realm of "metalloids" is not defined officially by the International Union of Pure and Applied Chemistry, the standardization entity which has been setting the rules for chemistry since 1919. However, boron, silicon, germanium, arsenic, and antimony are often categorized as such due to their particularly ambivalent physical properties.

Carbon, the most monumental of all bond-building elements, also occasionally gets lumped together with the metalloids because it has the most alarming array of *allotropes*—molecular arrangements of its atoms that give it vast variations of properties. Diamonds, graphite, and pure forms of coal are all allotropes of carbon that we are fairly familiar with. There are even more exotic forms of carbon that have gained prominence in recent years because of their fascinating applications for green technologies. The specific molecular order that they possess, in terms of the layering and topology of atoms, gives these materials extraordinary strength, conductivity, and flexibility, suggesting remarkable opportunities for a very different sort of carbon revolution. Carbon compounds also have the ability to form mirror images of each other that cannot be superimposed on one another—a property known as *chirality*. Human hands are the easiest corollary to this phenomenon, and hence chiral molecules are often labeled as "left-handed" and "right-handed." This three-dimensional differentiation or "stereochemistry" gives a high degree of specificity to carbon compounds in how they can be selected for fine-tuned action in living systems.

At this juncture it is also important to differentiate *topology* and *geometry* in understanding structural order at the level of molecules—or indeed the universe. While geometry is concerned with localized structure, such as the connection between dots in a plane, topology is concerned with more "global structures." Mathematicians use the term "modulus" and its plural "moduli" as a mark of any deformations in structure when they consider if there is rigidity or flexibility in a system. Topological and geometric systems can coexist if we consider different scales. On a curved Earth we can have localized flatness, and on a relatively flat universe we can have topological curvature in systems, like Mobius curves. Depending on whether we are "intrinsic" or extrinsic" to a system will determine whether geometrical or topological features are more salient. This insight also links to Einstein's *equivalence principle*, which linked special and general relativity by suggesting that "acceleration in flat spacetime is indistinguishable from gravity." *Special relativity* deals with flat space-time (geometric features), and *general relativity* considers curvature (topological features). Metaphorically, being on Earth's surface is thus similar to being on a spaceship, distant from any gravitational force and accelerating with its engines; hence the expression "spaceship Earth," coined by Buckminster Fuller. In curved spacetime, the line of motion of an inertial particle or a pulse of light is *functionally* straight in space *and* time; this is referred to as a *world line* (including time as a fourth dimension in addition to the three spatial dimensions). Such a world line is called a *geodesic*, which Fuller also used in his architectural representation of geodesic domes. Among the elements, in common discourse, the one which has the most structural similarity to such formulations is carbon—fantastically, also the core element of earthly life.

The New Carbonic Order?

The atomic similitude of the various allotropes of carbon also raises an intriguing proposition for us to consider. The most abundant source of fixed elemental carbon on the earth's surface is the rock that fueled the Industrial Revolution: coal. Since the carbon imbalance in the atmosphere leading to climate change is triggered largely by burning coal and is linked to human-induced climatic change, might we be able to convert that coal to other materially useful forms of carbon? Should we consider a future where coal mining jobs could be preserved because we have found a way to rearrange carbon atoms more effectively from coal to other noncombustible forms of carbon? Before getting too excited about this prospect, we need to first consider why coal and diamonds are so utterly different in physical properties. Both substances are largely made of carbon but are not necessarily pure carbon. Even in the case of a fairly pure diamond, some minor impurities—metals entering the lattice—can give diamonds slight colorations. In the case of coal, we have a noncrystalline or amorphous form of carbon which has numerous organic impurities that are the result of its origins as a decay product of complex plant molecules. The arrangement of carbon atoms in coal occurs through entropic processes which do not lend themselves to ordered outcomes. Thus, even though coal can be as much as 97% elemental fixed carbon in its composition, it is, for practical purposes, a very different substance than other forms of carbon.

Under conditions of high pressure and temperature, coal can transform to the more ordered carbonic form of graphite. Currently, synthetic graphite is largely produced from *coke*, which is a fuel made from heating coal in anoxic conditions. There are also other pathways to graphite formation, such as from cooling magma, in a way similar to how diamonds are created. Indeed, graphite can be converted to diamonds under high pressure, a process mastered by General Electric in the 1950s.[2] Diamonds have long been lionized for being the hardest industrially used materials in drill bits, and, until recently, most of the world's synthetic diamonds were produced for this purpose. Now, gem-quality diamonds are also being produced through processes involving chemical vapor deposition on a small diamond seed. The extraordinary hardness of diamonds emanates from their elemental order. The molecular arrangements of carbon in particular crystalline forms lends remarkable properties to the materials concerned. Diamonds have a tetrahedral-cubic structure which is very stable. Graphite is characterized by the hexagonal honeycomb structures of carbon molecules, reflecting a stable form of structural order observed in so many natural phenomena.

From graphite emanates a curious two-dimensional form of carbon called *graphene* that was only discovered in 2004 and has captivated a new generation of researchers and inventors. Graphene shares some similarities with an earlier discovery in the 1980s of compounds called *fullerenes*, which are characteristically "geodesic" and named after the designer Buckminster Fuller. The discovery of fullerenes involved a complex deposition of carbon vapors in a helium tank and won a trio of researchers the 1996 Nobel Prize in Chemistry. Similarly, the discoverers of graphene also received a Nobel Prize 14 years later, but their path to discovery was far more humble and shows how simple human interventions can harness natural order. In an attempt to clean a slab of graphite, Andre Geim and Konstantin Novoselov used simple adhesive tape to peel off dirt and dust and noticed that some of the graphite also peeled off. Knowing that the structure of graphite essentially comprised layers of two-dimensional hexagonal plates, the duo wondered if they might have isolated a thin atomic-scale layer of graphite.

All their previous learning in thermodynamics led them to conclude that two-dimensional matter was not stable at room temperatures and that kinetic forces would lead to its transition to three-dimensional forms. Before the discovery of graphene, the only two-dimensional structure which had been synthesized was an organic molecule with 222 carbon atoms linked together through 37 benzene rings, which was not particularly stable or useful. As Novoselov would later state in his Nobel lecture, against their initial impulses, they decided to wander into the quest for "materials in the flatland." The results of their venture into this uncharted territory of material science research were astounding. Graphene was able to defy the conventional wisdom against molecular stability as a two-dimensional crystal because its carbon bonding structure allowed the lattice to also be flexible like a wave. Note that this did not mean that the individual atoms of carbon were exhibiting a wave-particle duality analogous to quantum particles. Rather, the physical form of the structure was flexible, which gave it both stability and a unique set of superlative physical properties. The list of graphene superlatives stems from these unique properties of order. While being only an atomic layer thick, it is an excellent light absorber and hence visible to the human eye. Graphene has a greater tensile strength than our strongest steel by several orders of magnitude and thus is highly suitable for infrastructure construction. Heat and electric conductivity are so great that graphene-coated homes might not need wiring or thermal control.

So extensive were the list of the material's wonder properties that the European Union invested a billion euros to help develop graphene applications in 2013, in a program known as the Graphene Flagship. So far, the project has developed 17 commercialized products of graphene ranging from fast-charging and longer-lasting batteries to thermal control jackets. Figure 2.2 shows the range of products and the timeline that academic–industry partners envisage for this project. Many of these

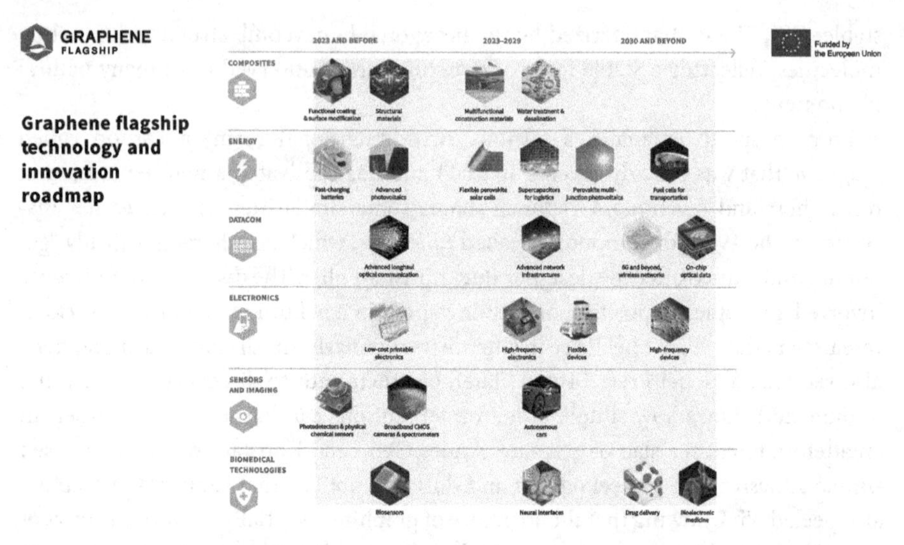

Figure 2.2 Graphene applications and timeline from the European Union's €1 billion Graphene Flagship project.

products are also complementing materials such as carbon fibers and composites that currently have petroleum-based source chemicals and are often not recyclable. In the case of graphene, the material may well be more durable as well as recyclable, and thus graphene brings an additional temporal order to our sustainability quest.

Graphene demonstrates another cautionary property of ordered systems that we can often neglect in our excitement in analyzing the amazing micro-state of a material. Even though the discovery of graphene was made possible by the most rudimentary mechanical means of using adhesive tape, to upscale such a process for commercial purposes is immensely difficult. What may work at the micro-scale might well be immensely difficult to undertake at a larger scale. The range of methods being developed to mass produce graphene is profuse and testing the limits of human ingenuity.[3] Carbon is ubiquitous on the planet, but we are confined by our ability to retrieve it from the bonds that bind it so firmly in other forms. Whether or not these efforts are able to reap the full potential of this super-ordered molecule may depend on how well we are able to find suitable ways of manufacturing graphene at low cost and with minimal environmental harm.

We do have a hopeful precedent for harnessing a vital molecule from an abundant element that seemed equally intractable. More than a century ago, Fritz Haber and Carl Bosch were able to find a way to use the most abundant element in our atmosphere, nitrogen, and combine each of its atoms with three hydrogen atoms to form the foul-smelling gas ammonia. Energy analyst Vaclav Smil has called this discovery the most consequential chemical achievement in human history because it allowed humans to "fix" nitrogen into fertilizers, which in turn allowed us to produce food at unprecedented rates.[4] The Haber-Bosch process was up-scaled successfully because it emulated a simple principle of chemical order identified by the French chemist Henry Le Chatelier, which can be stated as follows: when any system is at *equilibrium* (no change occurring independently in its state) and is subjected to a physical change, the system shifts to counteract the applied change. Such a process is often less pronounced in complex systems. Pressure and temperature are key physical change agents which can drive the reversibility of the reaction in favor of ammonia production from elemental nitrogen and hydrogen. While carbonic order is quite different from nitrogenous order, the broad insights from Le Chatelier's principle may well help us consider ways of graphene synthesis.

Beyond the allotropic forms of carbon, there is also a fascinating array of nuclear properties in carbon's isotopes (atoms with differing numbers of neutrons but with the same number of protons) that further our fascination with this element. While carbon generally has six protons and six neutrons in its most common isotope (hence called C^{12}), there are also forms with seven or eight neutrons (recall that the number of protons needs to be the same for the same element). C^{14} with eight neutrons is particularly important because it is radioactive; hence the nucleus is not in equilibrium but is slowly "decaying" in a manner whereby one of the extra neutrons is converted into a proton—and hence the carbon atom is converting into a nitrogen atom. As noted in Chapter 1, this process is driven by the weak nuclear force, and energy is released as well through the decay process.

Our planetary atmosphere is able to maintain a curious order of the ratio between C^{12} to C^{14}, which stays the same over extended periods of time because C^{14} is created in small amounts by cosmic rays interacting with our atmosphere and then slowly decays to nitrogen. Thus, even though C^{14} itself is not in equilibrium with its decay products, the ratio of C^{12} to C^{14} has stayed fairly constant over time. There are, however, some 10–15% shifts due to Earth's magnetic field, sunspot activity impacting cosmic rays, ocean absorption, and even possible nearby stellar explosions (supernovae). Tree ring and organic sediment analysis, which also have a "fixed" ratio with a known year of deposit, can give us a means of calibrating these past minor changes in the ratio as well. By observing the decay of C^{14}, we can ascertain the age of many carbon-containing materials through a simple calculation process linked to the speed of the material's decay. The American chemist Willard Libby was able to refine this technique that gave us the "radio-carbon dating revolution" and won him the Nobel Prize in 1960.

The ubiquity of carbon and the peculiarities of its elemental order can thus help us date an ancient parchment or a mummy in an Egyptian tomb. The key constraint, however, is that C^{14} decays beyond detection accuracy after about 50,000 years, and so this technique cannot be used beyond that. However, that is long enough for us to say with fair certainty that human remains dated by this process are at least 20,000 years old or older. This has led to a litany of challenges to the method's veracity by creationist scholars—all of which have been adequately addressed and refuted by the National Center for Science Education.[5] A more cogent concern of the technique's applicability and accuracy stems from the human generation of fossil fuels, which are contaminating the C^{12}:C^{14} ratio. During the Cold War, the reverse concern was true as nuclear weapons testing led to higher C^{14} generation due to radiation entering the atmosphere. Since the carbon in coal, oil, and gas comes from organic material millions of years old, the amount of C^{14} in this is negligibly small. Thus, through our burning of fossil fuels over the past two centuries, we have diluted the ratio in favor of C^{12} carbon. In the words of *Scientific American*, "a T-shirt made in 2050 could look exactly like one worn by William the Conqueror a thousand years earlier to someone using radiocarbon dating if emissions continue under a business-as-usual scenario."[6]

While C^{14} levels are decreasing in the atmosphere, they are increasing in another area of human activity—nuclear fission reactors. The graphite rods used in nuclear power plants are concentrating C^{14} carbon, and research conducted by the University of Bristol (UK) suggests an opportunity to manufacture a carbonic "beta-voltaic battery" from this decay energy. Carbon's unique allotropic properties could allow for the graphite to be converted to a diamond form that could better harness the beta particles/electrons coming from the decay while shielding unwanted penetrating radiation impacts. The so-called *diamond battery* has some potential for powering our future satellites, pacemakers, and precious electronics, though it has become a target of skeptical Vloggers as well.[7] Such is the promise and peril of all matters dealing with that heart of the atom, which has its very own order that deserves singular attention.

Nuclear Order

The decay of the C^{14} nucleus to nitrogen with the release of energy highlights a hallmark of elemental order that it is often neglected by scholars of sustainability. As we consider recycling metallic elements for more sustainable material consumption outcomes, let us keep in perspective that the elements we mine from the earth are inherently renewable unless subjected to a nuclear reaction. Thus, the copper, tin, iron, and aluminum we have mined over the ages from the Earth is largely still somewhere on Earth's surface. The iron in ancient metallic pots may have rusted and formed new bonds with oxygen, but its iron atoms themselves are very much still around from those materials. They have just been rendered more diffuse and distributed, and energy is needed to bring them back to functional order. On the other hand, if we disturb the nuclear order of an element, we can convert it to another elemental form, one that has a radically different array of consequences in terms of both material and energy balance. Furthermore, resource efficiency provides a mechanism whereby we can stretch our fortunes further even with a finite resource. The late cornucopian economist Julian Simon, who famously won a bet[8] with Neo-Malthusian ecologist Paul Ehrlich about copper supplies not being depleted by 1980, once noted that the length of a line segment is fixed, but the number of points along it are infinite. Simon's metaphor is functionally deceptive with most finite resource concerns but has greater salience with the particle-level energy harnessing potential of the atom.

The splitting of the nucleus, or "fission," releases the energy of the weak nuclear force, but it can also release additional neutrons that further split other nuclei and cause a "chain reaction" (as occurs in nuclear bombs). Nuclear *fission* of earthly elements is much easier to undertake as compared with nuclear *fusion*, which is the process that occurs in stars. Indeed, stars have been the nurseries where most elements came into existence in the cosmos through repeated nuclear fusion. The curious reality is that while nuclear *fusion reactors* are far more common at the universal scale, nuclear *fission reactors* are far more plausible at an earthly scale. Under specific conditions, fission reactions, similar to nuclear power plants, have even been evidenced to take place naturally in uranium deposits. The Oklo mine in Gabon is the only known occurrence of such a natural nuclear reactor whereby uranium ore, immersed in algae-rich groundwater, was able to sustain a chain reaction over hundreds of thousands of years. This mine has fascinated science fiction writers searching for long-lost ancient alien energy sources or mineral-rich African civilizations such as Marvel's vibranium-rich Wakanda. There is also more serious research ongoing at this site about whether the nuclear reaction that has been evidenced here for the past 1.8 billion years could give us some insights into any changes to the fine structure constant over time.[9]

No doubt, releasing nuclear energy through synthetic fission has been a blessing and a bane for human societies. The polarized contemporary discussion about the opportunities and obstructions for continuing uptake of nuclear energy could benefit from an "order-oriented" analysis of the problem. To harness the vast energy of the weak nuclear force, an element has to be *fissile*, meaning that its nucleus should be large and unstable enough to be easily split with a bombardment of relatively low-energy neutrons. Often the instability results from an odd number of neutrons in the atom. Thus, uranium 235 and U^{233} are both fissile whereas uranium 238 (the more commonly occurring isotope of the element) is not. However, uranium 238 is *fertile*, meaning that it can also absorb a neutron and convert to fissionable plutonium, which can be used for energy generation and, more ominously, for nuclear bombs as well. As current uranium-based nuclear power plants in the United States become economically unattractive due to high safety compliance costs, it may be time to consider the next generation of nuclear energy by going back to a basic understanding of how nuclear order can be most efficiently harnessed through fission.

First, an orderly analysis of nuclear energy has to harken back to the source of the raw material needed. Minerals containing fissionable or fertile elements have to be mined from the earth. Given the natural occurrence of fissionable uranium 235, albeit in very small quantities, it has been the preferred pathway for sourcing nuclear energy. However, if a more structured approach to considering the full systems input and output benefits of using a fertile but not fissile material such as thorium had been up-scaled, we might have ended up with a more promising future for nuclear power. Many of the safety challenges of current nuclear power plants stem from the need to maintain high pressures to sustain the nuclear reaction, with an accumulation of highly explosive hydrogen gas as a byproduct. Furthermore, there is a constant need to control temperature rise with a series of mitigation measures. Thus, we are trying to force order on a system which is hell-bent on releasing its entropic power. The explosion we observed at the Fukushima Daichi power plant following the tsunami of 2011 was actually hydrogen gas exploding because the safety pumps failed and led to an uncontrollable set of reactions. Even so, it is worth noting that not a single fatality has been linked to the actual nuclear radiation exposure at Fukushima, nor have any long-term health impacts been found 10 years hence. The fatalities were caused by the tsunami itself and not the nuclear meltdown. Thus, the actual quantitative risk to human morbidity or mortality from nuclear power remains much lower than almost every other form of energy production.[10]

The next generation of nuclear reactor design is considering these matters from a more orderly systems perspective than did the earlier nuclear power plants. Although the new generation of power plants could continue to use uranium as a fuel, there is good reason to reconsider thorium, which was neglected earlier because it was deemed to not be directly fissile. Reactors with thorium as the primary fuel would also generate materials that would have much less harmful radiation generation potential, owing to the shorter half-lives of the decay isotopes. Let us also not forget that nuclear radiation has played a role in our planetary order, from providing heat for circulation flows in the Earth's mantle to potentially generating the requisite cellular mutations upon which natural selection could act for evolution. The role of active nuclear elements, such as thorium, in driving us toward sustainability should thus not be dismissed. Thorium is also three times more abundant than uranium in the earth's crust and found far more widely in economically extractable deposits. The element is a byproduct of existing mines or is found in waste piles, and hence, for the foreseeable future, no new thorium mining would be required to source thorium reactors. Many of the rare earths mining and processing operations produce thorium as a byproduct, and it is labeled as a "waste" although it has much potential to be a feedstock for the next generation of reactors. This is where contemporary environmental activism needs to apply the same metrics of sustainability that they do for any material recycling efforts.

The stigma of nuclear "waste" as an imponderable issue has led to a range of errant policies that continue to haunt the future of sustainable energy supply from this vital resource. Applying our understanding of entropic systems, nuclear "waste" materials are actually fission byproducts. The great value of how these byproducts are generated lies in the low entropy state in which we can contain them. Thus, even the most radioactively toxic byproducts of nuclear power are highly contained, whereas the byproducts of other energy sources such as natural gas burning are dispersed widely through emissions. Even with nascent carbon capture and storage devices on emissions stacks, we are unable to contain many of the byproducts of such combustion. For other sources, like hydropower, wind, and solar energy, the entropic effects are not measured in terms of direct waste generation, but rather in the extent of land or sea area needed for their production and, of course, the material needs of producing and maintaining the infrastructure. Let me be clear: we cannot be glib about nuclear energy. There are still many technical hurdles and risk assurance issues to be addressed, but a dispassionate review of the next generation of reactors is in order.

The concept of *power density*—the rate of energy transfer per unit volume of fuel—is crucial in terms of understanding the value of nuclear order in providing sustainable energy as compared with other sources (Figure 2.3). This point has been admirably advocated by the Canadian-Czech energy analyst Vaclav Smil in his numerous books on the topic. Building on his work, Jesse Ausubel, from Rockefeller University, has calculated that uranium in a light water reactor is at least four orders of magnitude—10,000 times—denser than coals, oils, and hydrocarbon gases. A fast breeder reactor involving thorium would multiply the ratio another hundred times or more.[11] One may wonder why there is continuing controversy about the economic viability of nuclear power and even controversies over its carbon emissions. Much of the economic calculations have focused on current fissile reactors and their safety needs, which require massive and expensive concrete structures. *Molten salt reactors*, which do not require high pressures and also have a built-in disengagement mechanism to prevent accidents, could be far less expensive. Such reactors are tentatively being developed through a range of companies, including Terra-Power, which was

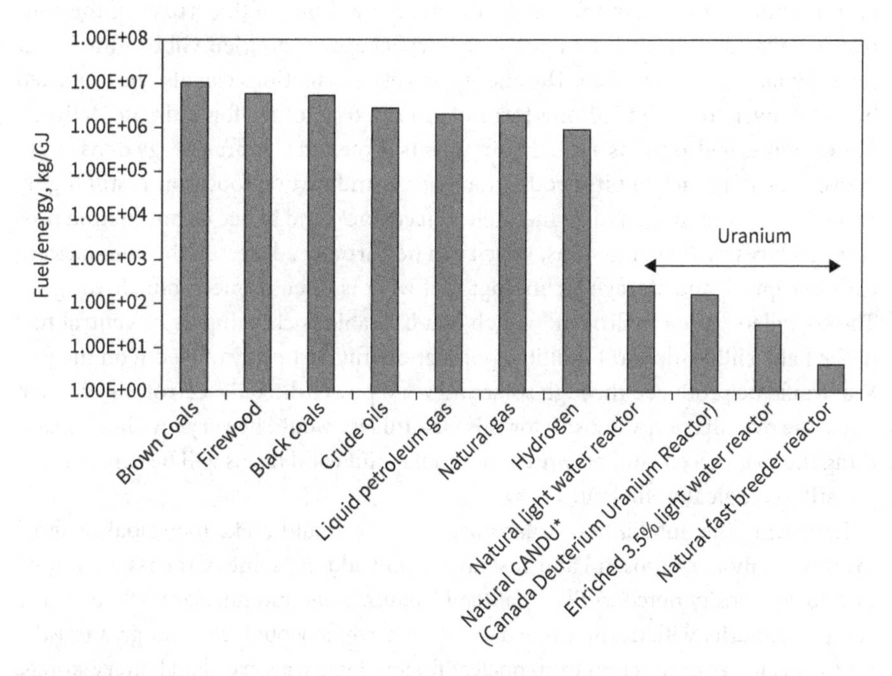

Figure 2.3 Fuel mass per energy production in gigajoules generated (y axis is logarithmic scale due to very high-energy density of nuclear compared with other sources). Wind, solar, and hydropower have even lower energy densities than brown coal.

From Ausubel 2015 (accessed July 10, 2020): https://phe.rockefeller.edu/docs/PowerDensity_Final120815.pdf.

initiated by Bill Gates, alongside *traveling wave reactors*, which use depleted uranium as a fuel. "Depleted" uranium connotes a byproduct of enrichment processes, and it is often used to make armor-penetrating bullets and counterweights in aircraft. Given the challenging regulatory environment around nuclear energy in the United States, many of the new nuclear ventures have been developing their prototype projects in partnership with China. This is where particular notions of political order may come into conflict with a quest for sustainability through nuclear order. Terra-Power reached an agreement with the China National Nuclear Corporation in 2017 to build a demonstration reactor south of Beijing. However, the US government imposed additional restrictions on joint ventures with China in 2019, which has prevented the company from moving forward with the plan as it seeks new partners.

Bill Gates and Elon Musk are in a power tussle over nuclear versus solar order for the future of energy. Musk has bet on solar, with the argument that even if solar's energy density is much lower than nuclear, it can be situated in areas with ample vacant land, such as in deserts. The bigger issue with regard to solar power's ascendance will not be the land area per se but the enormous material needs for building the infrastructure and the durability of the material. Currently, solar panels have a warranty of a 30-year life span before losing efficiency. The metrics on the actual age of solar and wind infrastructure is one of the cruxes of the controversy that surrounded the release of Michael Moore and Jeff Gibbs film *Planet of the Humans*, in April 2020. The energy density calculations can also be impacted by the conversion needs of one form of energy to another for ultimate delivery. For example, hydrogen as a fuel in aircrafts is three times more energy dense than conventional jet fuel, but its production energy and carbon footprint is still highly variable. "Gray hydrogen" uses methane as feedstock, and hence its production process produces carbon emissions, which can be partially addressed through nascent carbon capture and storage technologies of what is often termed "blue hydrogen." The so-called "green hydrogen," which Saudi Arabia is claiming as its central fuel in the half-trillion-dollar, 1-million-population, futuristic city of Neom on the Red Sea, would be produced through solar and wind power–linked electrolysis of water. Thus, the overall energy density for this wishful city would be very low, but considering the vast barren land on which these solar and wind farms will be constructed, it is still ecologically efficient.

Improving the durability of solar infrastructure would make the initial material investment more purposeful and sustainable and address some of the issues around low power density noted in *Planet of the Humans*. Solar and nuclear could also find complementarity with the new breed of reactors, which would allow for greater *valve control* of energy production from nuclear fission. Even with excellent battery storage infrastructure, solar power will always need some secondary backup supply to ensure high-quality delivery for particular uses. Natural gas or biofuels have the advantage of easy valve control (switching on and off) as compared with conventional nuclear fission power, where it takes considerable time to switch a reactor on and off. However, the new generation of molten salt reactors and other innovations will allow for more flexibility in this regard, thereby allowing for effective backup of solar and wind power. Nuclear power alongside other technologies that seem disruptive to human notions of a pristine order exemplify what Roger Kasperson has termed the "social amplification of risk." While this amplification may have been partially caused by past environmental injustices around nuclear supply chains, such as the impact of uranium mining on the Navajo Nation, there is less justification for its persistence with the next generation of nuclear technologies.

An ironic connection between solar and nuclear energy recently came through as America's newest and possibly last conventional fission reactor opened at Watts Barr, Tennessee, in 2017. Wacker Polysilicion came to the area almost as a direct response to Watts Barr. The company, which makes polysilicon for solar panels, started investing in Rhea County, Tennessee, in 2009, after construction had started on the new unit at the power plant. Gary Farlow, CEO of the Bradley/Cleveland County Chamber of Commerce, noted that "the biggest consideration for Wacker when they looked at our area was not only the capacity of the power system but also [it's] quality and reliability." The Wacker facility needs so much power that it requires 20–25% of the full capacity of a nuclear plant. The Wacker plant provides jobs for approximately 650 full-time employees and has invested a total of $2.5 billion into the local area since 2009. It is an intriguing example of the energetics of human technology even for competing sources of power.[12]

The most tangible connection between solar and nuclear lies within the sun itself. Our life-giving orb of plasma at the center of the solar system is a massive nuclear fusion reactor. The power density of nuclear fusion is dizzyingly enormous. The fusion of two hydrogen isotopes to form helium releases around four times as much energy as uranium fission and around 200 million times more energy than breaking hydrocarbon bonds. Energy released increases with square of the pressure. Igniting a fusion reaction is a classic example of reaching a "tipping point" in natural order, which, upon achievement, can release a bounty of desired outcomes. Since nuclei contain positively charged protons, they have a natural repulsion that comes from the same electromagnetic force that we observe in the repulsion of magnets. The high energy injection is needed to overcome this repulsion so that fusion can be forced to occur. In stars like our sun (which is a million times more massive than Earth), fusion is assisted by the immense gravitational mass of the stellar body, which is able to overcome the protonic repulsion. The major challenge with fusion technology on Earth's surface is to compensate for stellar gravitational and pressure conditions by increasing temperature in order to "ignite" the reaction. Thus, the energy needed to start the reaction currently ends up being more than what is generated. Indeed, fusion scientists note with some measure of satisfaction that their reactors are the hottest places in the solar system, as the ignition temperature to start a fusion reaction between hydrogen isotopes on Earth is a staggering 100 million degrees Kelvin—this is at least five to six times hotter than the temperature at the center of the sun! Human ingenuity has been able to concentrate energy in the highly ordered form of lasers to reach such mind-boggling temperatures.

Fusion's promise to supply relatively clean energy at stellar scales with relatively small inputs of hydrogen isotopes derived from water and perhaps small amounts of oceanic lithium is enough to motivate major investment in further research. For the past decade, a $20 billion multicountry project to investigate thermonuclear fusion is being constructed in Provence, Southern France (the International Thermonuclear Reactor [ITER]). This massive device will be a prototype for what scientists are hoping could lead to economically viable fusion power. We will revisit the unusual international alliance of the United States, India, China, Russia, South Korea, and France that brought forth this unusual collaboration in our discussion of political order in the final part of this book as well.

The win-win dream of getting limitless nuclear fusion energy at room temperature promised by the infamous experiment by Fleishman and Pons in the 1980s remains elusive. Fleischman died in 2012, but Pons moved from Utah to France, not far from where the ITER facility is being constructed. Although ITER will only focus on very hot fusion prospects, the case of cold fusion is not entirely closed either. In 2015, a team of researchers supported by Google published a review article in the journal *Nature* which laid forth some specific chemical conditions under which hydride compounds of precious metals like palladium may be able to most effectively harness fusion power at lower temperatures.[13] However, the most likely opportunity for nuclear fusion energy being extracted remains linked to electromagnetic manipulation at high temperatures.

The intersection of nuclear energy and electromagnetism is now widely used in machines such as magnetic resonance imaging (MRI) scanners. These revolutionary diagnostic tools make use of the tiny magnetic field created by the spinning of protons in a nucleus and their interactions with a strong external magnetic field. The difference in elemental response of the protons in different tissues to this magnetic field can create accurate images of internal body morphology. MRI machines also do not generate the dangerous ionizing radiation that computed tomography (CT) scanners or X-rays produce. The subtleties of nuclear order intersecting with magnetic order are gaining increasing attention as we attempt to understand much broader planetary processes and mechanisms.

Magnetic Order

In July 2020, physicists announced a remarkable discovery that an elusive set of "quasi particles," whose existence had been theorized for more 40 years, had been most probably evidenced in a two-dimensional experiment at Purdue University.[14] These purported *anyons* can display a quirky physical property: they can have one magnetic pole instead of the usual two. The observance came from a "braiding" of the particles, which was possible due to their interference with wave functions, between two thick layers of a complex array of prepared materials. Magnets in general have an inseparable spectrum from north to south poles, which is linked to fundamental states of energy stability and the inherent linkage between electrical and magnetic forces. Thus, in order for electric monopoles, such as negatively charged electrons and positively charged protons, to exist in isolation, magnetic dipoles become an existential necessity for a magnetic field to be established. The potential discovery of a monopole magnetic particle would be a momentous achievement similar in scale to the discovery of the Higgs boson and redefine our conception of magnetic order. Further experiments are still likely to be needed over coming years to confirm the anyon's existence and its monopole properties.

While electric charge can exist with a stationary particle, magnetism requires the motion of a particle. A magnetic field is generated when an electron moves laterally as well as when it spins, a process called *electromagnetism*. Magnetic order comes in seven flavors and is a result of how the alignment of these motions is able to leverage force in a particular direction. The reason why a metal like iron is able to display magnetic properties is because of unpaired electrons that are present due to its particular elemental configuration. Those lonely unpaired electrons are able to create alignment in a particular direction without canceling out the directionality effects of magnetic dipoles. In the list of the seven forms of magnetic order presented here, the first four are linked to having unpaired electrons around the atom but with different magnitudes and interactions with spins, leading to slightly different outcomes (Figure 2.4).

1. *Ferromagnetism*: Certain materials can display observable magnetic force even in the absence of an external magnetic field. This is due to a net directionality to the magnetic moment coming from the unpaired electrons in the system. Ferromagnetism may suggest a particular ordered arrangement of the magnetic domains analogous in some ways to the order of a crystal. When we call a material a "magnet," this is the order we are usually referring to.
2. *Paramagnetism*: In cases where there are some unpaired electrons and spins are favorable, materials can display attraction to magnetic fields through an induced effect, which is referred to as *paramagnetism*. Ferromagnetic substances can become paramagnetic beyond a certain *Curie temperature*, which varies for each material. At that point heat energy can break the unidirectional fixation of moments. However, when a magnetic field is nearby, they can still regain their magnetic order and be attracted therein.
3. *Anti-ferromagnetism*: As the name suggests, this magnetism counteracts positive magnetic effects by having the spin magnetic moments of adjacent atoms be equal and opposite in direction. Such alignment often happens at low temperatures, but there are compounds such as manganese oxide (used in batteries) which also display this property owing to an oxygen bond necessitating two electrons sharing a shell. However, this is only possible if they have opposite spins, as described by Nobel Laureate Wolfgang Pauli in his eponymous "exclusion principle" on particle wave-function stability.
4. *Ferrimagnetism*: When an anti-ferromagnetic substance has adjacent moments that are not equal in nature, the result is a substance which exhibits properties

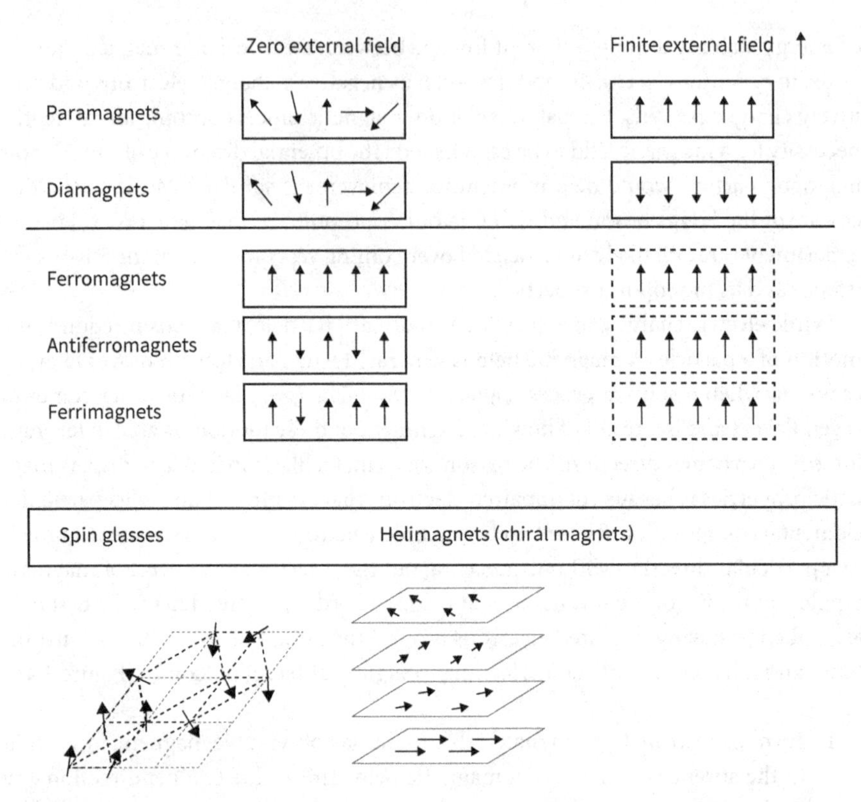

Figure 2.4 Magnetic order in atomic domains result in the different properties of each form of unpaired electron magnetism.

very similar to a ferromagnet. Magnetite, a naturally occurring material which displays natural attraction to metals, was initially thought of as a "ferromagnet." However, closer analysis revealed that it was actually a ferrimagnet, alerting us to the reality of how very different orders can exhibit similar effects. Ferrimagnetic substances are especially important in studying the history of Earth's magnetic field.

5. *Helimagnetism*: This is a specific magnetic ordering in a helical or spiral pattern; it is usually observed at very low temperatures but has recently been observed in iron-germanium alloys, which could provide a novel way for storing data in future computing devices.[15] Such memory storage uses of magnets have already been used in simpler form in commonplace applications like the magnetic stripes on credit cards or magnetic discs and tapes.
6. *Spin glass*: The magnetic domains in these substances are disordered even at low temperatures and analogous to a physical glass, where molecular structure is randomized and "quenched"—meaning the randomness is stuck due to the strong structural and molecular integrity of the substance. The unique molecular structure of spin glasses has also inspired a larger genre of complex systems

research in things as far afield as airline scheduling to vaccine development[16] and led to the 2021 Nobel Prize in physics being awarded to Italian theoretician Giorgio Parisi.

7. *Diamagnetism*: When there are no unpaired electrons in a substance and it weakly repels a magnetic force, it is referred to as *diamagnetic*. The human body is diamagnetic, but our repulsion is overcome by gravity. An experiment by Andre Geim used a frog's body to highlight diamagnetism through levitation against a magnetic field, for which he was awarded the "Ig-Nobel Prize" by the *Annals of Improbable Research*. Geim is the only such laureate to also win the actual Nobel Prize later for his discovery of graphene![17]

The coexistent linkage between electric current and magnetism is a manifestation of the electromagnetic force and has profound planetary implications. The existence of north and south magnetic poles, which are distinct from the geographic poles of the Earth, is as clear as a compass. Human exploration of the planet depended on this earthly magnetic order for navigation as early as the sixteenth century. However, there were numerous peculiarities about the field that baffled navigators. While rudimentary compasses made from lodestone can be traced back to the Han Dynasty in China around 20 BC, they were used for geomancy fortune-telling (Feng Shui) rather than instrumentally for navigation. There is some evidence that the ancient Olmecs of Meso-America may have used magnetic hematite for rituals even earlier, as far back as 1400 BC. Before the geographic value of magnetic materials was known, the celestial geometry of the stars, bird migration observation, and sampling of seafloor mud were key ways of navigating the Earth. However, around the twelfth and thirteenth centuries, there was a convergence of knowledge around using magnetic materials in a compass. This knowledge transfer happened from direct Chinese trade with Europe, as well as through Arab scientists who published interpretive texts on such phenomena. The first textual mention of a magnetic needle comes from the English scholar Alexander Neckam, in his book *De naturis rerum* (On the Natures of Things), written in 1190. However, the most momentous use of the compass, one that was to reconfigure world order, came from its use by Christopher Columbus in his voyages to America.

For much of the Colonial Era, Columbus was also credited with being the first to notice the divergence between geographic north and magnetic north.[18] Subsequent research has shown that the Chinese were aware of this *magnetic declination* many centuries earlier.[19] Regardless of who first observed the difference between geographic order and magnetic order, the impact of this discovery was transformational for it paved the way for understanding Earth's most inner secrets of order. A key insight from this understanding was the fluidity of Earth's magnetism and that the mere mass of the Earth's surface could not adequately explain magnetism. Various forms of inner Earth order were proposed, such as the "hollow Earth hypothesis" popularized in the early eighteenth century by Edmond Halley (also the discoverer of the famous comet that bears his name). This hypothesis itself proved to be totally hollow and was disproved a few decades later through an experiment supported by the Royal Society at a mountain in Scotland called Schiehallion. The experiment's goal of measuring the density of the planet was not fully achieved, but the results made it clear to scientists that the Earth could not be hollow! Even so, this fictional worldview was perhaps an inspiration for the dinosaurs of the inner realm that won literary acclaim in Jules Verne's 1867 novel *Journey to the Center of the Earth.*

The order of inner Earth did not reveal itself from visual observations but through smart readings of waves that traveled through the planet after major seismic events like volcanic eruptions and quakes. Magnetic order emanated from a fortuitous layering of solid and molten elements deep beneath our feet. The Danish Earth scientist Inge Lehmann can be credited for confirming through careful observations of seismic waves the nature of the Earth's interior that gives us the magnetic field. She discovered that the core of the Earth was likely solid metal. Subsequent calculations also revealed that, surrounding this solid orb of mostly iron, nickel, sulfur, and radioactive metals, was a fluid outer core which was in constant circulation. Adding to the complexity was the discovery many years later that the rotation of the inner core was in the opposite direction to the circulation of fluid metal in the outer core. The complex movement of liquid metal in the outer core is responsible for the "dynamo effect" that generates the magnetic field around the Earth. However, due to the fact that the source of the field is a highly dynamic liquid core, there are opportunities for the direction of the magnetic field to be reversed (whereby the current magnetic south pole could be reversed and find its place at our geographic north for several millenia). Such a reversal cannot be predicted but has been shown to have occurred several times in Earth's history. There have been 183 pole reversals over the past 83 million years – on average around once every 450,000 years. More frequent geomagnetic "excursions" also occur when a reversal is aborted because only the outer core's field reverses but not that of the inner core. The process of either form can take a few thousand years tp play out.

The value of the Earth's macro-magnetism is that it forms a massive force field which protects the planet from cosmic rays coming from outer space and, especially, the wrath of solar storms. Indeed, the rapid, complex evolution of life around 500 million years ago is coincidental with the solidification of the Earth's inner core. This elemental order at the core further strengthened the magnetic field of the planet, thereby protecting life's slow evolution from destructive radiation to the point where multicellularity could take root.[20] However, the impacts of the reversal of the Earth's magnetic field are still contested in the literature. In February 2021, a large team of geoscientists published a landmark study that considered multiple records of evidence from 42,000 years ago, when a magnetic excursion happened.[21] The climate variations coincident in the fossil record from ancient Kauri trees in New Zealand studied by the researchers were stark. Some researchers even posit that the disappearance of Neanderthals during this period may have been hastened by this transition. However, the scenarios presented in Hollywood movies such as *The Core*, in which a reduction in the Earth's magnetic field would lead to birds errantly crashing into buildings by losing their sense of navigation are quite implausible, although field reversal may lead to some disruptions in telecommunication and elevated exposure to harmful cosmic radiation. Experiments by astronauts are building further knowledge on these risks, and this remains a fertile research area for further inquiry.[22]

Our sun has its own complex magnetism, which is manifest in features such as sunspots. Due to its fluid nature, the reversal of the sun's magnetic field is somewhat more regular than that of the Earth's and occurs every 11 years or so. The most remarkable interaction between solar magnetism—induced "wind" of particles and our Earth's magnetic field is observed in polar regions in the form of spectacular aurora displays. These "polar lights" occur due to energetic interactions between ionized particles from the solar wind that are concentrated in polar magnetic regions releasing light energy. The colors are a manifestation of which elements in the ionosphere are colliding with the charged particles: oxygen atoms tend to give green displays, whereas nitrogen gives red and violet displays. A quick tangent at this point on the nature of "color" is in order. Colors as we know them exist only in our minds due to the light receptors in our eyes transcribing different wavelengths of light photons across a spectrum. Visible light interacts with objects by being absorbed or reflected. The color of an object is determined by the material properties that dictate which frequencies reflect into our eyes. The discovery of why the aurora effects occur is credited to a Nobel Laureate in physics, Hannes Alfven, who was a neglected figure in the pantheon of science even after his prestigious accolade. Part of the reason for this neglect was that Alfven often intuited an answer to a problem and then tried to find the appropriate physics to explain it as such. He was often right, but in many cases he was purely speculative. Ronald Merrill, author of *Our Magnetic Earth*, notes how magnetism has led to unconventional but successful discursive science, such as the work of Alfven. This has also sparked the wildly speculative popularity of magnetic bracelets which needs to be questioned, although there are legitimate "transcranial magnetic stimulation" procedures for certain neurological ailments.[23]

The inextricable linkage between electric charge and magnetism not only delights us with planetary phenomena such as the auroras; it is also an essential mechanism by which we convert electricity into motion. Magnets are the beating heart of motors, for it is through the attraction and repulsion dynamics of magnetism within a particular mechanical design that the rotation of an anvil can drive myriad devices. For very large rotational devices, such as wind power generators, we need particularly high-performance magnets made of elements like neodymium and dysprosium coupled with boron and iron. Finding that optimal mix of unpaired electron spins in alloys and composites to display high functional magnetic order continues to be a quest for mechanical engineers. Magnetic order confers important properties for effective and efficient energy delivery, but the source of these magnetic materials can also have important geopolitical implications. Magnets also have a physical "memory": their past magnetic exposure can impact their future magnetic moments. This property is termed *hysteresis* and is found in many natural systems: hysteresis is *the dependence of the state of a system on its history*. This is why magnets were key for memory storage on cassette tapes or disk drives in early computational hardware, and indirectly why they are still critically important for clean energy technologies. As we will see later in Chapter 4, trade in magnetic materials—because of these key properties—has important implications for economic order. If supplies are constrained due to geology or regulatory restrictions, there may be a motivation to find alternatives. Nevertheless, all such alternatives will still be inherently subservient to certain limits imposed by magnetic order.

The elements and the range of properties which they exhibit across our planet, from chemical bonding to nuclear energy to magnetism, are the functional edifice of natural systems. However, it is not just in their composition that natural order is maintained and understood. There are pathways which elemental interactions take in ecological processes. Just as the lines of a magnetic field have directionality and consequence, natural cycles and periods have fundamental impact on natural order as well. Such cycles can develop into adaptive pathways that follow *panarchy* processes in multiple simultaneous feedback loops as an antidote to traditional *hierarchy* analyses. The "Pan" in panarchy comes from the Greek god of wild nature, who is shown as a half-goat humanoid with capricious instincts that also have given us the word "panic." These cycles, which were first conceptualized as three-tiered hierarchical loops by the ecologist C. S. Holling, move from disruption to a series of adaptive feedback loops. Often shown as attractive infinity-style looping cycles, the patterned feature of these cycles have less value than their underlying meaning. What are the mechanisms by which such cycles emerge, and what are the limits of the order they can confer? Examining the nuanced responses to this question will be the next chapter's goalpost.

Notes

1. Among the finest books on this topic is Kean, S. 2011. *The Disappearing Spoon: And Other True Tales of Madness, Love, and the History of the World from the Periodic Table of the Elements*. Back Bay Books.
2. An excellent book on the early history of synthetic diamonds is Hazen, Robert. 1999. *The Diamond Makers* (revised edition). Cambridge University Press. In the past decade, there have been extensive advancements in gem-quality diamond formation from chemical vapor deposition on a seed gem.
3. Johnson, Les, and Joseph E. Meany. 2018. *Graphene: The Superstrong, Superthin, and Superversatile Material That Will Revolutionize the World*. Prometheus.
4. Smil, Vaclav. 2004. *Enriching the Earth: Fritz Haber, Carl Bosch, and the Transformation of World Food Production*. MIT Press.

5. For a good compendium of responses to each of the creationist critiques of C14 dating, refer to the National Center for Science Education https://ncse.ngo/answers-creationist-attacks-carbon-14-dating (accessed July 1, 2020).
6. This was Kaenel, Camille von. July 21, 2015. "Fossil Fuel Burning Obscures Radiocarbon Dates." *Scientific American*. https://www.scientificamerican.com/article/fossil-fuel-burning-obscures-radiocarbon-dates/.
7. For details on the prototype of the battery Amosov, V. N., V. N. Babichev, N. A. Dyatko, S. A. Meshchaninov, A. F. Pal', N. B. Rodionov, A. N. Ryabinkin, A. N. Starostin, and A. V. Filippov. 2018. "Experimental Simulation of a Diamond Betavoltaic Battery." *Technical Physics Letters* 44 (8): 697–699. https://doi.org/10.1134/S1063785018080023. Refer to the video challenging the promotional material from the University of Bristol in https://www.youtube.com/watch?v = JDFlV0OEK5E.
8. Sabin, Paul. 2013. *The Bet: Paul Ehrlich, Julian Simon, and Our Gamble over Earth's Future* (1st edition). Yale University Press.
9. Davis, E. D. (2019). "The Oklo Natural Fission Reactors and Improved Limits on the Variation in the Fine Structure Constant." *AIP Conference Proceedings* 2160 (1): 070012. https://doi.org/10.1063/1.5127735.
10. Montgomery, Scott L., and Thomas Graham Jr. 2017. *Seeing the Light: The Case for Nuclear Power in the 21st Century* (1st edition). Cambridge University Press.
11. Ausubel, Jesse, 2015. "Power Density and Nuclear Opportunity." Rockefeller University. https://phe.rockefeller.edu/docs/PowerDensity_Final120815.pdf.
12. I am indebted to my former student Brian Capobianco (currently pursuing his doctorate at Ohio State University) for alerting me to this case. Singer, Mitch. September 12, 2016. "Watts Bar 2: Economic Catalyst for the Tennessee Valley." Rhea County Economic and Community Development. http://www.rheaecd.com/watts-bar-2-economic-catalyst-tennessee-valley/.
13. Berlinguette, Curtis P., Yet-Ming Chiang, Jeremy N. Munday, Thomas Schenkel, David K. Fork, Ross Koningstein, and Matthew D. Trevithick. 2019. "Revisiting the Cold Case of Cold Fusion." *Nature* 570 (7759): 45–51. https://doi.org/10.1038/s41586-019-1256-6.
14. Castelvecchi, Davide. 2020. "Welcome Anyons! Physicists Find Best Evidence Yet for Long-Sought 2D Structures." *Nature* 583 (7815): 176–177. https://doi.org/10.1038/d41586-020-01988-0.
15. Zhang, S. L., I. Stasinopoulos, T. Lancaster, F. Xiao, A. Bauer, F. Rucker, A. A. Baker, et al. 2017. "Room-Temperature Helimagnetism in FeGe Thin Films." *Scientific Reports* 7 (1): 123. https://doi.org/10.1038/s41598-017-00201-z..
16. Stein, Daniel L., and Charles M. Newman. 2013. *Spin Glasses and Complexity* (1st edition). Princeton University Press.
17. Annals of Improbable Research: https://www.improbable.com/2010/10/05/geim-becomes-first-nobel-ig-nobel-winner/.
18. Heathcote, N. H. de. Vaudrey. 1932. "Christopher Columbus and the Discovery of Magnetic Variation." *Science Progress in the Twentieth Century (1919–1933)* 27 (105): 82–103.
19. Smith, Peter J., and Joseph Needham. 1967. "Magnetic Declination in Mediaeval China." *Nature* 214 (5094): 1213–1214. https://doi.org/10.1038/2141213b0.
20. Doglioni, Carlo, Johannes Pignatti, and Max Coleman. 2016. "Why Did Life Develop on the Surface of the Earth in the Cambrian?" *Geoscience Frontiers* 7 (6): 865–873. https://doi.org/10.1016/j.gsf.2016.02.001.

21. Cooper, A., Turney, C. S. M., Palmer, et al. 2021. "A Global Environmental Crisis 42,000 Years Ago." *Science* 371 (6531): 811–818. https://doi.org/10.1126/science.abb8677.
22. Chancellor, Jeffery C., Rebecca S. Blue, Keith A. Cengel, Serena M. Auñón-Chancellor, Kathleen H. Rubins, Helmut G. Katzgraber, and Ann R. Kennedy. 2018. "Limitations in Predicting the Space Radiation Health Risk for Exploration Astronauts." *NPJ Microgravity* 4 (1): 1–11. https://doi.org/10.1038/s41526-018-0043-2..
23. Merrill, Ronald T. 2011. *Our Magnetic Earth: The Science of Geomagnetism* (reprint edition). University of Chicago Press.

3 Circularity, Cyclicality, and Sustainability

> The eye is the first circle; the horizon which it forms is the second; and throughout nature this primary figure is repeated without end. It is the highest emblem in the cipher of the world.
>
> **—Ralph Waldo Emerson**

Circles are clichéd metaphors for the mystique of life and a sense of ordered closure. They symbolize an infinite loop of recurrence without a beginning or an end. In natural systems circularity is synonymous with renewal and hence renewability. Functionally, we think of natural processes in cycles, which are an elaboration of circularity. But cycles do emerge from linear and perhaps stochastic processes as well. The most pivotal elements on Earth go through a series of cycles, which we have come to take for granted, but they also had their origin at some point. Even catastrophic disruptions can bring forth conditions for order which in turn leads to cycles of sustainability. In contemporary environmental conversations, "circularity" tends to lead us down the path to discuss the fabled "3Rs"—Reduce, Reuse, and Recycle—which have recently been rebranded in the context of a "circular economy." Recycling, in particular, has been put forward by some industries, specially the plastics sector, as a panacea through a highly stylized illusion of order, as noted with plastic waste categories discussed in the previous chapter. We need to go deeper into structural geological order to better understand the opportunities and constraints of employing circularity in earthly order.

Consider the concentrations of the elements of carbon, sulfur, nitrogen, and hydrogen on our planet. Recent research at Rice University has shown that the most likely explanation for these elements and their resulting life-giving cycles was the most catastrophic event in the history of our planet.[1] Around 4.5 billion years ago, when Earth was a mere geological embryo, a Mars-sized proto-planet named Theia is believed to have struck our planet with such impact that the debris generated formed the moon and set it in orbit. This collision was immensely consequential in terms of the elemental concentrations needed for life, but also in the creation of our planet's protective magnetic field. The collision also gave us the clockwise and slightly angular rotational speed for both Earth and the moon. Comet collisions brought forth water

Earthly Order. Saleem H. Ali, Oxford University Press. © Oxford University Press 2022.
DOI: 10.1093/oso/9780197640272.003.0004

and eventually gave us a hydrological cycle. The moon's rotational and orbital synch around the Earth in turn gave us the tidal cycles and a cascading story of the evolution of life.

In contrast, Venus experienced some other enigmatic episodes in its early history which led to its rotational axis being reversed and also slowing its speed dramatically. A day on Venus is 243 Earth days long, which means that there is very little dynamo effect and essentially no functional magnetic field. The slow circulating Venus also does not have any moons, and much of the gravitational influence exerted on its geosphere comes from the sun. Without a functional magnetic field, the planet was also not protected from solar radiation, and volcanism led to a dense atmosphere of carbon dioxide that has rendered the planet the archetype of a greenhouse hell in the solar system. Disruptive forces can thus spark virtuous cycles or vicious spirals toward doom. Material size and orbital scale have played an important role in how we classify planetary bodies. The demotion of Pluto from a planet to a "dwarf planet," that is part of a vast field of icy rock orbiting celestial bodies, was met with street protests in 2006. As a circular orbiting entity, there was emotional appeal to keeping it in the interplanetary order of the solar system. Yet the discovery of the vast number of other similar smaller bodies in the Kuiper Belt made its demotion appropriate.

The allure of the circle was most influentially exemplified by Barry Commoner's landmark 1971 book *The Closing Circle*. This book was partially meant to consider underlying structural aspects of ecological order as more important than linear variables such as human population growth. A few years earlier, the "Club of Rome" had been formed and their landmark book *Limits to Growth* had been published. While partially agreeing with the concerns about unlimited economic growth, Commoner's work was concerned about the ecological system's resilience and vulnerability. Commoner thus engaged in a series of debates with population biologist Paul Ehrlich and John Holdren (who was to later become Science Advisor to President Obama). As a result of these engagements, a notable heuristic equation came through which is now a staple of any conversation on planetary sustainability:

$$\text{Ecological Impact} = \text{Population} \times \text{Affluence} \times \text{Technology}$$

The equation is now simply called the IPAT equation and has been the subject of much further debate and controversy over the years.[2] Population and affluence (a corollary variable for consumption of resources per capita) are fairly straightforward as contributing factors, but technology is more complex. Some technologies can increase impact on the environment through resource consumption while others can mitigate impacts through efficiency or pollution control. Mathematicians are often uncomfortable with such heuristic equations, which can give the allure of exactitude because of their arithmetic form. Indeed, the mathematician Neil Koblitz has called such equations "propaganda."[3] A revised form of the equation has also been developed using more mathematical rigor and termed the Stochastic Impacts by Regression on Population, Affluence, and Technology (STIRPAT). This modified version posits the measure of "ecological elasticity," which has more policy appeal as well since it incorporates metrics of resilience and adaptive opportunities.[4]

Apart from the population–affluence debate, the Closing Circle also gave us the "Four Laws of Ecology," summarized as follows:

1. *Everything is connected to everything else.* There is one ecosphere for all living organisms, and what affects one affects all.
2. *Everything must go somewhere.* There is no "waste" in nature and there is no "away" to which things can be thrown.
3. *Nature knows best.* Humankind has fashioned technology to improve upon nature, but such change in a natural system is, says Commoner, "likely to be detrimental to that system."
4. *There is no such thing as a free lunch.* Exploitation of nature will inevitably involve the conversion of resources from useful to useless forms.

The wisdom of circularity embodied in these four simple statements leads us to consider how such ecological mechanisms operate in natural systems. Circularity leads to cycles and periods over space and time that give us the systems of survival for life. These cycles are generally elemental in nature, such as the carbon, nitrogen, and phosphorus cycles, and are mediated by specialized plants and bacteria. Consider the animate and inanimate aspects of these cycles, clunkily termed "biogeochemical cycles." Such an integrated approach to considering elemental cycling can be traced back to the pioneering work of the Russian mineralogist Vladimir Vernadsky, whose 1926 book *The Biosphere*, illuminated the interlinkages between animate and inanimate cycling. The term "biosphere" (sphere of life) had been coined by the Austrian geologist Eduard Suess alongside the term "lithosphere" (sphere of rock) to differentiate the animate versus inanimate parts of planetary systems. However, Vernadsky was remarkably astute in seeking the connections between the mineral/nonliving aspects of the Earth and the biotic/living parts of the planet. The writer Dorion Sagan, who wrote a book by the same title 80 years later, credited him for making us realize these connections: "Life is not life but rock rearranging itself under the sun.... In a sense Vernadsky did for biological space what Darwin did for biological time." A series of diagrams and models developed in *The Biosphere* by Vernadsky, although contrived, were key to understanding natural ecological cycles (Figure 3.1) and led to the field of biogeochemistry.

Several decades later, in the middle of the Arizona desert, there was an attempt made by humanity to replicate a self-sustaining biosphere through mechanical means and labeled *Biosphere 2*. This experiment was inspired by the "ecovillage movement" for self-sustaining communities and Buckminister Fuller's concept of a "spaceship Earth" as *Biosphere 1*. Systems ecologist John Allen, who started his career at the Synergia Ranch ecovillage, partnered with billionaire oil tycoon Ed Bass to provide $150 million for the construction of the facility to test human abilities to recreate earthly order. The prototype was unfortunately not successful in its ability

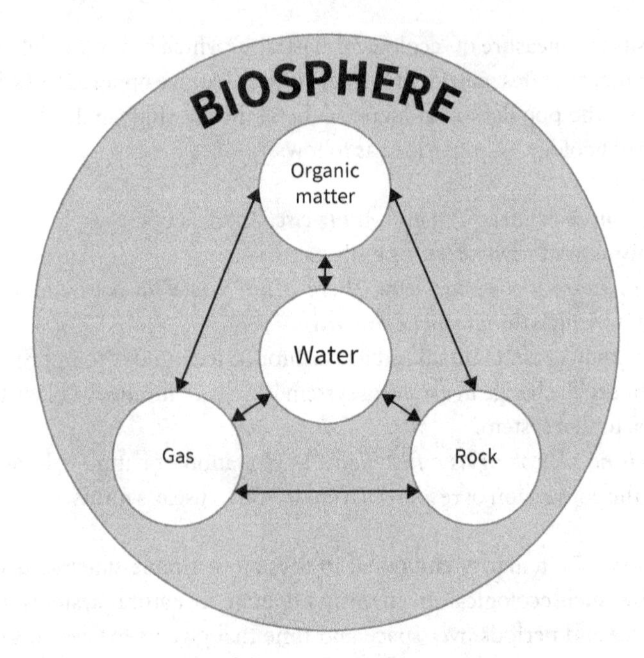

Figure 3.1 Vernadsky's diagram of the biosphere from 1926.

to sustain either the biotic life introduced nor the eight humans who inhabited the massive facility from 1991 to 1993. Apart from the technical challenges, the social interactions within this community were also fraught—a phenomenon that is observed in other "intentional communities" as well. While there are success stories of ecovillages using techniques such as permaculture to live in greater harmony with the environment, the upscaling of such efforts remains a challenge until we can fully understand underlying structures and hierarchies of earthly order.[5] *Biosphere 2* is now a glorified greenhouse research facility and conference center run by the University of Arizona—a monument to human ambition and consequent humility in mimicking earthly order. While worthwhile environmental education is carried forth at the facility, it is clearly not serving the ambitious goals for which it was originally designed.

Within Vernadsky's schematic of the biosphere, water is central to understanding the cyclicality of elements and materials in the biosphere. It was thus quite astounding that *Biosphere 2* was constructed in one of the most parched ecologies on Earth—albeit the goal was to isolate the inside hydrology from the external environment. Simplistic as the heuristic in Figure 3.1 may be, there is little doubt that the hydrological cycle is a vascular imperative for sustaining life on the planet. In some of his work, Vernadsky invested considerable effort to present water as a mineral. He was referring back to a three-point eighteenth-century classification of planetary substances as *animals*, *vegetables*, and *minerals* developed by the taxonomic polymath Carl Linnaeus. These mineralized attributes of water took on confounding complexity as he considered how this essential liquid cycled through the veins of the earth. In his treatise on the *History of Natural Waters*, he classified three broad hydrographic types: freshwater, saline ocean water, and groundwater. However, based on the mineral content, he also delineated 531 "species" of water but noted there could be as many as 1,500![6] As this vast specialization of "water types" suggests, the attributes of the hydrological cycle have often been misunderstood and require more in-depth analysis that builds on Vernadsky's pioneering insights.

Hydrological Order

The circulation of water through the environment was first presented as a diagram by American hydrologist Robert Horton in a paper read at a meeting of the American Geophysical Union in 1931. This diagram (Figure 3.2) was not drawn to scale and

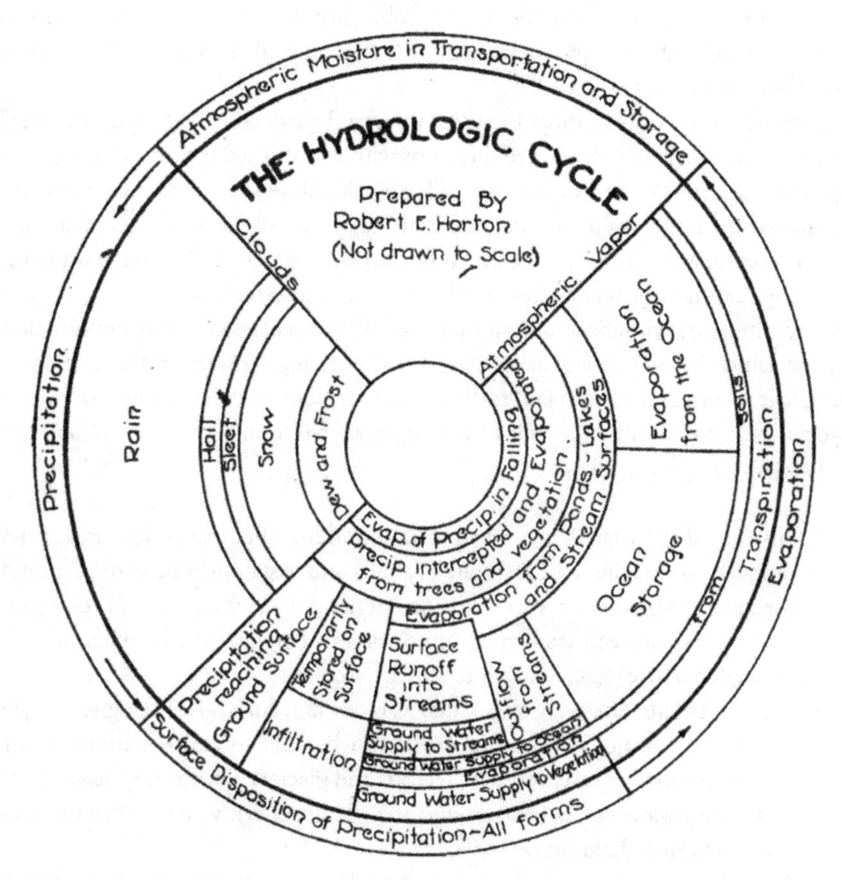

Figure 3.2 First illustration of hydrological order from 1931.
Prepared by Robert Horton for the American Geophysical Union (used with permission).

showed circulation at the circumference of a circle between four key fields: (a) atmospheric water, (b) precipitation, (c) deposition, and (d) evaporation. Radiating inward from these four dimensions were various reservoirs of each category and then a vacuous hole in the middle of the diagram. Horton was also insistent on developing a mathematical precision to what he termed a "water balance" and formulated the following rather simplistic equation: Rainfall = Evaporation + Runoff. The origins of water on the planet remain a mystery, with competing hypotheses related to asteroids, comets, and even the giant collision of the proto-planet Theia which created the moon (as discussed earlier in this chapter). Most recently, it has been proposed that water in the mantle may trace its origins to a primordial solar nebula—a great gaseous nursery for our own Sun—which led to the emergence of Earth and other planets. Regardless, the total amount of water on the planet (on surface and internally) has been relatively constant for millennia. While hydrogen escapes from the planet through atmospheric processes, the impact of this on the water balance is functionally negligible at present, although extreme climate change may alter that prospect. Planetary cycling of water for most practical purposes is thus fairly tight. South African biologist Lewis Wolpert humorously stated this reality as such: "probability theory suggests that when we drink a glass of water, we are imbibing at least one molecule that has passed through the bladder of Oliver Cromwell!"[7]

Subsequent to Horton's diagram, great interest developed in hydrology as a field, and a series of increasingly refined diagrams with visuals showing landscapes sprang up across the world. Such cycles of flow have held allure since antiquity. Heraclitus developed the notion of *panta rhei* ("everything flows") in his treatise *On Nature* in the fifth century BC. There was clearly some intuitive logic to the cycling of water in a closed system like our planet. Yet the diagrams also unsettled some regional experts who saw immense variations in how their own water access issues were not functionally explained by such macro-level diagrams. Reflecting on some of these critiques, the geographer Jamie Linton laid forth a cogent critique of the conventional view of hydrological order under the following categories, synthesizing work by other physical and human geographers[8]:

1. *Humid fallacy*: There has been a presumption that the temperate climatic view of water flows is the "natural" state of affairs and that deserts need to be humidified by human agency. Vast variations in natural water patterns in ecoregions are ignored and can lead to an imposition of engineered hydrological order through massive water transfer or storage projects.
2. *Blue water bias*: Surface waters, such as rivers and lakes, are often given higher salience in hydrological cycle descriptions, whereas they constitute only around 1.2% of freshwater reserves, with icecaps and glaciers constituting around 68% and groundwater comprising around 30%. Atmospheric water is often underestimated in such diagrams as well.
3. *Vegetal interventions in hydrological order*: The impact of forests and vegetation can be highly variable in terms of the kind of climate being considered. Thus,

forests in temperate areas can regulate the water flows depicted in hydrological cycles, but in tropical areas evapo-transpiration effects from plants can be much greater. This can lead to errant prescriptions on forest policy predicated on hydrological presumptions.

All of these critiques pertain to a realization that hydrological systems may well not be in natural equilibrium as the cycle may suggest. There could be a natural dynamic flux which impacts evolutionary processes of species in the "ecosystem"—a term coined in 1935 by the British botanist Sir Arthur Tansley to consider the interactions among plants, animals, and the nonliving environment. Water was seminal to the functioning of ecosystems but also to defining the template of the planet's physical form, at least in Earth's crust, but perhaps even deeper. Recently, it has also been discovered that there is more water in the upper mantle than the entire volume of water in the oceans of the world, suggesting a sync or cycling at larger geological timescales.[9] The slow seepage of water from the earth's oceans to the mantle has also been observed in this context, though the speed of this is much less than polar melting that could happen due to climate change. The vastness of ocean systems gives a certain asymmetry to the hydrological cycle and provides an example of *relative homeostasis*—a dynamic equilibrium—when it comes to salinity levels. For the past 1.5 billion years or so, the salinity levels of ocean water have remained relatively constant at around 3.5% by weight. Of course, within seas and other more confined arms of the oceans, there is further variation in this level. However, the relatively constant salinity in the open ocean is due to *salt sinks*, which allow for the accumulation of salts in oceans to be mineralized and taken out of the hydrological system. This happens at approximately the same rate as they are brought into the oceans through terrestrial mineral flows from rivers and groundwater seepage, while the vastness of the oceans is able to buffer and normalize any sporadic variations.

The removal of salt from the oceans to maintain the relative homeostasis is facilitated by numerous processes ranging from mineral precipitation to shell formation by crustaceans. Thus, the life-giving properties of oceans are linked to salts, but the salinity balance is also an essential part of how the climate is regulated by the oceans. Oceanic currents are moderated by temperature and salinity in what is collectively referred to as *thermohaline regulation*. Salinity of the water increases as we move closer to colder latitudes since freezing selectively takes out fresh, non-saline water from the ocean. At the same time, colder water is also denser, and hence both these factors lead it to sink. As more warm water moves to take its place, the dense, cooler water moves toward the tropics and traverses a long journey to the other polar region and then goes through a long loop along the planet. This "oceanic conveyer belt" is a seminal feature of how water is circulated across the planet. However, this movement is much slower than surface currents generated by wind and tidal forces. The oceanic conveyor belt takes approximately 1,000 years to complete a cycle. This movement is essential for nutrient flows and determines fish movements and hence food supply for millions of humans. There are also regional circulation currents, such as the Atlantic thermohaline circulation, which play an essential role in creating relatively comfortable, habitable climates for major cities in the Northern hemisphere. One of the key concerns related to climate change's impact is that the acceleration of Arctic melting may lead to more fresh water entering the oceans, which could disrupt this circulation. Such a disruption would not only impact nutrient flows but also impact atmospheric phenomena such as the Jet Stream, which gives milder winters to Western Europe. More extreme weather would thus be likely as a consequence.

The ability of water to act as a solvent adds a life-giving complexity to hydrological order. It is also an example of the input-output frameworks by which so many aspects of natural order are manifest. The "budget" of water on the earth is regulated to a large part by the great "oceanic bank" that also holds immense saline reserves of the elements in solution. Carrying our monetary metaphor further, the ocean is a remarkably "solvent" bank with plenty of liquid assets that living organisms capitalize on. However, despite its immensity, human activities are now beginning to impact this massive reservoir. The input of water and salts from land has been impeded, on the one hand, by engineered systems such as dams; by one recent estimate, only 37% of the world's rivers longer than 1,000 kilometers are now "free-flowing." On the other hand, we are exacerbating inflow of water through climate change due to polar sea level rise as well. Over geological time, we see how the hydrological order of the oceans has shifted dramatically depending on stochastic events.

Among the most significant drivers of hydrological order shifting in the oceans have been very large lava flows from deep within the earth's outer core that have made their way to the mantle, resulting in "flood basalts." These events, which have occurred at least 10 times in Earth's history, lead to massive releases of carbon dioxide gas into the atmosphere, resulting in major temperature rise. The geological history of the earth suggests that, at 1,000 parts per million (ppm) of carbon dioxide, no polar ice caps existed, and, during these periods, the oceans would essentially have uniform temperatures and there would be no oceanic conveyor belt (which requires a temperature differential to work). In its most apocalyptic version, this would lead to a build-up of sulfur-generating bacteria which would make the ocean purple with hydrogen sulfide, which is highly toxic to life and also diminishes the ozone layer. This scenario is believed to have played out in the most serious mass extinction episode in Earth's history around 250 million years ago, when 96% of marine life and 70% of terrestrial vertebrate species became extinct. The Permian-Triassic extinction, as it is called, is one of the five major extinction episodes in Earth's history. Except for the asteroid impact extinction which affected the dinosaurs, all the others are believed to have been connected at least proximately to the relationship between the oceans and atmospheric levels of key gases. The inextricable relationship therein between air and water in earthly order takes us next on an intellectual quest steeped in contending metaphors from Greek mythology which still enchant our narratives of life, death, and rebirth.

Orders of Gaia and Medea

The ancient Greek poet Hesiod is believed to be among the earliest sources of mythological order that now continues to grace the name of space missions or video games. The endurance of Greek mythology stems in part from his masterful poem *Theogony*, in which the origin entity in the universe is termed *Chaos*, who abstractly gives birth to the deified embodiment of Earth: *Gaia*. Hence begins earthly order: the various progenies of Gaia, known as the 12 *Titans*, set forth the Olympian ruling order over the planet. The allure of Gaia as a being has captivated many cultures and traditions but, in its contemporary scientific context, is linked to a hypothesis ascribed to the centenarian British systems scientist and engineer James Lovelock.

While working at NASA's Jet Propulsion Laboratory in the early 1970s, Lovelock was asked to consider if there was likelihood for life on Mars based on an analysis of the planet's atmosphere and surface geochemistry. Lovelock surmised that no life was likely since the Mars atmosphere and surface appeared to be in chemical equilibrium, whereas the presence of life is often associated with non-equilibrium processes. Thus, if you took away life, within a short span of geological time, all the life-giving oxygen would be locked up in stable chemical compounds such as nitrates, carbonates, iron oxides, and silicates. Instead, we have oxygen levels which are well-suited for the kind of life that we have present today. What this means is that life creates conditions for a series of negative feedback loops wherein there occur constant fluxes of inputs and outputs. Such a dynamic equilibrium occurs because of the set of negative feedback loops set forth by living systems, particularly microbes and plants, in regulating the earth's biosphere. No such fluxes are occurring on Mars, which led Lovelock to the next bold conjecture: that the animate and the inanimate features of Earth had created certain optimal conditions for life to exist and thrive. This is partly in congruence with the notion of *biogenesis* put forward by Eric Smith and Harold Morowitz, in which they suggest that the biosphere is essentially a fourth geosphere, with the first three being (a) the lithosphere, (b) the hydrosphere, and (c) the atmosphere. Thus, life is an extension of geological reality with an organized state of activity beyond fixed structure. Energy flows within the planet, coupled with the sun's perennial radiation, create a giant "battery" through which hydrogen loss from earth's atmosphere due to solar and ionic neutralization processes creates an "energy channel." On the surface of the earth, there are particular points at which energy from the earth's interior is similarly released (such as hydrothermal vents), and a series of geochemical processes ensue therein that create an opportunity for the emergence of life, as we will see a little later in this chapter.

Lovelock was particularly focused on the atmosphere and how, in the absence of life, the composition of the environment in its current form would be highly improbable. He was well aware of the period of geological history when the composition of the atmosphere was very different. Some of the variations were due to geological processes and some due to the planet's changing orbital and axial rotations, or changes in solar activity, which could also lead to changing climatic cycles. With even slightly higher oxygen levels around 300 million years ago, giant insects roamed the planet because of their particular respiration morphology. Yet only one variable hindered their dominance: birds, which coevolved around this period, started to prey on these larger insects. Then, around 150 million years ago, insect size began to decline even though oxygen levels rose.[10] Such variations on a theme led to a cascade of other prognostications about the pivotal role organisms were playing in maintaining conditions for life to continue to exist. In essence, just as organisms themselves have a process of homeostasis that allows them to thrive within their environment, Earth was also maintaining such a homeostasis. The next metaphorical leap made by Lovelock, perhaps for public effect, was to call this panoply of homeostasis observations on Earth the *Gaia hypothesis*. The invocation of the primordial Greek goddess was an invitation for criticism because it implied agency and personification of the planet. Lovelock also took some poetic license in his earlier writings by referring to Earth as a "single living entity" or by suggesting that "the quest for Gaia is an attempt to find the largest living creature."[11] At an organismic level, the most compelling metaphor for Gaia is perhaps an egg: encased in its shell, amid a delicate order of liquids and membranes, a full-fledged living organism develops. The shell must not be too thin lest it break with minor impact, nor too thick lest the hatchling not be able to emerge on maturity. An amazing harmony of calcium chemistry is enacted by birds as they extract the element from their own bones or by eating shells of mollusks to generate eggshells. The developing chick then consumes some of the calcium in the shell to slowly erode the shell's strength and build its own calcified beak which can break through the weakened shell. As naturalist Lord David Attenborough has noted, "the egg is perhaps nature's most perfect life support system."

The planet, by this metaphorical extension of flows, was in Lovelock's view behaving like a super-organism and hence deserved the emblem "Gaia." This view was immediately criticized by biologists as pseudo-science, but Lovelock ascribed that perspective to his use of a mythic deity as a descriptor. There was considerable debate about whether the metaphor was even apt. Was Gaia a model, a mechanism, or merely a metaphor? From a purely functional perspective, the Gaia hypothesis had a major impact, for which even its detractors give it credit. By considering planetary processes as being interlinked, dynamic mechanisms, this worldview ushered in the advent of *earth systems science*. There were many permutations of the Gaia hypothesis and subsequent development of models as well to consider how to take the hypothesis to the next step of having a theory of planetary order associated with it. However, the detractors of the hypothesis were also noting many notable exceptions to the regulatory aspects of Gaia and pushed back on any suggestions that organisms were creating some sort of optimal order for their survival. Instead, a more plausible way of putting this forward is that there has been a close association between natural selection of organisms for particular environments and the perpetuation of those ecologies through feedback mechanisms set forth by those organisms. There is thus a proposed "tight coupling" (close impactful relationship) which provides this order over geological time scales, but they are by no means immune to disruption.

The notion of *coupling* is a keen feature of natural order also linked to the "rhythms" and synchronization features in many natural systems. For example, the seventeenth-century Dutch engineer, Christian Huygens, inventor of the first pendulum clock, noted that his clocks' pendulums would automatically become synchronized when they were linked with a plank. However, when they were not connected with a plank, this synchronization did not happen. Later, the impact of synchronization was also observed with metronomes on a plank becoming synchronized. This is a key feature of "coupled" natural systems that leads to pulsating behavior across nature, from firefly flashing to the celestial orbits of planets and their moons. The implications of this observation were also linked to how crowded bridges can lead to synchronous pedestrian behavior. This in turn can actually destabilize the bridge because energy can become concentrated in lateral vibrations that would otherwise be canceled out if the movement of the pedestrians was asynchronous. Soldiers are thus asked to eschew their usual marching patterns when crossing bridges to avoid creating instability in these structures. This phenomenon was observed when the Millennium Bridge in London was opened in 2000, and a record number of pedestrians started crossing and the crowding led them to walk synchronously. The bridge started to vibrate horizontally and had to be closed for repairs. What was also noted was that the impact of this synchronization effect manifests itself rapidly at a tipping point, similar to a phase transition. Coupling effects at a planetary level are also observed for the entire planet. For example, seismologists have observed that the entire planet goes through a small earthquake pulse every 26 seconds. The exact nature of this coupling that may cause the Gaia corpus to "beat" rhythmically is not yet fully understood.

One of the most significant insights from Gaia theory has been to explain why the sun's increasing luminosity and heat over the course of Earth's history has not led to more extreme variations in temperature in congruence with this change. The late astronomer and public intellectual Carl Sagan met with Lovelock during his time at NASA and gave him the challenge of trying to explain such a phenomenon. This led to the development of a computational model in which Lovelock considered how a simple and appealing floral organism—a daisy—might develop a self-regulating temperature mechanism. Lovelock considered various properties of the flowers and feedback loops in their interactions with the planet to develop various permutations of the "Daisyworld" model. This model was further refined by academic protégés of Lovelock to explain this planetary thermostat in terms of continental plate tectonics and resulting volcanism, the carbon cycle, and the carbon silicate cycle.

Large-scale variations in temperature, such as the "Snowball Earth" phenomenon that occurred twice in Earth's early history during life's development, were likely caused by rapid weathering and sequestering of carbon dioxide in the world's oceans, leading to massive cooling. In the later episode of this phenomenon, plants may well have also played a role as voracious consumers of carbon dioxide, thanks to a marvelous class of ever-evolving enzymes called *rubisco* (ribulose-1,5-bisphosphate carboxylase-oxygenase), which continue to be a major point of research for climate science.[12] Volcanism was, however, the major geochemical mechanism that broke this cycle and appears to be a key driving force of both extinction and procreation events, depending on its scale and time of occurrence. Plate tectonics and the occurrence of geological resources through planetary processes were no doubt an instrumental factor in determining where humanity evolved and settled in early times, as lyrically depicted in Lewis Dartnell's book *Origins: How the Earth Made Us.*[13] Such agency on the part of planetary forces is powerful but need not be integrated through some particular set of feedback loops to sustain life itself.

The romanticism and quasi-determinism of the Gaia hypothesis, which gives primacy to organisms for acting as collective "geo-engineers," can also be challenged. Indeed, there are plenty of organisms which produce a host of noxious materials that can also inhibit life, and Earth's history presents us with some examples of that through four of the five mass extinctions. The paleontologist James Ward has led some of the discourse in this arena by extending the metaphor of Greek myth and suggesting a counterhypothesis named after a nemesis of Gaia. Further down the divine lineage, Medea, wife of the Argonaut Jason, also appears in Hesiod's *Theogony* as a vengeful and malevolent force. Ward built on critiques of the Gaia hypothesis presented by biogeochemist James Kirschner to present the *Medea hypothesis.* He did so in a book published by Princeton University Press as part of their "Science Essentials" series, which at the time required a supportive sponsor who was a member of the US National Academy of Sciences. The book thus clearly had scientific currency, but his presentation of an alternative hypothesis was not an endorsement of such. Indeed, Ward's main point was to raise the specter that the Gaia and Medea hypotheses held some important insights about positive and negative feedback loops in nature but countered each other's functional veracity. Lovelock had earlier acknowledged that the balance could be tipped toward more deleterious feedback loops in his work with cloud formation from chemicals released by phytoplankton (see Figure 3, which shows the "CLAW and "anti-CLAW hypotheses"—the acronym stemming from the initials of the authors' surnames: Charleson, Lovelock, Andrea, and Warren).[14]

The biogeochemical cycles that Vernadsky described resonated with this view of the planet's life-giving feedback loops as a system, as well as with the earlier work of the brilliant naturalist Alexander von Humboldt who had noted that "nature is a web." Through aristocratic expeditions between the late eighteenth and early nineteenth centuries, von Humboldt gained insights on natural order reminiscent of Lovelock. His book, *Cosmos,* was initially going to be named *Gaia,* but a more universal title seemed more appealing. A recent biography of von Humboldt has an apt title—*The Invention of Nature*—which highlights the ambivalence of such well-intentioned and often well-reasoned harmonious orders. Often such orders are also dynamic and true at certain moments in time. Von Humboldt's views on natural agency were influential on Darwin's development of natural selection, but while the

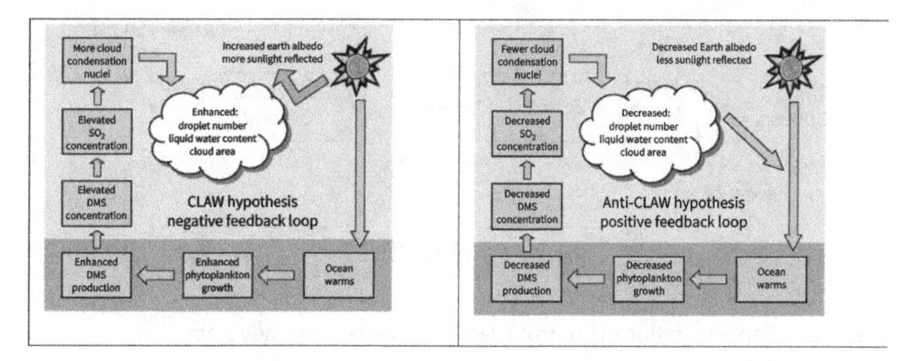

Figure 3.3 The Gaian and Medean aspects of negative and positive feedback loops leading to positive and negative biotic outcomes in the "CLAW" and anti-CLAW hypotheses
Adapted from diagrams on Wikimedia Commons.

former was focused on systems interactions, the latter was more concerned with the fitness for individual survival. Eleanor Jones Harvey, who curated a major exhibition on von Humboldt's and Darwin's ideas in 2019, has noted that "both views were at two points in the swing cycle of a pendulum, having salience at particular moments and intrinsically connected."[15] The Gaia hypothesis has also been presented in a similar light as a process of "sequential selection" by one of Lovelock's most erudite protégés, Tim Lenton.[16] There are clearly many appealing features of Gaia theory, and some of its specific insights are indeed true. But then taking the next step of macroscopic causality, elegant as it may be, is perhaps reminiscent of the words of polymath T. H. Huxley: "the great tragedy of science—the slaying of a beautiful hypothesis by an ugly fact."

Perhaps the most structurally important critique of the Gaia theory pertains to the organisms themselves. If each organism is to create an optimal environment for itself, it will continue to make those influences and grow: this would be a positive feedback loop but ultimately a challenge to sustain. Such growth would itself be antithetical to one of the key underlying theories to the Gaia hypothesis—Control theory—which was developed by the physicist James Clark Maxwell and is shown in Figure 3.3. An expanded version of this theory, when applied to complex systems, gives us the field of *Cybernetics,* which gained ascendance in the mid-twentieth century thanks to its coinage by the systems scientist Norbert Wiener. In the botanical realm, these interactions have even been observed between trees in a forest, mediated by mycorrhizal fungi that create a nutrient exchange network. The German forester and nature writer Peter Wohlleben has controversially suggested that there is a kind of passive intelligence and social connection between trees in a forest.[17] In order to maintain some steady state, there needs to be a mechanism whereby the environment will become less conducive for growth of the organisms. While this is observed in nature through coevolution and predation patterns in natural systems, the fundamental notion of

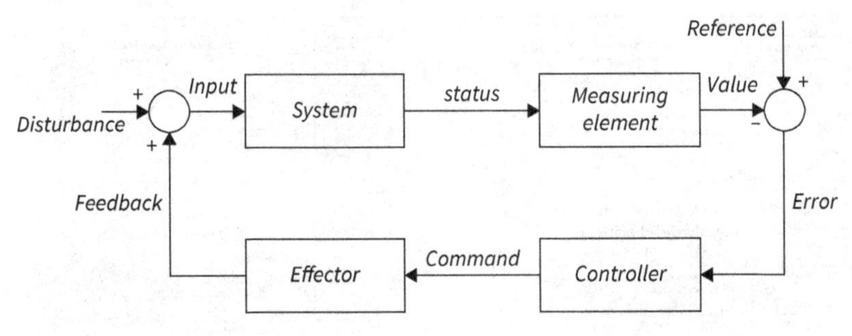

Figure 3.4 Representation of Control Theory to maintain a steady state.
Figure by Anish Devasia.

such a process can be a challenge to reconcile with the original tenets of Gaia theory (Figure 3.4).

As James Kirschner has noted with reference to the violations of such a mechanism: "the Gaian notion of environment-enhancing negative feedbacks is, from the standpoint of control theory, a contradiction in terms." Perhaps the best way forward with understanding "Gaian order," if we may call it such, is then to consider specific controlling mechanisms on the planet. Their functionality is what matters regardless of any agency on their past development. Contemporary earth systems scientists are more interested in such mechanisms in studying these complex interactions for their own functionality. Thus, when the University of Exeter held a centenary symposium in 2019 to celebrate James Lovelock's 100th birthday, most of the presentations were at the level of a specific phenomenon such as glacial albedo, forest fires, Saharan dust, and Amazonian fertility. There were even discussions of how Earth's context compares to the search for exoplanets with possible life, depending on a range of astrobiological and astrogeological variables such as the mass, elemental composition of the surface, magnetic potential of the core, and proximity to the parent star.[18]

An entire field of astrobiology has taken root in this regard, motivated by such a "systems vision" to consider the best variation of conditions chemically, gravitationally, and thermally for life to exist in other worlds. My only exposure to such lofty ideals of translating our understanding of earthly order to extraterrestrial order was the good fortune of being on a panel with Jill Tarter, the director of the Search for Extra-Terrestrial Intelligence (SETI) Institute. Dr. Tarter was inspiration for Carl Sagan's novel *Contact* and the subsequent movie in which her character was played by Academy Award-winning actress Jodi Foster. We were invited by the University of Wisconsin at Madison as part of their "Big Ideas Festival," and, during two days together, I learned about the "Drake equation" that has been an important driver in charting a pathway toward finding intelligent life elsewhere. In our conversations we discussed the visionary work of Dr. Lynn Margulis on scientific elaborations of Gaia theory and how she had been a role model to so many women scientists alongside Dr. Tarter. Married to Carl Sagan at the age of 19, Margulis had persisted with her research through motherhood and a rocky spousal relationship and defined her own unique space in the annals of earth systems science. Yet, at her core, Dr. Margulis was a biologist, and some of her most significant contributions came from seeing patterns of order at the cellular level of organisms.

Organismic Order

Peering under a microscope for years and considering the various organelles within eukaryotic cells—the kind that most plants and animals are made of—Lynn Margulis saw unions of earlier life forms with contemporary life. While traditional biology had framed much of interspecies relations as competition, she saw prospects of harmonization and cooperation. She was just as adept at seeing patterns and inferring trajectories of evolutionary order in organisms as she was in conjecturing the macro-level planetary order of Gaia theory. However, unlike Gaia theory, her cellular research was much more testable and verifiable—albeit here, too, scientific orthodoxy was miserably inertial. Margulis's paper, which presented the theory of *endosymbiosis*, was rejected by 15 journals before the *Journal of Theoretical Biology* finally accepted it in 1966. In this compelling theory, she proposed that primitive prokaryotic life (single-celled organisms with no nucleus in their cell) had embodied itself in eukaryotic cells and formed key components—mitochondria, chloroplasts, and flagella—in the larger cells. The initial insights came from Margulis's interest in *symbiotic order*, the way in which organisms cooperate with each other for mutual benefit and the antithesis of parasitism in which one organism takes advantage of another. The ancient prokaryotic cells which infiltrated early eukaryotic cells found nutrients in the cellular interior (cytoplasm) and in turn provided energy generation and use capacity to the cells. Chloroplasts provided the ability to conduct photosynthesis and harness the energy of the sun for metabolism. Mitochondria were able to provide and conserve chemical energy through a mechanism known as the *citric acid cycle* (or the *Krebs cycle*, named after the Nobel Laureate who discovered it). Flagella provided maneuverability and sensory properties to the cell. Metabolic processes across a wide range of biological traits and various biological systems also follow scaling laws, which were identified by the Danish biologist Max Kleiber. Thus, within a class of animals like mammals, you could essentially plot out dimensions of the size of an organism versus a range of variables and see a defined scaling pattern. The number four also holds a particular salience across organisms. Physicist Geoffrey West calculated the growth rates for a range of biotic metrics and found a nonlinear but regular relationship in scaling that had four in the denominator fraction of the exponent. Thus for example a one-order increase in the mass of an aorta between the hearts of different sized mammals would lead to only a ¾ order increase in its cross-sectional area and this would hold true across a full range of sizes (hence the exponent of a function). The same nonlinear scaling or "power law" of ¾ growth would be observed for the cross-sectional area change for all tree trunks as well. In some cases the proportionality may be inverse but still applicable. For example as size increases heart rates, mitochondrial densities, rates of evolution, and diffusion rates across membranes decrease by a power law factor of negative ¼.[19]

Evolutionary processes due to ecological pressures can allow for slight variations from patterns, but clearly the speed and scale of life's diversity has benefited from such structures. Consequently, the discovery of endosymbiosis by Margulis took place at a time when the temporal order of evolution was being debated because the advent of advanced genetics research led us to question organismic order. The compelling metaphor of a "tree of life" was tempting and had even resonated with some of the research in complex systems science. Yet vertical evolution and the transfer of genetic material from parent to progeny was only part of the story of how such a tree took shape. The work of Carl Woes and a host of other trail-blazing associates led to the discovery of *horizontal gene transfer* between organisms. Endosymbiosis was a mechanism whereby such horizontal gene transfer could occur, and later genetic analysis confirmed Margulis's assertions of the cellular migration between cellular species which had allowed this to happen. In the eponymous title of David Quammen's book on the topic, biologists were now confronted with a more "tangled tree." Genetic analysis also allowed for greater complexity to be accorded to the taxonomic lineage, and higher levels of species differentiation was observed.

These advances in phylogenetics led to the delineation of a separate "domain" of prokaryotic organisms, termed "archaea," found in extreme environments like hydrothermal vents and geysers. These were particularly special ecosystems because they existed in the absence of sunlight, which had previously been considered essential for life to evolve. The discovery of archaea solved a major puzzle around the energetics of the origin of life. The biochemistry needed to develop chloroplasts that can harness sunlight is itself highly energy-intensive and complex, perplexing scientists and forcing them to ask, "How could such complexity spontaneously arise?" Furthermore, oxygenated ecosystems need a flow of light energy to keep them functional, whereas reductive ecosystems, such as in the deep sea where autotrophs like archaea reside (without needing other life forms to feed them), use chemical energy from geological processes and reactions. The realm of hydrothermal vent archaea solved the problem, since, in these ecosystems, unlike in oxygenated systems where energetics favor disorder, the energy gradient in reductive ecosystems favors a drive toward complex self-organization of molecules. It is this remarkable anoxic ecosystem which is now considered to be the location where complex life originated through the formation of the aforementioned citric acid cycle, which itself formed through a geochemical energy gradient. This cycle is able to focus chemical energy to subsequently form a wide range of other complex organic molecules through a metabolic molecular engine of sorts. The biologist Everett Shock summarizes this extraordinary process which occurs in these anoxic ecosystems as "a free lunch you are paid to eat!" Through this process, the tree of life was generated from the humblest of origins. As more complex chemical structures were broken within metabolic cycles, more "events" of biotic creation became ordered.[20]

Within the highly ordered field of organismic taxonomy, restructuring the "tree of life" was a massive undertaking. There were vitriolic arguments between scientists at conferences on whether the tree of life should have two domains of prokaryotes and eukaryotes or an additional third trunk of archaea. Eventually genetics became the touchstone for defining taxonomic order.[21] The branches, or *klados* in Greek, became a point of immense contention among a group of scientists who defined themselves as "cladists." A key defining feature of their categorization process for a branch was that a legitimate evolutionary group must include all descendants of that group. By this measure, mammals are a legitimate branch, but fish are not because the defining features we use for fish can be found in many different branches. At a sociological level, we can use fish as a category for defining marine economic activity, but at a biological level it is not functionally useful. However, cladism had its own limitations of functionality when considering horizontal gene transfer. Thus, the most useful categorization of the tree of life needs to take us back to genetic roots. With the establishment of endosymbiosis, the only controversy that remains is whether eukaryotic cells emerged on their own from a common primordial cellular ancestor and then absorbed primitive bacteria into their organelles, or whether a form of archaea consumed bacterial cells to form eukaryotes (Figure 3.5).

As recently as 2019, this debate took a contentious turn again with the discovery of peculiar archaea from the depths of the ocean near Greenland. Given their location in the oceanic playground of the Vikings and their many shipwrecks, scientists named the new phylum Asgard, after the ancient realm of the Norse gods. Some of the specific archaea are named after Loki, the mysterious trouble-making brother of Thor, the god of thunder. The genetics of "Lokiarchaeota" appear to favor a two-domain ordering, but the debate rages on as there are still so many undiscovered "missing links." The good news is that we are getting better at identifying ancient organisms, and even ancient viruses, which might shed further light on the tree of life. Viruses are still officially not on the tree because they are dependent on host cells to remain active. However, in 2017, the journal *Science* published an article reporting the discovery of large Klosneuviruses which had protein translation genetics.[22] This has again raised the specter of where in the order of life we can place these complex molecular agents. Much of the search for the definitive order of viruses and living cells will require far more discovery of ancient organisms in the paleontological record or perhaps even directly in ice cores where viruses as old as 30,000 years have been revived. Bacteria as old as 100 million years from the deep Pacific seabed were revived in 2020 by Japanese researchers as well.[23] The concept of a temporal order of "life spans" is thus lost in some remote niches of the microbial realm. It is also important to note that both viruses and bacteria can play important roles in sustaining symbiotic relations with other organisms. The digestive bacterial balance for good health is well known and advertised through yoghurt and kefir cultures, but some viral agents are also showing similar salubrious effects. An example is a virus that infects

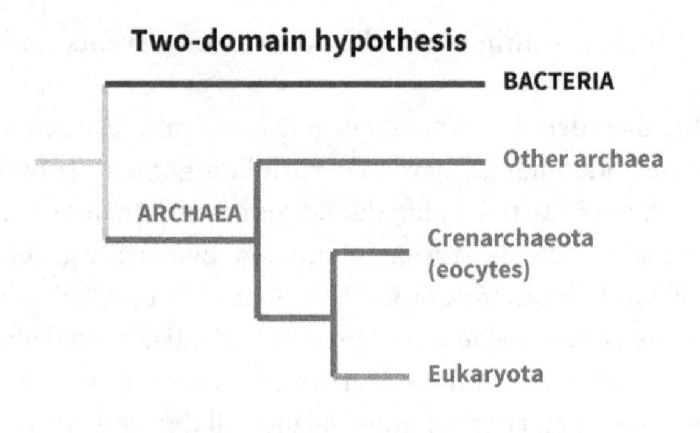

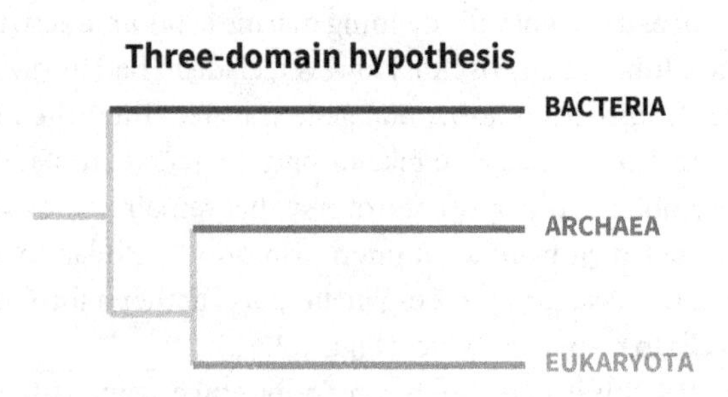

Figure 3.5 Domain hypotheses for the tree of life.

a fungal endophyte that colonizes a specific grass in the extreme geothermal soils of Yellowstone National Park. All three—virus, fungus, and plant—are required for survival in soils with temperatures higher than 50°C.[24]

Despite their redeeming qualities, there are indeed many microbes which kill life. Yet even these pathogens, predators, and many agents of death and distress among organisms are considered either harbingers of balance or as brute beacons of renewal. The "circle of life" symphonic cadence or the "law of the jungle" clearly has appeal beyond just Disney musicals and Kipling tales. However, the trajectories of organismal dominance on the planet have many forks of possibilities that should be noted as yet another limit to predicting outcomes from existing natural order. For example, the competition between reptiles and mammals for dominating evolutionary speed has been linked to levels of oxygen in the atmosphere. Peter Ward (who popularized the aforementioned Medea hypothesis), alongside other prominent paleontologists, believes that reptiles dominated the earth at a crucial period of the Jurassic because their bodies were more efficient in lower oxygen environments. Mammals that coexisted at the time with reptiles remained relatively small during the age of the dinosaurs. Had the oxygen levels been higher, mammals may have evolved faster, and then we humans could have potentially evolved several million years earlier—which would have potentially led to very different technological advances by this time! Indeed, with the oxygen levels and the dinosaurs being as they were, humans might not have even had a chance to evolve had it not been for the asteroid impact which destroyed many of the largest predatory dinosaurs.

Another peculiar asymmetry in organismic order pertains to how multicellular organisms evolved almost exclusively with eukaryotic cells. Prokaryotic cellular organisms form colonies, which indicates some level of orderly cooperation but not of the same level of functional coherence that multicellular organisms such as ourselves possess. How this initial order arose from unified cells to multicellularity is still widely debated. The size of individual cells is constrained by biophysical laws of diffusion by which nutrients can be absorbed efficiently and that is why all single-celled prokaryotes are so small. However, multicellular organisms can avoid this limitation on size and grow substantively in size and organization. Collagen is considered the key miracle molecule that allowed for cells to adhere and form increasingly larger mechanisms of cooperative existence. Collagen is a molecule which needs many oxygen bonds; it arose to prominence around 600 million years ago. No wonder it has acquired a cult following now among health supplement enthusiasts.

The benefits of such cellular cooperation were also encoded within the genes that later transferred this order to subsequent generations of cells. A malfunction in these genes, or of the cooperative biophysical mechanisms between these cells, is the pathological phenomenon we call *cancer*. Essentially, cancerous cells lose their ability to cooperate and share resources in an orderly way. Instead, they become selfish and proliferate for their own sake rather than for keeping order within their host organism. The chance of such a pathology developing stems from mutations which may occur during cellular reproduction. Mutations are, of course, the template with which natural selection can operate and species can evolve. Yet each mutation also causes a slightly higher chance of an aberrant cell to also form and lead to cancer. Thus, the benefits of mutation in providing innovations are constantly at odds with the potential for problematic cells arising. The process of cellular aging is also believed to have linkages to mutation rate changes in cells which may occur as a way to counter other chemical processes of cellular decline. Ultimately the "life span" of an organism is defined by how well such processes of decline can be abated through maintaining cellular order. This varies from species to species based on their respective metabolisms and cellular composition. Death is, of course, the essential part of circularity in the organismic order.

Yet the predator–prey relationship of humans with other animals has created a synthetic order of factory farming that is extraordinary in terms of its long-term impact. No doubt the evolution of the *Homo sapiens* brain is attributed to our transition to a carnivorous diet that some ancestoral apes in the species tree may have made. However, our current consumption patterns of meat are rightly questioned by environmentalists as a disruption of original predator–prey relationships. The upscaling of meat production in areas where plant-based diets had previously been dominant is a cause for alarm. As a counterweight to this trend, absolute vegetarianism and veganism trends are also rising—only humans are capable of creating such antidote orders that are predicated on morality. Adding to the complexity is also a potential win-win opportunity for growth in plant-based meat mimicry products as well as lab-grown biological meat products. Ultimately, the human predator–prey relationship has also been desensitized between the consumer and the producer of meat products, which is why we need to consider its impact at a planetary scale.

Bounded Natural Order

The life and death of organisms is a parable of circularity, but within our finite orbiting planet, what these organisms do in their existence leads us to a final coda for functional natural order: the notion of *planetary boundaries*. While the term itself is relatively new, the concept of some physical constraints on inanimate resources, such as raw materials and food for biotic energy and human development, has been around for centuries. Aristotle divided the world into "animals, vegetables, and minerals" with the presumed dominance of animals consuming vegetables in turn consuming minerals through the soil. Humans had the ingenuity to also consume minerals directly, too: hence our monikers of metallic ages that define human development as well as our proclivity for salt. There were also odd exceptions like Venus flytraps, carnivorous plants which reversed the predation hierarchies. Peculiar creatures like sponges, polyps, and coral also confounded the bounded order of animal, vegetable, or mineral. Working with polyps and hydras led the eighteenth-century naturalist Abraham Tremblay to explore the realm of "stem cells" without fully knowing their remarkable abilities to transform into almost any other cell in an organism. Corals were clearly among the most prolific mineral producers on the planet, forming hard exoskeletons of calcium carbonate. They have thus been instrumental in sequestering so much of Earth's carbon that reef systems are an integral part of any contemporary discussion of planetary boundaries for sustaining the diversity of life.

Naturalists for much of the Enlightenment period wrestled with natural order at a planetary level and tended to be biased in favor of animal properties in defining the hierarchy of life. As historian of science Suzannah Gibson has observed in her history of defining planetary order during this time, the animal bias would later become enshrined in the "Mrs. Gren" definition of life, articulated through these seven properties: *Movement, Respiration, Sensation, Growth, Reproduction, Excretion, Nutrition.*[25] Not all cellular life carries the totality of these properties, and clearly we now define organismic order through genetics rather than through any such properties. Yet the notion of some planetary boundary conditions also harkens back to a human bias in defining the ideal conditions for the diversity of life that favors our own existence. What level of species diversity is optimal in a particular ecology? This is a confounding question for ecologists. Ecologist Gene Evelyn Hutchinson popularized this question in his 1959 paper, written while visiting an Italian monastery and eponymously titled "Homage to Santa Rosalia or Why There Are So Many Kinds of Species?" Core to his analysis, as well as to subsequent work in this arena, has been an understanding of energy flows through *trophic order*—food chains that are best adapted to achieving a dynamic equilibrium for an ecosystem. The size of the organisms also matters in terms of the energy balance. Tardigrades, the smallest of animals (micro-animals) at less than a millimeter in length, are among the most resilient creatures because they have such low sustenance needs and thus survived all five great extinction events and even the hostile environment of outer space.

Yet resilience appears to be augmented when particular organisms are part of an ordered ecosystem. The Soviet-era Russian biologist M. M. Kamshilov conducted a remarkable series of experiments in the Murmansk region of his country to test the toxicity of phenolic acids, a common set of naturally occurring organic toxins. He exposed four different model ecosystems to the acids: first consisted of bacteria; second, bacteria and aquatic plants; third, bacteria, aquatic plants, and mollusks; and fourth had all constituents of the preceding three plus fish. Although, the chemical breakdown of phenolic acid is only possible by bacteria, the fourth system with the highest diversity of organisms deactivated the toxin the fastest.

Food chains for all creatures great and small are ultimately dependent on nutrient cycles, and this is where the notion of *carrying capacity* in ecology finds basic chemical resonance. Ultimately, all life has some basic nutrient needs, and if a system is unable to cycle nutrients effectively, that capacity to sustain life is lost. The Australian ecologist Will Steffen and Swedish systems scientist Johan Rockström analyzed such cycles and human interventions around the turn of the millennium and, in 2009, formulated an extension of the carrying capacity concept to delineate a bounding order to sustain life as we know it. They defined nine "environmental limits within which humanity can safely operate" and termed them "planetary boundaries." Later these nine boundaries were also complemented with 12 *social floors* by the economist Kate Raworth to constitute what she referred to as a "doughnut's space" within which we could sustainably reside (Figure 3.6). This heuristic donut has charmed the world ever since and made appearances at august gatherings such as the World Economic Forum in Davos, Switzerland, and the British Parliament.

Sustainability is bounded by the donut, and anything that exceeds those boundaries starts a cascade of positive feedback loops that could lead to eventual ruin, though the authors are adamant that this does not necessarily imply an irrevocable "tipping point." Recovery is possible, and the boundaries are set forth before thresholds are reached for irreversible harm. Further refinements also built resilience and uncertainties into this framework. The United Nations has given an endorsement to this framework by forming an Earth Commission, in 2019, to further define these boundaries and to develop ways of monitoring humanity's performance in staying within these bounds through "science-based targets." Whether or not the Earth Commission's edicts are able to keep humanity within this ostensibly safe operative space remains to be seen. Although setting forth such boundaries is unsettling to some scholars, the

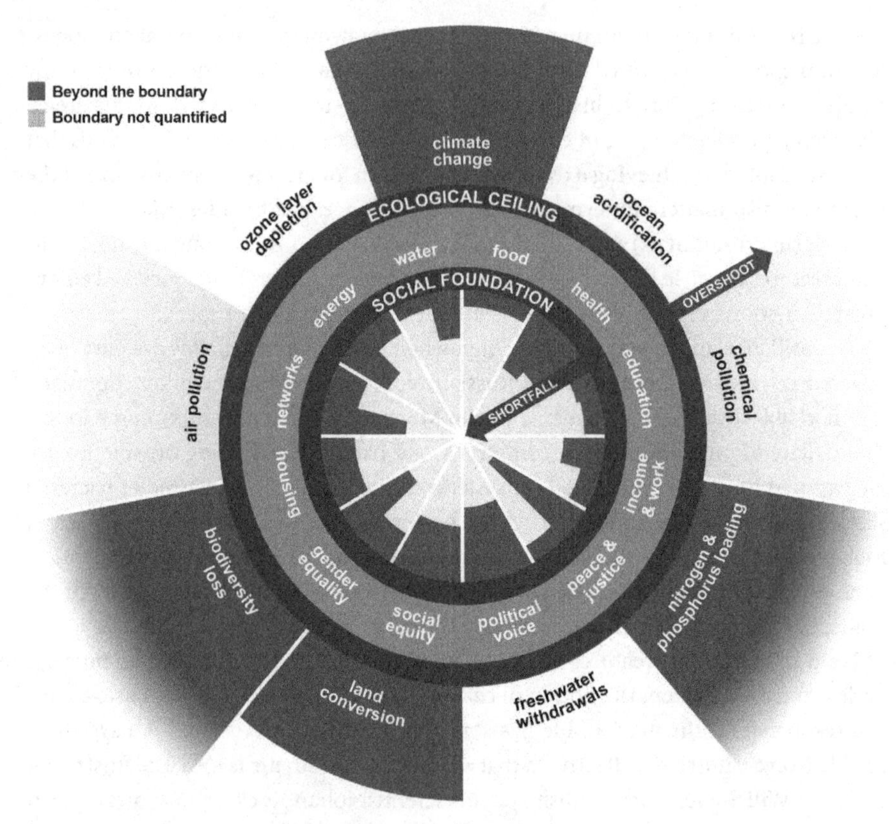

Figure 3.6 Planetary boundaries, ecological ceilings, and social foundations.
From Raworth, Kate. 2017. "A Doughnut for the Anthropocene: Humanity's Compass in the 21st Century." *The* Lancet Planetary Health *1* (2): e48–49. https://doi.org/10.1016/S2542-5196(17)30028-1.

geological reality that *H. sapiens* (or perhaps now *H.* "*geosapiens*"[26]) has been blessed to live in the interglacial Holocene epoch for the past 12,000 years cannot be underestimated. This epoch has provided us a proverbial "goldilocks" zone, and the safe operating space being proposed is to ensure that we do not push ourselves over key tipping points toward a much more uncomfortable environmental reality. Just as the universe has key parameters that allowed existence to emerge in its current form (as discussed in Chapter 2), so, too, does our planet's functional scale. The discoverers of the ozone hole gave us an insight into how close humanity came to reaching a planetary tipping point with synthetic interventions. Had we developed compounds with bromine rather than chlorine as their elemental basis, the damage to the stratosphere could well have been irreversible, dooming humanity to a diabolically cancerous future. Since humans even lived in the Ice Age, there is little doubt that we would survive in some form, but it would be a far more unsafe and uncomfortable existence with greater resource constraints.

The attraction of bounded order is an essential aspect of environmentalism. The notions of circularity with which this chapter began are finite in space but potentially infinite in process. Round and round we go, so long as our motion is within that bounded space, just as with celestial orbits. Yet if we did not have the exact same systems of planetary operation, it would simply mean that we would have had different trajectories—some more biologically consequential than others. If the Saharan dust had not balanced the phosphates leaching out of the Amazon due to rain, the flora and fauna may have been less biodiverse but adaptive in other ways. If the same dust did not slow down many hurricanes on its path across the Atlantic, we may have had more frequent violent storms and perhaps not have built as many coastal cities. Similarly, trees in the Amazon counterintuitively transpire more moisture during dry times, and the massive moisture released forms an aerial water vapor flow that eventually condenses back as rain to foliate more parched parts of neighboring ecosystems. Eventually the rains return to the giving forest again. Insect migrations also form such remarkable systems of flows. Dragonflies from India can migrate across the Indian Ocean on monsoon vapor currents and breed in the swamps of East Africa. Their nymphs prey on mosquito larvae and help to reduce disease vector transmission. Ultimately, the different possible trajectories that come out of these natural processes reflect how we can manage order out of natural resource constraints. Many of the systems we have developed for managing resources on a day-to-day basis are often not in sync with the ebbs of and flows of planetary processes. Moving from the natural order of *ecos* (our planetary home) to the dominant functional order of *H. economicus* in the marketplace is the next phase in our quest.

Notes

1. Grewal, Damanveer S., Rajdeep Dasgupta, Chenguang Sun, Kyusei Tsuno, and Gelu Costin. 2019. "Delivery of Carbon, Nitrogen, and Sulfur to the Silicate Earth by a Giant Impact." *Science Advances* 5 (1): eaau3669. https://doi.org/10.1126/sciadv.aau3669.
2. Chertow, Marian R. 2000. "The IPAT Equation and Its Variants." *Journal of Industrial Ecology* 4 (4): 13–29. https://doi.org/10.1162/10881980052541927.
3. Koblitz, Neal. 1981. "Mathematics as Propaganda." In *Mathematics Tomorrow*, edited by Lynn Arthur Steen, 111–120. Springer. https://doi.org/10.1007/978-1-4613-8127-3_12.
4. York, Richard, Eugene A. Rosa, and Thomas Dietz. October 1, 2003. "STIRPAT, IPAT and ImPACT: Analytic Tools for Unpacking the Driving Forces of Environmental Impacts." *Ecological Economics* 46 (3): 351–365. https://doi.org/10.1016/S0921-8009(03)00188-5.
5. A good discussion of ecovillages and "ecotopia" experiments can be found in Lockyer, J., and J. R. Veteto, eds. 2015. *Environmental Anthropology Engaging Ecotopia: Bioregionalism, Permaculture, and Ecovillages* (1st edition). Berghahn Books.
6. Edmunds, W. M., and A. A. Bogush. 2012. "Geochemistry of Natural Waters: The Legacy of V. I. Vernadsky and His Students." *Applied Geochemistry* 27 (10): 1871–1886. https://doi.org/10.1016/j.apgeochem.2012.07.005.

7. Quoted in Dawkins, R. October 5, 2006. "Strange Science in the Middle World." Beatty Lecture, McGill University. http://www.reporter-archive.mcgill.ca/39/05/dawkins/index.html.
8. Linton, John. 2008. "Is the Hydrological Cycle Sustainable: A Historical-Geographical Critique of a Modern Concept." *Annals of the Association of American Geographers* 98 (3): 630–649.
9. Fei, Hongzhan, Daisuke Yamazaki, Moe Sakurai, Nobuyoshi Miyajima, Hiroaki Ohfuji, Tomoo Katsura, and Takafumi Yamamoto. 2017. "A Nearly Water-Saturated Mantle Transition Zone Inferred from Mineral Viscosity." *Science Advances* 3 (6): e1603024. https://doi.org/10.1126/sciadv.1603024.
10. Clapham, M. E., and J. A. Karr. 2012. "Environmental and Biotic Controls on the Evolutionary History of Insect Body Size." *Proceedings of the National Academy of Sciences* 109 (27): 10927–10930. doi:10.1073/pnas.1204026109.
11. Lovelock, J. 1979. *Gaia: A New Look at Life on Earth*. Oxford University Press.
12. Zhang, Zhiyuan, Renduo Zhang, Alessandro Cescatti, Georg Wohlfahrt, Nina Buchmann, Juan Zhu, Guanhong Chen, et al. 2017. "Effect of Climate Warming on the Annual Terrestrial Net Ecosystem CO_2 Exchange Globally in the Boreal and Temperate Regions." *Scientific Reports* 7 (1): 3108. https://doi.org/10.1038/s41598-017-03386-5.
13. Dartnell, L. 2020. *Origins: How the Earth Made Us*. Vintage.
14. Charlson, R. J., J. E. Lovelock, M. O. Andreae, and S. G. Warren. 1987. "Oceanic Phytoplankton, Atmospheric Sulphur, Cloud Albedo and Climate." *Nature* 326 (6114): 655–661.
15. Presentation at the C. V. Starr Center for the Study of the American Experience at Washington College, October 13, 2015. Video of presentation available on Youtube.
16. Lenton, Timothy M., Stuart J. Daines, James G. Dyke, Arwen E. Nicholson, David M. Wilkinson, and Hywel T. P. Williams. 2018. "Selection for Gaia across Multiple Scales." *Trends in Ecology & Evolution* 33 (8): 633–645. https://doi.org/10.1016/j.tree.2018.05.006.
17. Wohlleben, P., T. Flannery, and S. Simard. 2016. *The Hidden Life of Trees: What They Feel, How They Communicate: Discoveries from A Secret World* (J. Billinghurst, Trans.; illustrated edition). Greystone Books.
18. The Z-pulse power facility at Sandia National labs is simulating exogeological environments to see where else in our galaxy such a Goldilocks zone of planets exists.
19. West, Geoffrey. 2017. *Scale: The Universal Laws of Life, Growth and Death in Organisms, Cities and Companies*. Penguin Press.
20. A marvelous analysis of these processes can be found in Smith, E. 2016. *The Origin and Nature of Life on Earth* (1st edition). Cambridge University Press.
21. For a highly readable account of the development of taxonomy, see Yoon, Carol Kaesuk. 2009. *Naming Nature: The Clash Between Instinct and Science*. W. W. Norton & Company.
22. Schulz, Frederik, Natalya Yutin, Natalia N. Ivanova, Davi R. Ortega, Tae Kwon Lee, Julia Vierheilig, Holger Daims, et al. 2017. "Giant Viruses with an Expanded Complement of Translation System Components." *Science* 356 (6333): 82–85. https://doi.org/10.1126/science.aal4657.
23. Morono, Yuki, Motoo Ito, Tatsuhiko Hoshino, Takeshi Terada, Tomoyuki Hori, Minoru Ikehara, Steven D'Hondt, and Fumio Inagaki. 2020. "Aerobic Microbial Life Persists in Oxic Marine Sediment as Old as 101.5 Million Years." *Nature Communications* 11 (1): 3626. https://doi.org/10.1038/s41467-020-17330-1.

24. Roossinck, Marilyn J. 2015. "Move Over, Bacteria! Viruses Make Their Mark as Mutualistic Microbial Symbionts." *Journal of Virology* 89 (13): 6532–6535. https://doi.org/10.1128/JVI.02974-14.
25. Gibson, Susannah. 2015. *Animal, Vegetable, Mineral?: How Eighteenth-Century Science Disrupted the Natural Order*. Oxford University Press.
26. An intriguing suggested change to our species title. Wolfgang Lucht in his review of *The Human Planet: How We Created the Anthropocene. Nature* 558: 7.

PART II

ECONOMIC AND SOCIAL ORDER

It is better to be roughly accurate than to be precisely wrong.

—Alan Greenspan, Former Chair of the US Federal Reserve

The only laws of matter are those which our minds must fabricate
And the only laws of mind are fabricated for it by matter.

—James Clerk Maxwell

4
The Orders of Economic Harmony

> The aim of science is to seek the simplest explanations of complex facts. We are apt to fall into the error of thinking that the facts are simple because simplicity is the goal of our quest. The guiding motto in the life of every natural philosopher should be, "Seek simplicity and distrust it."
>
> **—Alfred North Whitehead, *Concept of Nature,* 1919**

To many readers, moving from natural order to economic and social order may imply easier reading territory. Humanity often feels more intimately connected to social and economic narratives than to primers on physical and biological science. Social and economic stories capture the headlines of newspapers because they seem to be more proximate and consequential forces of order in our daily lives. Yet the irony of learning is such that natural science and engineering, while more incomprehensible to the average public, are actually more predictable and less complex in analytical terms. Mathematics, the language of natural order, may seem esoteric and convoluted, but it is inherently structural in its exposition. Even in the realm of probability, there are confidence intervals and boundary structures of accuracy. Some of the most landmark recent developments in mathematics have been to understand ostensibly random or "stochastic" processes. The Austrian-British mathematician Sir Martin Hairer, who has assembled most of the accolades in mathematical thought from the Field Medal (2011) to the $3 million Breakthrough Prize (2020), is at the forefront of such efforts. His novel application of differential equations has been applied to finding "regularity structures" in tea-stirring, water droplet diffusion in a tissue, and forest fires. In the social and economic domains, such notions of regularity structures become more challenging to impose even as we find the narratives more intelligible. There remain core challenging problems in mathematics, even with the rise in data processing abilities. Machine learning has to contend with phenomena such as "the curse of dimensionality"—a term used to describe the need for vast amounts of data to get reliable inferences if we increase the features or "dimensions" of a circumstance. This is also related to the notion of a "combinatorial explosion," whereby the complexity of a problem rises rapidly, as observed in Sudoku puzzles. Functionally, Hairer is less concerned about tracking the structure of such numerical "explosions" and hopes that mathematics can advance to the point of understanding brain function

Earthly Order. Saleem H. Ali, Oxford University Press. © Oxford University Press 2022.
DOI: 10.1093/oso/9780197640272.003.0005

processes, which could perhaps give us more precise insights on social behavior as well.

In their aspirational goals of structure, many social scientists have successfully branded themselves with ostensible mathematical exactitude. Economists, whose original goals have been to understand human responses to resource scarcity, negotiate the vagaries of human consumption patterns through pricing mechanisms. Statistics and mathematics are required courses in all economics programs. Many anthropologists have also blended much of their study of human societies and cultures with behavioral determinism that is predicated in natural science methods and discrete datasets. Yet such methodological hybridity has not always been framed around the hierarchies of order in the context of achieving some optimal level of planetary biotic sustainability for human habitation. Part of the challenge remains the hubris with which we consider our grasp of universal knowledge and the ability with which we can comprehend complexity. Physicist John Barrow confronted this challenge in his landmark book titled *Impossibility: The Limits of Science and the Science of Limits*, in which he applied the process of scientific inquiry to the broadest range of learning.[1] In this volume, Barrow presented a diagram which considered complexity and uncertainty orthogonally to each other (Figure 4.1). This diagram lays bare the stark reality that social and economic order is both far more complex and far more uncertain than we thought regarding our equations that describe the phenomena in question.

Barrow also hits upon an important aspect of scientific learning which has to do with the need for humility in the face of uncertainty. The notion of having universal

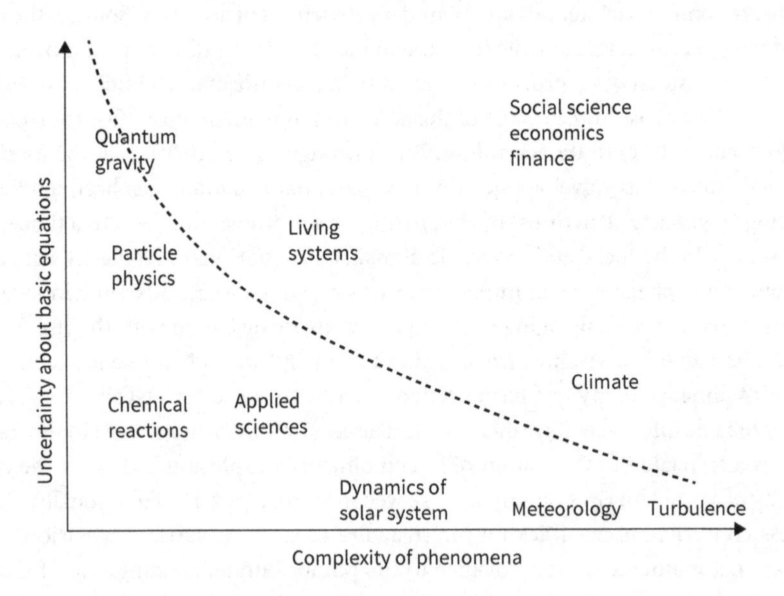

Figure 4.1 John Barrow's placement of economic and social order on a complexity-measurement Cartesian plane.

"natural laws" was not common in antiquity[2] and was largely a product of the Renaissance, when there was a more advanced convergence of mathematics and observational science. Key to this development was also a debate about whether natural systems had exhibited uniformity over space and time—the principle of *uniformatarianism*. Geology accepted this principle most readily with John Hutton's treatise *Theory of the Earth*, with reference to the origins of landscapes on the planet. Though questioned in various forms by the basic sciences, for functional purposes, many of the key features of uniformitarianism hold true. Natural laws that can be expressed as equations can offer a "deep simplicity" in the words of astrophysicist John Gribbin,[3] but they can also unravel and present a range of new dimensions of research just when we thought order was upon us. Simplicity is alluring; nuance is irritating. This is why teachers all too often harken to the thirteenth-century maxim of the Franciscan friar William of Ockham, which simply stated that "entities should not be multiplied without necessity," or, in other words, simple explanations are the best. Multitudinous principles of parsimony in the philosophy of science have emerged from "Ockham's razor." While natural systems in a proximate sense take paths of least resistance and work on pathways that optimize energy flow, the ultimate outcome might not be simple. We are beset by inexplicability and a lack of knowledge or understanding when it comes to many macro-scale problems of human inquiry. It is that wonderment of ignorance which drives the scientific enterprise but also, more broadly, the philosophical quest to find meaning in other ways. Biologist Stuart Firestein celebrates the importance of "ignorance" in a course he teaches by this title at Columbia University. In an accompanying book with the same title, he notes George Bernard Shaw's famous quotation: "Science is always wrong. It never solves a problem without creating 10 more." Using this observation, he goes on to elaborate with literary flair:

> Isn't that glorious? Science produces ignorance, possibly at a faster rate than it produces knowledge. Science, then is not like the onion in the often-used analogy of stripping away the layer after layer to get at some core, central fundamental truth. Rather it's like the magic well: no matter how many buckets of water you remove, there's always another one to be had. Or even better, it's like the widening ripples on the surface of a pond, the ever-larger circumference in touch with more and more of what's outside the circle, the unknown.[4]

Considering this view of science, it is intriguing how the social science fields, particularly economic sciences, have evolved to consider human behavior. Economics has gained ascendance in social science because of our fundamental measure of markets and money as the touchstones of human welfare. Invoking the fabled metaphor of "the invisible hand" that guides market wisdom, Adam Smith's spirit still permeates much of the implicit views of economic order. Markets, one may argue through this view, provide the means by which economic order can emerge out of chaotic individualism. However, the functioning of markets is predicated on certain norms, and Smith was well aware of issues of market failure. During the mid to late eighteenth century, the East India Company was in its heyday and Smith denounced the multinational colonial enterprise as a "blood-stained monopoly." Yet still his views were used, or perhaps misused, to give money and markets an ascendancy in determining economic order. *Laissez faire* became the maxim of economic order for much of the Industrial Revolution of the nineteenth century soon after his demise in 1790. This was also a convulsive time in Europe, with the advent of the French Revolution highlighting the violent whiplash of a fundamentally unequal economic order.

In this milieu, an oft-neglected French scholar named Charles Fourier came to the stage with a concept of "economic harmony," presented as an antidote to the social upheaval of the times. Today, his work is often dismissed as "Utopian Socialism," but many of his insights about conspicuous consumption in capitalist order resonate with a contemporary quest for ecological sustainability. Fourier suggested that economic order should be characterized by the greatest possible consumption of varieties of food and the smallest possible consumption of varieties of clothing and furniture. He was thus the father of economic minimalism at a time of immense largesse. Fourier's concern for economic order arose from the asymmetries between human need and greed. Yet, as population grew, subsistence economies where people simply could survive by growing their own food became obsolescent. Greed of the rich became part of a long supply chain of seemingly useless products whose manufacture still provided needed income to the poor. This form of supply chains that have few alternative paths for delivering components of complex machines can themselves lead to a more fragile order.[5]

Thus, the mercantile economic order became a survival trap in the views of Fourier and other critics of the purported order of "the invisible hand." Fourier was keen to notice neglected ingredients in assumed social orders. He is also credited for coining the word "feminism" to highlight the neglected role of women in society's productivity. Similarly, the contribution of natural resources in providing the template for all economic order was largely taken for granted by economists of his time. Yet his work inspired some intentional communities to be formed in the mid-nineteenth century which attempted to showcase "economic harmony." Villages, towns, and cities became the microcosms for a range of design experiments for social and economic order and heralded the advent of "planning" as the locus for considering sustainability.

Urbanism and Resilience in Socioeconomic Order

The original survivalist impulses of humans and the constant struggle for subsistence necessitated nomadic habitation. The lifestyles of our hunter and gatherer ancestors, as well as those who continue to flourish in such socioeconomic environments, have been linked to a fluidity which, at its surface, defies any order. Yet these societies timed their movements across the Earth with the rhythm of seasons and hence mimicked natural order more closely than settled societies. Still, humans were not always able to keep their ostensible order within ecological limits, and we have examples of extinctions spurred by human hunting in traditional subsistence societies. While the causality of many extinctions in the Americas is contentious, the Maori in New Zealand recognize their part in the extinction of the Moa, a giant flightless bird, as well as its main nonhuman predator, a giant eagle, the Pouakai.[6] The economic order of hunters and gatherers was focused on what some contemporary economists term "spontaneous order" and what great economic thinkers like Adam Ferguson and Michael Polanyi explained as the result of human action but not of human design.

Polanyi's starting point was a critique of positivist science and reductionism in his seminal essay *Life's Irreducible Structure* (1968), in which he suggested that the genetic information in DNA had chemical functionality by being bound to higher-level ordering principles or "boundaries." For systems in which humans have designed machines or cities, the higher-order boundary may appear to be the design, but the functional limitations will be dependent on physical laws of nature and emergence. He further expanded this insight to experimentation and engineering: "the experimenter imposes restrictions on nature in order to observe its behavior under these restrictions, while the constructor of a machine restricts nature in order to harness its workings." In human social systems, we may vary in our attention to the two orders of boundary conditions. In an experimental crucible, the boundary of the vessel is of lesser interest than the reaction occurring within its boundary. However, in a chess game, the boundaries of the rules of chess and their interaction with strategy are of more consequence than the physical structural boundaries of the chess pieces themselves. Spontaneous order in social systems is the result of this interplay of natural and anthropogenic functionalities that have hierarchies but that can develop order from the bottom up as well as from the top down. This insight can be traced back philosophically to the Taoist Chinse philosopher Zhuangzi (369–286 BC), who considered the emergence of social systems from the grassroots rather than from the technocratic authoritarianism for which Confucius had argued in his writings.

Once human societies were able to develop agricultural mechanisms (often termed the Neolithic Revolution) and settle in more permanent locales, the concept of spontaneous order (or emergence, as discussed earlier) became manifest in the emergence of cities. Recent archaeological research as evidenced in Graeber and Wengrow's bestselling work *The Dawn of Everything* (2021) reveals a more meandering path towards hierarchical settlements and complex social order among hunter-gatherers. Nevertheless, the upscaling of cities no doubt happened more rapidly after the advent of agriculture. Cities also helped us operationalize the notion of *capital*—an asset that delivers some functional use for human systems. The concept of civilization as a human development construct became associated with cities. Even the root of the words "city" and "civilization" can be traced back to the same Latin *civitas*, which is also where were get the origin of the word "citizen." The inhabitants of cities were producers, merchants, and consumers whose cohabitation and proximity provided the means of wealth exchange. Innovation in harnessing natural resources and transforming them into other forms of capital provided the mechanism by which cities also grew. Some of the earliest centers of economic power were cities and the fulcrums of major civilizations we recall today. The ancient Sumerian settlement of Ur, founded around 3800 BC, was perhaps the primordial example of a successful city whose name might be the root for the word "urban." The Sumerians of Mesopotamia are considered among the earliest forces of centralized political power in the ancient world. Cities thus also formed a fundamental linkage between economic, social, and political order, a concept we will revisit in Part III of this book.

The economist and writer Paul Krugman, in one of his lesser-known books *The Self-Organizing Economy* (1996), used the example of cities as emergent phenomena in a mercantile economy. The book echoes some of the much broader insights to be found in Eric Jantsch's book *The Self-Organizing Universe* (1980). This monumental work was published during a proliferation of research in systems science as part of a series of books published by Pergamon Press under the title "Systems Science and World Order Library." A key feature of these books was to use complex systems in operationalizing order for human habitation. Cities often figured prominently in such discourse as the most concentrated locus of human habitation, even though there were grand discussions as well of physical laws and international relations in world order. Cities were emblematic of human cooperative impulses beyond kith and kin to deliver more for humanity than the relatively self-absorbed subsistence of nomadic societies. Urban ommerce at once provided the incentive for individualistic wealth generation, but also innovative impulses that could benefit society at scale.

The logic of urban development after the Neolithic Revolution was such that cities did not need to be "invented." They simply developed in various parts of the world as human societies saw their value to trade and commerce. Such an organic process of urbanization happened in some ways similar to cooperative impulses that lead to the formation of termite mounds, beehives, and ant colonies. However, the functionality of such concentrated insect communities is still survivalist rather than intentionally focused on innovation or the qualitative growth of a society. The concept of trade is itself part of the human experience, and hence any resultant economic order is not possible in animal societies. Adam Smith himself noted trade as a key differentiating feature of humans within the broader spectrum of animal societies in his magnum opus *The Wealth of Nations*. Even studies of advanced primates such as chimpanzees suggest that while bartering may be induced in chimpanzee communities through human intervention, it is not a spontaneously generated order, as it is in human societies.

Trade existed in hunter gatherer societies as well, which have been simplistically termed by Western philosophers like Jean Jacques Rousseau to exist in a "state of nature." Yet its significance was likely far less central to survival in comparison with broader cultural attributes and social connections.[7] There is evidence that even without a "social contract" for human relations, trade existed. As the zoologist and public intellectual Matt Ridley states in *The Origins of Virtue*, "exchange for human benefit has been part of the human condition for at least as long as Homo sapiens has been a species. It is not a modern invention." However, the refinement of trade as a means of exercising what was later termed by David Ricardo as "the law of comparative advantage" was amplified through the advent of cities. As human societies clustered around particular natural resources or developed particular skill sets and innovations, they saw the value of their productive enterprises. Cities scaled up such specialization and thus became hubs of trade, which may be considered the most primal and defining form of human cooperation.

Yet the paradox of the rise of urbanism has been that while economists have argued that the historic emergence of cities were bottom-up phenomena determined by markets, much of the field of "planning" has arisen in the context of cities. Most academic programs in planning are labeled "urban studies and planning," and there is a long history of externally ordained designs on urban order. Among the most celebrated examples of such developments in the context of sustainability thinking is the concept of "garden cities," which has been credited to the itinerant entrepreneur Ebenezer Howard. Along with many other Europeans of his time in the nineteenth century, Howard saw the rapid rise of industrialization in Europe lead to economies of pollution and inequality. Industrialization created orders of "natural monopoly," with developments such as railroads and pipelines that led to an ever-greater accumulation of wealth. During his travels in America, Howard was inspired by the writings of novelists such as Edward Bellamy, who had considered the opportunities for a more equitable economic order.

The Gilded Age in America also led to massive accumulations of wealth in the hands of industrialists, inspiring writers like Bellamy, alongside other progressive thinkers of the age, to reimagine the future through mechanisms like cooperatives and sharing. Interestingly enough, many of the principles of our current-day "shared economy" manifest in Uber and Airbnb are reminiscent of the economic order envisaged therein. Whether or not such economies could lead to ecological sustainability would ultimately depend on aggregate levels of consumption and impact reduction. A shared economy, without an eye on distortions of incentives for equitable access, can lead to negative impacts on sustainability indicators as well. For example, the rise of Airbnb in tourist cities led to urban rental scarcity as many owners asked long-term tenants to vacate their homes in favor of higher paying short-term rentals. Despite such pitfalls, the garden city movement had at its core an economic order which would intersect with natural systems, at least in spatial terms. Figure 4.2 shows the archetypal diagram of a network of garden cities envisaged by Ebenezer in his book *Garden Cities of To-Morrow* in 1902.

The "slumless, smokeless" Utopia envisaged by Howard considers interdependence and trade between the satellite cities as well and a core economic capital. There is also a clear sense of diversification within the model, with agrarian and industrial manufacturing sectors coexisting within the network. This physical network, with its series of nodes and "hubs" (with multiple intersections), could also give rise to a social network between the various nodes based on resource availability. A few decades later, researchers in social psychology such as Stanley Milgram would develop the "small world" experiment by which such hubs and spokes are able to connect vast social

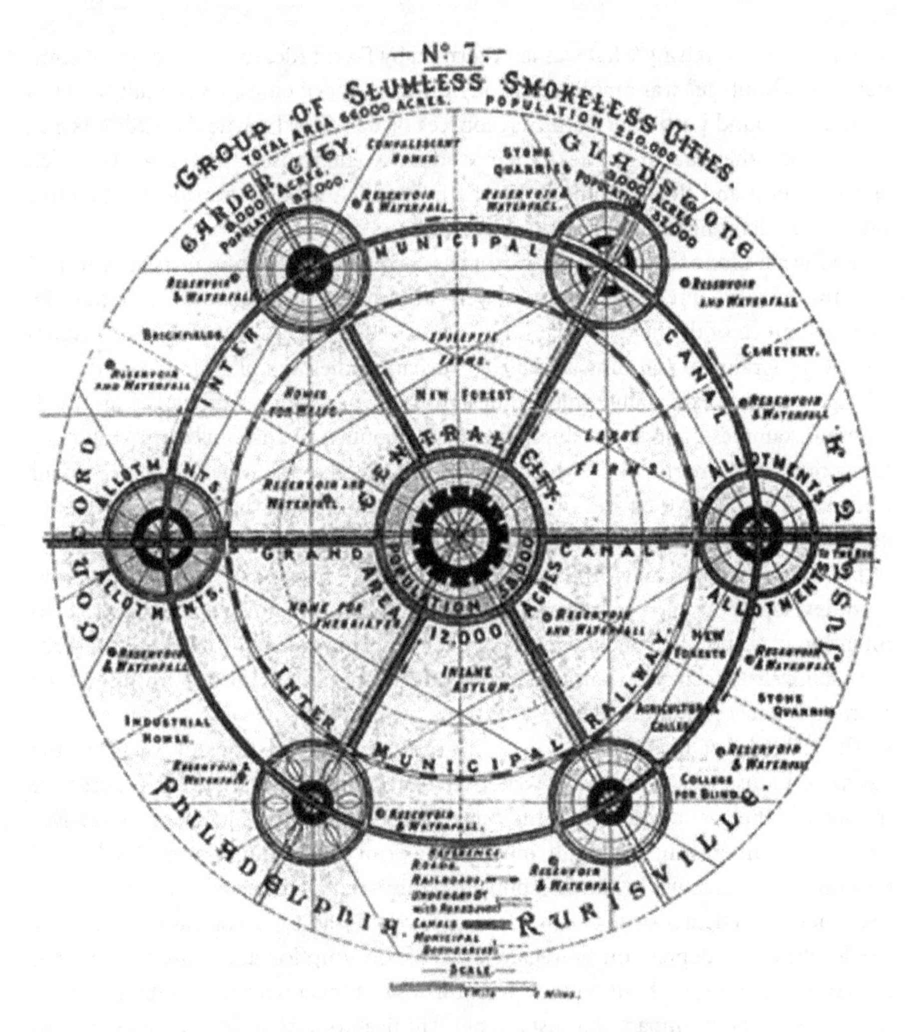

Figure 4.2 The Garden City Network of Ebenezer Howard (1898).

entities alongside economic interests as well. Remarkably, network theorists such as Jon Kleinberg[8] have since shown that the proverbial "six degrees of separation" aphorism that Milgram suggested between any nodes (or humans, for that matter), no matter how distant, is indeed a plausible rubric for most complex networks. "It's a small world after all" is far more than just a Disney song: it has fundamental meaning in the short paths toward computational connectivity between a vast range of natural and social phenomena. A globalized world has made such a realization even more ascendant, as noted by political scientist David Grewal in his book *Network Power*, in which he suggests that such "power" depends on three key attributes: compatibility, availability, and malleability. Network nodes have to be "compatible" to communicate with each other, they need to have the capacity to absorb information and hence be "available," and they need to accommodate revisions to their code and hence be "malleable."[9] The networks implicit in the garden city model and their modern social counterparts have remarkable positive potential in wielding power, but they also inexorably move us to a collectively self-inflicted conformity that can constrain choice.

Although Utopian at the level of design, the garden city framework recognizes social maladies and human pathologies. Stated in the insensitive terms of its times, there are places on the plan for psychiatric treatment and addiction rehabilitation programs as well. Unlike the fictional roots of *Utopia*, from the celebrated sixteenth-century satirical book by Sir Thomas Moore, who coined the term, garden cities were envisaged to be more grounded in reality (Utopia has its roots in the Greek meaning "no place," though Moore also noted that it sounds like *Eutopia*, which happily means "good place"). Although considered paragons of spatial planning, garden cities also raised fundamental questions about how economic order could arise organically or if it needed more central orchestration. Before Howard, the influential landscape architect Frederick Law Olmstead had also considered the structural elegance of city forms with large parks and boulevards, which are manifest in his design of Washington D.C. and of New York's Central Park. The sheer size of some of these cities as they became capitals and financial centers required a rethink of how economic activity could find harmony with natural and social order. Key to figuring out whether or not bottom-up or top-down economic order could arise were the essential notions of *scale* and *speed*. Historian Niall Ferguson's book *The Square and the Tower* (2016) considers the historical power of networks and hierarchies as being a recurring structural theme. Markets with complex trading orders in the "square" with a looming tower of hierarchical governance alongside have constituted both the literal and figurative economic geography of the world.

Scales and Speeds of Economic Order

The garden cities concept, attractive and elegant as it was in its heyday, led to major criticism from contemporary ecological planners. The suburbanization of America is often blamed on this concept because it can be used as a justification for low-density sprawl and gated communities with manicured "gardens." Yet such a manifestation of the concept is a caricature of scale and was unfortunately captured by the American architectural imagination in the post-World War II boom period. Among those culpable of this distortion of the garden city concept toward individualized socioeconomic order was the legendary architect Frank Lloyd Wright, whose cluster of works are now listed as World Heritage Sites. The allure of blending natural systems within the human built environment was compelling to Wright, but he was keen to individualize the experience for residents of his habitats. Thus, his love of embracing nature within his design often neglected the aggregate impact of the cities and economies which would result from such individualized bubbles of natural habitation. The worldwide population growth in the middle of the twentieth century led to a spurt of creative activity in formulating new forms of urban economic order. Urban sociologist Jane Jacobs noted the importance of what is termed "social capital" generation alongside any aesthetic developments of cities. The hustle-bustle of people in urban cores was embraced by Jacobs as "intricate minglings of different uses in cities." They "are not a form of chaos. On the contrary, they represent a complex and highly developed form of order."[10]

As the size of cities continued to grow as centers of economic activity, the Greek planner Constantinos Doxiadis developed the concept of *ekistics*. The organized upscaling of systems of life was deliberately articulated from Anthropos (individual) to Ecumenopolis (global city of up to 50 billion people, that was fictionally represented as Trantor in Asimov's *Galactic Empire* novel) with 15 hierarchical levels represented by density of human settlements. A pivotal feature of the ekistics approach to upscaling was the development of networks. This is now a core element of social order at multiple levels and is termed the "Fourth Industrial Revolution" by the World Economic Forum (coal-powered steam engines, electricity, and computers comprising the first three revolutions). The vision of Doxiadis is personally significant for me because he designed the new capital of Pakistan, Islamabad. The city, where many of my family members continue to reside, was planned to grow along a structured polycentric grid wherein each growth unit would have its own market center with a park and mixed-density residential areas intermingled. It remains one of the more livable cities in the developing world despite having increased in population by almost tenfold in 50 years. Around the same time, Italian architect Paolo Soleri considered urban density and nature in more synthetic vertical environments through the field of *arcology*, which he presented as low spatial impact, high-density, economically self-reliant architecture. A prototype of an arcological town was created by Soleri in a dry plateau in Arizona and called Arcosanti. His ideas were also adopted in science fiction works, like William Gibson's novel *Neuromancer* (1984) and Peter Hamilton's *Neutronian Alchemist* (1997). Perhaps the most enduring principles of scaling sustainable economies with natural order came from the Scottish-American landscape architect Ian McHarg, articulated in his 1965 book titled *Design with Nature*, in which he put forward a vision of design that adapted to nature rather than synthetically crafting it within our preferred environs. He exhorted us to dismiss "the self-mutilation which has been our way and give expression to the potential harmony of man-nature."[11] McHarg trained a generation of urban planners who have managed to also build vibrant economies into their plans across the world.

The original concept of a garden city was thus functionally meant to exist with small population sizes of 32,000, with all the key amenities of economic activity and habitation at a human scale of access. The network of cities and the sharing of larger resource productivity was also to be linked by public transport, and the concept by no means envisaged individualized automobile transport. Thus, in its original form, the economic order of garden cities was in sync with a "small is beautiful" ethos that was championed by the renegade economist E. F. Schumacher. The writings of Schumacher later operationalized economic orders of scale and linked them to cities and intentional communities. Yet the advantages of the small-scale natural order approach were not just limited to urban economies. Some of the same ideas were also applied by entrepreneurs to develop companies as a foil against the dominance of large monopolies in economic systems. Apple and Microsoft, for example, grew to battle it out with IBM, which had almost complete dominance of the computer sector at the middle of the twentieth century. Yet these companies themselves achieved dominance over the years and became targets of the same conflicts of scale in which they had originally pitched themselves as protagonists. The same trajectory has happened with many niche businesses, such as Ben and Jerry's, Body Shop, Tom's of Maine, and others that eventually were acquired by bigger companies (Unilever, Loreal, and Colgate Palmolive, respectively, in the case of these examples) and/or themselves became enormous (like Walmart, which disrupted Sears, Macy's, and JC Penny). This leads us to the question: Is there a certain inherent order in scale as economies grow, and is such an order, which may well be efficient in narrow financial terms (giving high returns on investment), also ecologically efficient in terms of natural resource usage?

Given the resistance of conventional economics to consider environmental constraints directly, a parallel field of *ecological economics* had to develop, led by a few rebel researchers who strayed outside their disciplinary domains to consider a more panoramic view of resource security. Notable among these rebellious scholars was Frederick Soddy who coined the term "isotope" and won the Nobel Prize in chemistry in 1921. The science fiction writer H. G. Wells was inspired by his work to consider how radioactive isotopies could be used for energy in his 1914 novel *The World Set Free*. Wells dedicated this novel to Soddy as a mark of his admiration in what was among his most prescient works. Later in life Soddy turned his attention to applying physical and chemical observations to economic systems and wrote several treatises on economics. In the reductionist cadence of that era, despite his Nobel accolade, he was dismissed as a crank for going beyond his own disciplinary bounds. However, he posthumously received acclaim for linking natural order to economic order. The Canadian journalist Andrew Nikiforuk gave him the following tribute, which encapsulates his prescient journey from chemistry to economics:

> After working with a pound of radioactive rock that contained more concentrated power than 150 tonnes of coal, Soddy realized that society had entered a new era. Moreover, he saw that the transition from wind, slaves and sunshine to hydrocarbons and their fuel-fed machines ultimately gave an elite group of people unprecedented power over everyone else's lives and labor.[12]

Soddy's mantle was carried forward within economic thought a few years later by the Romanian-American economist Nicholas Georgescu-Roegen, who had been a protégé of the great economist Joseph Schumpeter. While Schumpeter had noted the biological evolutionary order for capitalism as "the process of creative destruction," Georgescu-Roegen focused more on the physical and thermodynamic limits of economic growth. His seminal book, *The Entropy Law and the Economic Process* (1971), was the first treatise to consider physical constraints on capitalism. The systems economist Kenneth Boulding was also instrumental in further linking such an approach to social and ecological systems. Subsequently such work in academia was also applied in policy discourse by a few strident and visionary professionals such as former World Bank executive Herman Daly who suggested that steady-state economics was eventually needed to ensure environmental sustainability. Daly was concerned not only with scale but also with the speed of economic growth and whether the assumed "win-win" outcomes of a rising tide lifting all ships could be sustained with finite natural resources. In sync with natural order, Daly envisaged a new field of ecological economics which would be subservient to natural order and move toward a "steady-state." Yet a core critique of such an approach from mainstream economists was twofold. First, there was the challenge of poverty alleviation and inequality, which required economic growth to alleviate, at least in developing countries. Second, a lack of economic growth could imply a stagnation in human innovation.

The arguments made by ecological economists were fundamentally linked to broader critiques of growth at the planetary systems level that were most notably synthesized by the Club of Rome in the early 1970s. This famed intellectual club, which is still in existence, had its origins with Italian industrialist Aurelio Peccei and Alexander King, the erstwhile head of science at the Organisation for Economic Cooperation and Development (OECD). With the help of long-range computational forecasters and modelers from MIT, led by Jay Forrester, Dennis Meadows, and Donella Meadows, the group of 100 experts began to consider long-term growth trajectories for humanity within an underlying premise of exponential changes in most parameters. Supported by the Volkswagen Foundation, the Club published their famous report, *Limits to Growth*, in 1972, when the first United Nations Conference on the Human Environment was also held in Stockholm, Sweden. More than 30 million copies of this report have been sold since then, and a 30-year update was also published in 2004. A subsequent book, *2052: Forecast for the Next Fifty Years*, was also published in 2012. Many of the predictions about resource scarcity in the forecasts have not been accurate, but the overall messaging around the flaws within the assumed virtues of economic growth have gained currency with resource systems scientists.[13] A pragmatic and more measured path forward between these debates on growth lies with the field of *complexity economics*, which finds trenchant synthesis in Eric Beinhocker's book *The Origins of Wealth*. His key insight is that, just as complexity in natural systems has led to myriad forms of evolutionary products, so, too, can wealth be a product of complexity acting on social systems.[14] How such wealth operates in sync with constraints of natural systems should be our key question rather than agonizing over growth, per se.

Economics is essentially a science of managing resource scarcity. As we know all too well, natural resources and human populations are unevenly distributed across the planet. Global trade has allowed for some exchange, but for many people who are living in the most desperate of conditions technological change has also played an important role. The rapid rise in agricultural productivity in the mid-twentieth century is often referred to as the "Green Revolution," made possible by Nobel Peace Prize Laureate Norman Borlaug. This viosnary Norwegian-American from Iowa used his agronomic training as a seed breeder to develop high-yield seed varieties for wheat in Mexico in the 1940s and the 1950s. His subsequent collaborations with agronomist M. S. Swaminathan in India eventually led to successful replication of the techniques that were originally developed in Mexico. However, this progress, similar to the economic tunnel vision on economic growth, also led to a discounting of many ecological concerns, such as heavy fertilizer and pesticide usage.

So how do we consider such progress trajectories analytically? The Russian-American economist Simon Kuznets considered human development trajectories and their connection to inequality and suggested that there was parabolic relationship between the two (Figure 4.3). As countries become richer, they initially become more unequal since wealth is generated by the high-risk, high-reward entrepreneurs. Over time the wealth leads to more sustained development across the society: the rising tide proverbially lifting all ships. The same approach has also been suggested by some economists in terms of environmental harm becoming worse initially along a development path due to a rush to develop and then slowly diminishing as communities realize the need to regulate pollution. Such a trend happened to some degree for inequality and environmental harm as well, but it can be argued that the curve can be reversed as people begin to exploit the system or other ecological pressures increase. Furthermore, some irrevocable limits or carrying capacities for environmental

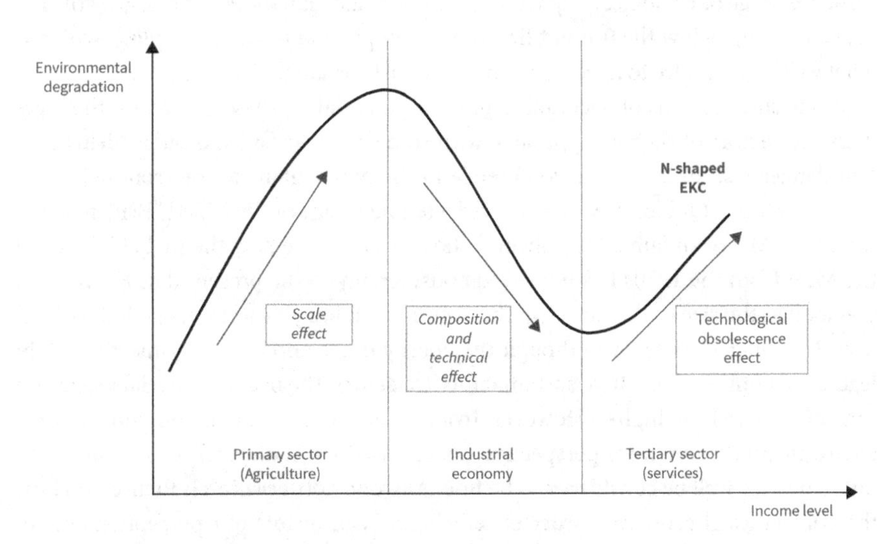

Figure 4.3 The N-shaped environmental Kuznets curve.

From Lorente, Daniel Balsalobre, and Agustín Álvarez-Herranz. 2016. "Economic Growth and Energy Regulation in the Environmental Kuznets Curve." Environmental Science and Pollution Research *23* (16): 16478–94. https://doi.org/10.1007/s11356-016-6773-3.

systems could be exceeded, which would have irreversible impacts, such as the extinction of species.

Perhaps it is time to consider *ecological security* as the lens through which we approach economic regulation as well, since ultimately resources are always predicated in ecological systems. Such an approach would require regulating the scale of consumption in developed countries while creating incentives for constructive consumption and trade in developing countries for poverty alleviation. We need to use measures of economic performance of countries not just in terms of economic indicators or even human development indicators, but also environmental indicators. Thus, country decision-makers should note indices such as Yale University's Environmental Performance Index that ranks countries by their environmental performance alongside conventional United Nations indices such as the Human Development Index. It's high time we have a more nuanced and "naturalized" approach to economic growth that acknowledges the resilience as well as the constraints of natural systems.

A key element of such a nuanced approach would be to consider whether there are win-win opportunities for having economic growth while reducing ecological impact as well as resource depletion. Such a prospect is termed "decoupling" in systems science parlance, meaning that the concomitant change in wealth is not linked to either irreversible resource decline or environmental impact through pollution. Such an approach has been embraced by bodies such as the United Nations International Resource Panel as an aspirational goal for economic and social order. Figure 4.4 shows how the negative divergence in percentage change of depletion or impacts from percentage economic growth rate is a mark of decoupling. All those curves above the 0-point line are an example of *relative decoupling*, meaning that there is still a rise in resource usage or ecological impact but at a lower rate than the economic growth. The curve heading below the 0-point line is an example of *absolute decoupling*, which is what we should aspire to if we want a more robust sustainability outcome.

While such notions of decoupling present potential win-win outcomes, there are many detractors of such an approach, with some going as far as to call it "delusional and dangerous." In 2018, at a conference in the Norwegian city of Trondheim, the material scientist Julian Allwood showed a telling image of the fabled "British Mini," the car of Mr. Bean fame. Dr. Allwood showed on a slide how the fuel efficiency of the Mini from the 1970s has improved considerably to the present day. However, if you look at the size of the car itself, the "Mini" is no longer diminutive—indeed, it is now 30% bigger! Thus, even though the energy usage and efficiency metrics might lead us to consider a positive sustainability trajectory, the overall material usage and impact may well be higher. However, from an economic growth and conventional environmental economics perspective (as opposed to "ecological" economics), the outcome may well be considered to be fine. Allwood's observation is focused on how the conventional economic order of "efficiency" can distort our perceptions of actual impact on sustainability outcomes. The observation of efficiency can still lead to overall higher consumption and is referred to as the *rebound effect*. Such an effect was noted as early as the nineteenth century by coal economist William Stanley Jevons in

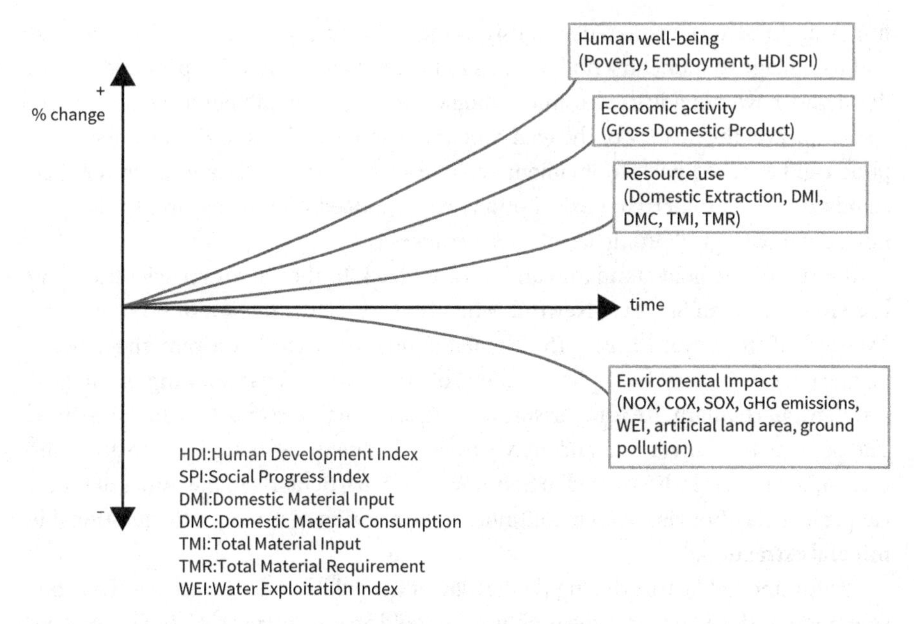

Figure 4.4 Various forms of resource and environmental decoupling.
From Scheel, Carlos, Eduardo Aguiñaga, and Bernardo Bello. 2020. "Decoupling Economic Development from the Consumption of Finite Resources Using Circular Economy. A Model for Developing Countries." Sustainability *12* (4): 1291. https://doi.org/10.3390/su12041291.

the context of efficiency improvements in steam engines (hence, it is also known as the *Jevons paradox*). One of the core reasons for our lack of realization of this paradox is that our monetary system, by which we consider pricing triggers in changing economic behavior, cannot directly incorporate resource scarcity. Thus, efficiency may need to be redefined in terms of broader sustainability criteria than just output per unit input. Contemporary economic order is largely linked to a system of trust in particular mediums of exchange that we call "currency." There were times when our currency was connected to particular natural resources, most notably to precious metals. Might a return to such a "raw material" order for our economic system also prompt us to be more in sync with natural order?

Currencies of Sustainable Economic Order

The nature of gold—an inert metal that stubbornly eludes most chemical reactions—has made it a symbol of wealth and power throughout history. Its resistance to oxidation and its wide cultural acceptance made it particularly appealing as a direct medium of exchange or as a surety for other monetary indicators. Why did gold acquire such prominence in economic policy during the age of scientific discovery and industrialization? Why was gold then abandoned as a monetary anchor, and why is it now again being considered seriously as the ultimate asset? The discourse on gold and its linkage to monetary policy tends to be highly polarized. Yet proponents and detractors have not adverted to the ecological aspects of gold's potential comeback. As we unravel the history of the *gold standard*, some insights about monetary discipline can be traced back to incipient ecological constraints. Variations on the gold standard and its underlying assumptions could indeed offer a solution to the economic *and* ecological profligacy of *Homo economicus*.

The rise of the gold standard can be traced back to the work of a scientist of no less eminence than Sir Isaac Newton, who also had the distinction of serving as the "Master" of the Royal Mint.[15] The elemental quality of gold as a rare and durable primary resource may have played some role in Newton's championing of the gold standard, particularly in comparison to its more reactive competing metal—silver. The popularity of silver as a currency was largely due to the abundant stocks of the metal that Spaniards discovered in South America, particularly in the mines of Potosi (in present-day Bolivia), which continues to be a region of ecologically questionable mineral extraction.[16]

Environmental factors during the heyday of the gold standard were hardly a consideration. Indeed, to many, the proclivity for gold was as "natural" as evolution, and, during the height of the nineteenth-century gold rushes, the imagery of evolution was often used to promote the gold standard as opposed to other metals and materials. As the debate over paper currency was being conducted, the symbol of gold resonated with many Christian theologians who had come to the New World. Historian Kathryn Morse notes in her study of the Klondike gold rush that, "the argument about values, both monetary and moral, was rooted in the symbol of gold, and in nature, a nature created with certain purposes by a Protestant God."[17] In contrast, those who opposed the gold standard also tried to "denature" the image of gold by showing that it was only representative of perceptions of value and did not have any inherently useful quality. The critics of the gold standard represented gold as sterile compared to, say, a seed planted by a farmer that could accumulate value as labor and nutrients gave it intrinsic worth as a vital form of nourishment. Yet the primacy of gold and the rush to find it endured even this line of critique.

The great chronicler of the Klondike gold rush, Jack London, conceded in his back-of-the-envelope cost benefit analysis that the miners had probably invested more than $220 million dollars to build infrastructure and sustain themselves in order to dig up around $22 million worth of gold. To him, this calculation at the turn of the twentieth century was still weighted in favor of the yellow metal's worth. Nonetheless, London believed that this monumental effort was still of "inestimable benefit to the Yukon country" because "natural obstacles will be cleared away or surmounted, primitive methods abandoned, and hardship of toil and travel reduced to the smallest possible minimum."[18]

Gold was almost universally valued, but there were some rare exceptions to the rule. In highly resource-scarce communities, such as the desert tribes of the Sahara, the mineral commodity of choice was not an inert metal like gold, but a compound formed by one of the most reactive metals in the periodic table—sodium. So reactive is this light metal that it is never found alone in nature; rather, it combines with numerous other elements, the most common of which is chlorine, thereby forming that essential ingredient for the human palate—salt.

Common salt was a deity for many tribal groups, including Native Americans such as the Navajo, the Zuni, and the Hopi, who prized the simple sodium chloride mineral despite their many other cultural differences. Peter Bernstein describes the comparison between gold and salt in the eyes of the Saharan tribesmen: "What must those poor diggers have thought of the funny people from the north country who swapped inestimable salt for stuff whose only role was to give men pride and pleasure by letting them see its luster."[19]

Unlike salt, gold was not a physical mineral necessity for human survival. Nevertheless, its durability made it highly attractive as a standard for monetary exchange. Like silver, gold coins were at times used as currency, but it was impractical to use them on a large scale with various denominations. Hence the move toward some form of centralized acquisition of gold at banks. Following Newton's earlier interest in gold, there was a period of uncertainty worldwide until huge reserves of gold were discovered in southern Africa toward the end of the nineteenth century. The relationship between gold production in South Africa and the accumulation of gold in Europe, particularly in Great Britain, is also stark and poses interesting ethical questions about responsibility for the contemporary environmental and social issues of gold mining. In his landmark study of the relationship between the gold producers in South Africa and the Bank of England, Russell Ally reveals that much of the prominence of gold in the global financial system at the turn of the twentieth century arose from this relationship.[20] The study also reveals that London controlled the minting of gold coins and bullion that could have provided added value to the African economy, and consumption patterns across the Empire were managed largely from the British capital. Thus, the United Kingdom spurred the demand for gold, and responsibility for the impact of gold mining can be ascribed to British policies.

The United States instituted the gold standard during a similar period of gold obsession. From 1834 onward, gold was considered alongside silver as a standard reserve metal in monetary policy, with the fixed price of gold set at $20.67 per ounce—which remained in force until 1933. During this period, there was a global resurgence in the primacy of gold, particularly toward the end of the nineteenth century, due to gold rushes worldwide. These rushes fueled phenomenal growth in various industrial sectors and paved the way for many other sectors of the economy to develop. Gold rushes created capital flows, which allowed for investment in many other sectors of the economy. Economic historians Paul David and Gavin Wright also contend that, in the larger scheme of things, the rush toward minerals laid the foundations for lasting development and diversification in ways that are often neglected in contemporary research: "Rapid resource extraction in America was also associated with an ongoing process of learning, investment, technological progress and cost reduction, generating a many-fold expansion rather than depletion of the nation's resource base."[21]

Nevertheless, the "non-use" of gold sitting in bank reserves is still unsettling. In the words of Robert Triffin, a Belgian economist who was a major critic of centralized gold reserves: "Nobody could ever have conceived of a more absurd waste of human resources than to dig gold in the distant corners of Earth for the sole purpose of transporting it and reburying it in other deep holes, especially excavated to receive and heavily guarded to protect it."[22] Economic historians have also debated the efficacy of the gold standard. Several economists, including Milton Friedman and former chairman of the Federal Reserve Ben Bernanke, have blamed the Great Depression on the gold standard.[23] Cover and Pecorino's detailed econometric analysis comparing the growth of the US economy and the "taming" of the business cycle suggested that, in fact, the rapid rise in the US economy could be traced back to the elimination of the fixed-price gold standard in 1933.[24] The arguments against the gold standard have been predicated on how it limits the range of policy tools dealing with extenuating circumstances, such as the needs of a wartime economy. Following World War II, the *Bretton Woods agreement*, which created the World Bank and related financial institutions, recognized the salience of gold by keeping a reserve currency system, albeit using a variable price of gold.

Following the Vietnam War, President Nixon withdrew the United States from the international gold exchange standard completely. The US dollar itself became trustworthy enough in the eyes of the international community for it to be liberated from the shackles of gold. However, these shackles are precisely what many current proponents of the gold standard consider so compelling. The gold standard has the potential to instill discipline in monetary policy and can prevent governments from wantonly printing money and causing inflation. The standard can also prevent governments from overspending and creating huge deficits, which is the reason it has attracted much attention recently from Tea Party stalwarts such as Ron Paul. However, it is not just those who are afraid of according power to centralized government banks who are championing the gold standard again. In May 2011, the Mexican government announced that it was buying $4 billion worth of gold to boost its reserves and provide economic security. Such international confidence has raised the price of gold to an all-time high, and the potential for a comeback of the gold standard in some form deserves serious revisiting.

From an ecological perspective, the gold standard has the attraction of linking economic growth to natural resource constraints, a linkage that has been a recurring theme in much of ecological economics discourse.[25] However, the environmental impact of mining gold is so intense that any support for resurrecting the gold standard is summarily dismissed by many activists. Indeed, some activists have argued that national gold reserves should be tapped in order to meet consumer demand for gold, thereby limiting the ecological impact of mining.[26] For example, each ounce of gold produces 30 tons of waste, and the US Environmental Protection Agency estimates the cost of cleanup for existing metal mines to be around $54 billion.[27] However, it is important not to conflate gold mining with the gold standard in terms of our discussion of sustainability. Given the durability of gold and its ease of recycling, the gold standard can in principle be maintained without having to mine more gold. Furthermore, if gold is being used as a reserve in and of itself, the main issue of consequence is property rights over the gold. If there were international consensus on the global gold reserves still in the ground, then trade and ownership of such reserves could also be handled through an international treaty system that regulates ownership of gold reserves, rather than through physical extraction. For example, if the world's total gold deposits could be centrally certified and shares issued for buying these reserves, the same purpose could be served as stockpiling gold in a vault. In other words, a nation's gold reserves could remain "stored" in their natural underground state, rather than being mined, purified, and deposited in a Fort Knox-like vault. Some discounting factor could be added to account for the accessibility of certain deposits versus others, and countries which have gold on their land could get preferred purchase rights to the shares, similar to how company founders or employees have preferred stock options. While this would be environmentally preferable to the present system of mining and storage, ecological economists would be right to ask: Why not simply anchor currency to some other metric of planetary carrying capacity which could be considered in the same way? Perhaps a "leave it in the earth" approach to gold could be a first step on the road to considering more ecologically benign means of storing and measuring "wealth."

For nonrenewable resources, such an approach was indirectly suggested by the Club of Rome in their 20-year retrospective book *Beyond the Limits.* A system dynamic computer model called "World 3" was put forward that considered nonrenewable resource stocks and flows, as shown in Figure 4.5. The model considers the capital investment of the resource base, the temporal dimension of resource usage, technological variables, and conservation. If we were able to obtain all these data and have an efficient policy apparatus to convert these material flows and related variables into economic decision-making, it would the optimal path forward. However, in the absence of such policy consensus, as well as measurement changes, the gold standard presents an intriguing, albeit coarse, calibrator for resource constraints.

The gold standard has a checkered history in terms of its overall efficacy in economic development but there is little doubt that we need to instill some discipline within our financial institutions. Historically, gold mining and the gold standard have contributed to a great deal of ecological devastation. However, they have also contributed to the great enterprise and economic growth of the modern era. The gold standard unwittingly acknowledged the connection between monetary power and natural resource reserves which constitutes the core of many contemporary environmental ideologies. While the earlier proponents of the standard did not envisage this connection, the "discipline" they sought was the result of the natural limits of gold extraction and the underlying finite gold reserves. Current conversations about resurrecting the gold standard should focus on this underlying presumption of the standard. Indeed, such an approach may help to also bridge the perceived tension between environmentalists and conservative politicians and economists. The take-home message remains that the salience of a gold standard for financial discipline

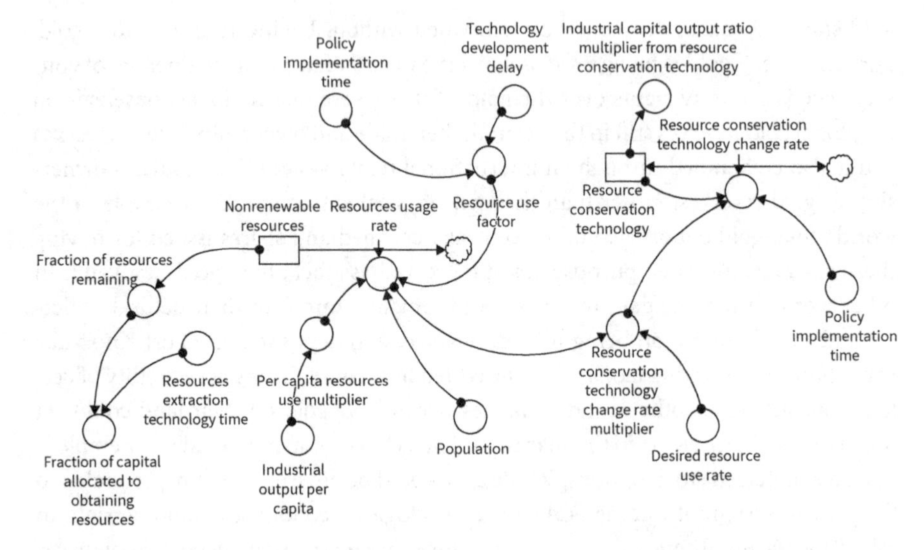

Figure 4.5 The World 3 Systems Dynamic Model nodes of analysis for nonrenewable resources in an economy.
Reprinted from *Beyond The Limits*, copyright 1992 by Donnella H. Meadows, Dennis L. Meadows, and Jørgen Randers. Used with permission from Chelsea Green Publishing (www.chelseagreen.com).

is due not only to the nature of gold itself but also is a product of the fact that gold is a relatively scarce natural resource with the persistent allure of durability. More fundamentally, gold, or any other natural element, provides an indelible means of connecting our economic order to natural constraints. A key underlying rationale for an ecological application of the gold standard stems from a fear that prices are inadequate to stimulate behavioral change under natural resource constraints. Signals of scarcity are thus not always efficiently transmitted through the economy. While such a return to an elemental currency standard is unlikely to occur in the short term, there may well be better ways in which signaling of scarcity can be incorporated in economic systems. Toward that end, we need to better understand the laws of supply and demand and equilibria that they present to us in our quest for sustainability outcomes.

Notes

1. Barrow, John D. 1998. *Impossibility: The Limits of Science and the Science of Limits*. Oxford University Press.
2. Zilsel, E. 1942. “The Genesis of the Concept of Physical Law.” *Philosophical Review* 51(3), 245–279. https://doi.org/10.2307/2180906.
3. Gribbin, J. 2004. *Deep Simplicity: Bringing Order to Chaos and Complexity*. Random House.
4. Firestein, S. 2012. *Ignorance: How It Drives Science*. Oxford University Press.

5. The OECD has considered this matter of late in the context of the COVID-19 pandemic by comparing pyramidical versus diamond-shaped supply chains. The latter involve an intermediary core for multiple sectors which, if impacted by a crisis, can have a spillover impact that is more consequential than pyramidical supply chains. See "Covid-19 and Global Value Chains: Policy Options to Build More Resilient Production Networks." n.d. OECD. Retrieved January 2, 2021, from http://www.oecd.org/coronavirus/policy-responses/covid-19-and-global-value-chains-policy-options-to-build-more-resilient-production-networks-04934ef4/.
6. Wehi, Priscilla M., Murray P. Cox, Tom Roa, and Hēmi Whaanga. 2018. "Human Perceptions of Megafaunal Extinction Events Revealed by Linguistic Analysis of Indigenous Oral Traditions." *Human Ecology* 46 (4): 461–470.
7. Henrich, Joseph. 2017. *The Secret of Our Success: How Culture Is Driving Human Evolution, Domesticating Our Species, and Making Us Smarter* (reprint edition). Princeton University Press.
8. Kleinberg, J. M. 2000. "Navigation in a Small World." *Nature* 406 (6798): 845–845. https://doi.org/10.1038/35022643. Kleinberg's work builds on the models developed in Watts, D. J., and S. H. Strogatz. 1998. "Collective Dynamics of 'Small-World' Networks." *Nature* 393 (6684): 440–442. https://doi.org/10.1038/30918.
9. Grewal, David Singh. 2008. *Network Power: The Social Dynamics of Globalization*. Yale University Press.
10. Jacobs, J. 1992. *The Death and Life of Great American Cities* (reissue edition). Vintage.
11. McHarg, Ian L. 1969/1995. *Design with Nature*. Wiley.
12. Nikiforuk, Andrew. https://www.questia.com/magazine/1G1-296841805/the-vision-of-frederick-soddy-in-the-economy-of-energy (accessed September 29, 2020).
13. For an excellent review of growth as a structural concept in nature and society, see Smil, V. 2019. *Growth: From Microorganisms to Megacities*. MIT Press.
14. Beinhocker, Eric D. 2007. *The Origin of Wealth: The Radical Remaking of Economics and What It Means for Business and Society* (1st edition). Harvard Business Review Press.
15. Newton was quite taken by the old alchemist's desire to turn lead into gold. His interests in this regard and their connection to his job at the Royal Mint are documented in Dobbs, B. J. T. 1983. *The Foundations of Newton's Alchemy*. Cambridge University Press.
16. An excellent ethnography of the Potosi mines in historical context can be found in Nash, June. 1993. *We Eat the Mines and the Mines Eat Us: Dependency and Exploitation in Bolivian Tin Mines*. Columbia University Press.
17. Morse, Kathryn. 2010. *The Nature of Gold: An Environmental History of the Klondike Gold Rush*. University of Washington Press.
18. Quoted in Kathryn Taylor Morse, *The Nature of Gold*, pp. 194–195. This theme is also found in the writings of Frederick Jackson Turner on the significance of the frontier in American history. See Turner, Frederick Jackson. 1999. *Rereading Frederick Jackson Turner: "The Significance of the Frontier in American History" and Other Essays*. Yale University Press.
19. Berstein, Peter L. 2004. *The Power of Gold: The History of an Obsession*. Wiley.
20. Ally, R. 1994. *Gold and Empire: The Bank of England and South Africa's Gold Producers*. Witwatersrand University Press.
21. Paul, David, and Gavin Wright. 1998. "Increasing Returns and the Genesis of American Resource Abundance." *Industrial and Corporate Change* 6 (2): 223.

22. Quoted in Manning, Richard. 1998. *One Round River: The Curse of Gold and the Fight for the Big Blackfoot*. Henry Holt and Co., 140.
23. Friedman had estimated at one time that the gold standard would cost the United States 2.5% of its gross domestic product (GDP) in terms of indirect costs, including social and environmental burdens borne by mining communities. For Bernanke's views on the standard, see Bernanke, Ben, and Harold James. October 1990. "The Gold Standard, Deflation, and Financial Crisis in the Great Depression: An International Comparison." *National Bureau of Economic Research Working Paper Series* no. 3488. http://www.nber.org/papers/w3488.
24. Cover, James P., and Paul Pecorino. 2005. "The Length of U.S. Business Expansions: When Did the Break in the Data Occur?" *Journal of Macroeconomics* 27: 452–471. Critical analysis of the methodology used in this research and the inferences which we can draw has also been challenged to some degree.
25. See, for example, Daly, Herman E. 1997. *Beyond Growth: The Economics of Sustainable Development*. Beacon Press.
26. Young, John E. 2000. *Gold at What Price: The Fate of the National Gold Reserves*. Report prepared for prepared for: Mineral Policy Center Project Underground Western Organization of Resource Councils.
27. Perlez, Jane, and Kirk Johnson. "Behind Gold's Glitter: Torn Lands and Pointed Questions." June 14, 2010. *New York Times*. http://www.nytimes.com/2005/10/24/international/24GOLD.html.

5
Elusive Orders of Economic Equilibrium

> The equilibrium between supply and demand is achieved only through a reaction against the upsetting of the equilibrium.
>
> —**David Harvey, *The Limits of Capital*, 2006**

There are two protagonists that maintain order in conventional economics—producers and consumers. It may be tempting to compare these protagonists to a natural system where there is a predator who consumes its prey. Yet such a metaphor would be deceptive since the relationship between a consumer and a producer in an economic system is transactional and likely to be mutually beneficial. Perhaps a more apt natural comparison would be the phenomenon of *symbiosis*, whereby two organisms gain from each other through a process of mutually beneficial exchanges and reach an "equilibrium" of costs and benefits for such interactions. Algae and their coral hosts are a case in point whereby algal cells use their photosynthetic prowess to produce excess sugars for coral polyps to consume while carbon dioxide produced by the coral is consumed by the algae. If the equilibrium in this symbiosis is disrupted, both organisms suffer. There are times, though, when the relationship is more unilateral, in which the consumer does not provide adequate returns to the producer due to structural asymmetries of power or capacity. If such an organism is consuming excess material without harming the producer, the relationship is *epiphytic*, similar to orchids growing on a tree trunk. However, if the consumption of resources adversely impacts the producer, the consumer becomes *parasitic* and the relationship between the individual entities falls out of equilibrium. A strangler fig tree which slowly consumes the nutrients of its host trunk would be a corollary example of a parasitic relationship.

In 2019, the Academy Awards (Oscars) gave a surprise win to a Korean film called *Parasite*, which astutely applied the biological origins of the title in a socioeconomic context. The story laid out how asymmetries in economic power led to a spiral of destructive and disastrous actions by both the poor protagonists and the affluent antagonists in the film, with a reversal of culpability in the plot. The film challenged us to consider how, unlike in natural systems, knowing who a parasite is within the complex functioning of economic markets is far more difficult. Those humans with a wealth of resources may well be exploiting the minions who work to add scaled value to the apex of "top consumers." But at the household level, small acts of theft by the poor may be considered parasitic as well. The film makes us consider the scales of

Earthly Order. Saleem H. Ali, Oxford University Press. © Oxford University Press 2022.
DOI: 10.1093/oso/9780197640272.003.0006

analyzing perceptions of parasitism in the context of global socioeconomic inequality and also challenges us to consider if the status quo or equilibrium which we may perceive is sustainable. Yet we can also zoom out further at the level of consumption on a planetary scale and consider if humans as a species are parasitic on natural systems and how an equilibrium may be reached between our consumption and production systems.

Producers and consumers are in turn are either suppliers or seekers of resources that they functionalize through various "factors of production." Land, labor, and capital have been the traditional trinity of factors that producers need to provide goods and services. Materials and energy are generated from these three primary factors. Technology and entrepreneurial talent can refine the value proposition offered as productive output and may be considered additional factors to note. Capital represents a "stock" of resource wealth that is often further divided into various forms depending on the scale of analysis. Geographers have used a "five-capital" framework harkening back to work done by the British government on how livelihoods can be generated and sustained. A refined version of this framework prepared by the erstwhile engineering consulting firm Halcrow is shown in Figure 5.1.

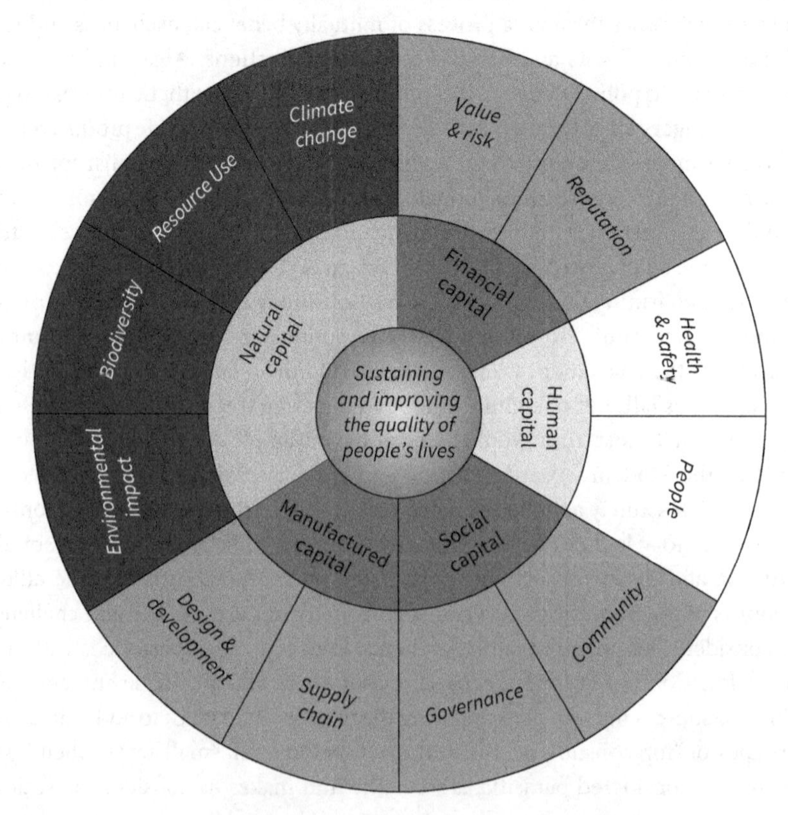

Figure 5.1 The Five Capitals framework and its resultant areas of influence.

The outermost circle in the diagram delineates various manifestations of capital and the pressure points on its quality and currency. There are other ways of unpacking capital, which can include cultural, intellectual, experiential, and even spiritual forms. Permaculture and other holistic living movements have particularly capitalized on such a broader range of human wealth indicators. Economists often assume that financial capital can be a proxy for other capitals due to the power of money as a medium of exchange. All the factors of production interact with each other at some level through monetary mechanisms in that most fundamental of equilibrating forces in an economy—the market. Human captivation with markets stems from how they can be a locus for all the various forms of capital to intersect with each other. Yet markets do not have any moral order and the value of exchange that occurs is not always commensurate with what humans may value for well-being. The key measure of value that economic forces are driven by is the price that markets ascribe as an indicator of value. Hidden beneath every price for a producer is an underlying layer of value that is termed the "cost of production." This cost of production is in turn determined by prices paid by the producer for the materials and energy to manufacture a product or deliver a service. The difference between the cost of production and the price a consumer pays determines the profit which spurs all commerce. Price is thus a dynamic entity which bridges the consumer and the producer and forms the fulcrum of economic equilibrium. Markets can be made "moral" either through the necessity of exchange and trust in creating economic order or through external oversight of ethical norms.[1] Ultimately, a confluence of norms between the consumer and producer on value in price versus value in content will determine how such morality becomes institutionalized.

Orders of Price and Quantity

If you open any economic textbook, the most ubiquitous graphic will be a set of intersecting lines and curves. Our eyes naturally travel to the point of intersection, and we focus more intently on the coordinates of convergence. The x-axis of these graphs is usually "quantity" or "Q," and the y-axis is the "price" or "P." The intersection is a point of equilibrium where the supply of a producer meets the demand of the consumer. These *supply/demand curves* are the bread and butter of economics. Economic order rests its case of emergent harmony on the notion that supply and demand for resources eventually create an equilibrium price. Core assumptions in constructing such an order come from our views on competition and trust between buyers and sellers. Can sellers develop products and find a market for them among buyers, or is there some signaling needed from buyers to produce a product that would be marketable? This is in some ways a classic "chicken-and-egg" question: Which comes first? In social science we call such quandaries *endogeneity problems*. Generally speaking, the lower the price, the higher the demand for the good, but there are exceptions to this sense of order when it comes to particular prestige goods like diamonds or branded accessories. There can be a bending back of the demand curve in such cases where, beyond a certain point of price increase, signaling of quality or prestige can lead to increased demand. Such goods were identified by economist Thorstein Veblen as an example of "conspicuous consumption," which can have serious implications for environmental sustainability. Related to this phenomenon is also the "diamond–water" paradox, which was noted by Adam Smith, wherein diamonds are useless for sustenance of humans and yet very expensive while water is essential and yet very cheap under most conditions. The order of markets is thus not necessarily aligned with biological order.

In an 1801 treatise, the French economist Jean Baptiste Say proposed that "a product is no sooner created, than it, from that instant, affords a market for other products to the full extent of its own value."[2] In other words, Say suggested a certain wisdom of the producer in generating products that would find consumers. Indeed, producers can produce without a consumer, but a consumer cannot consume without production. Say's proposition became known as a "law" because it underpinned the ability of markets to reach equilibrium through supply creating its own demand. Enrooted in Say's law is the germination of laissez faire economics and the presumption that, through a modulation of quantity and price, equilibria emerge in markets. Yet episodes such as recessions and depressions show us that signaling between production and consumption can go astray, and we can have a glut of production as well. Demand may need to be created in such circumstances, and governments have taken on this role. Furthermore, Say's "law," if followed by the producer, can be in conflict with constraints of natural order in terms of sustainable resource usage.

Following the Great Depression, British economist John Maynard Keynes became the most vocal and influential proponent of demand-driven order in the economic system and dismantling "Say's fallacy." He witnessed the collapse of markets during the Great Depression and saw the need for government action in creating demand through a range of macroeconomic policies. Keynes famously said that "markets can remain irrational longer than they can remain solvent." He did not consider equilibrium an inevitability, but rather saw loss of consumer demand as a potential death spiral for markets. Consumer demand was dependent on trust in products and in future purchasing power, both of which were less forthcoming in times of scarcity and uncertainty. The notion of "government stimuli" in times of crisis, which we have observed during calamities such as the 9/11 terrorist attacks or the COVID-19 pandemic, are in essence based on Keynesian notions of the state driving demand. Government could drive demand by injecting disposable cash to consumers or by purchasing and hiring on its own. Large infrastructure projects or defense expenditures have often been the most active locus of such activity by governments. Defense expenditure is particularly seductive in this regard because security imperatives can use the order of fear which even President Eisenhower, a former general himself, warned us of in his famous words: "we must guard against the acquisition of unwarranted influence, whether sought or unsought, by the military industrial complex."[3]

Keynes's most vocal critic was the Austrian-British economist Friedrich Hayek, who was heavily influenced by the economic fallout of the Bolshevik revolution in Russia in the early twentieth century. The rise of a highly centralized state and communism provided Hayek enough empirical ammunition to assault a wide range of government interventions. For Hayek, the wisdom of Adam Smith's "invisible hand" was manifest in the primacy of pricing mechanisms which calibrated an "extended order," similar in cadence to the notion of "spontaneous order" we discussed in the previous chapter. Government intervention and socialist worldviews to Hayek were a *fatal conceit* (the title of one of his co-authored books as well). In place of enforced egalitarianism, Hayek's "extended order" was "a framework of institutions—economic, legal, and moral—into which we fit ourselves by obeying certain rules of conduct that *we never made*, and which *we have never understood* in the sense of which we understand how the things that we manufacture function." At the heart of Hayek's view on extended order is a firm belief that prices have the power to drive innovation and process information more efficiently than does central planning. Furthermore, Hayek believed that government-driven demand would also lead us down a *Road to Serfdom* (the title of his most famous book). Despite their mutual antagonism, Keynes and Hayek shared a common detachment from environmental concerns. Keynesian spending through government purchases or cash-injected consumerism is just as ecologically problematic as the profligate primacy of the market. In the words of the erudite writer Eric Zencey, "free markets operated on infinite planet principles are just the other road to serfdom."[4] Having won a Guggenheim Fellowship for writing, Zencey had the reputational currency to get op-eds published in the *New York Times* while also holding academic appointments. I had the pleasure of interacting with Zencey during my years in Vermont, where he spent summers and frequently corresponded with me about his writings. He diagnosed a key feature of the dominant economic paradigm as one of "cornucopian winning" that surrounds short-term thinking while neglecting long-term loss.

The equilibrium being sought by the conservative and liberal economists alike neglected the inability of markets to consider natural order. This complete inability of both extremes of the conventional economic spectrum led to increasingly greater frustration within the environmental science community. Relative abundance of natural resources at a functional level for much of the twentieth century led to the sanguine belief among economists that pricing signals would ultimately provide the warning signal on depletion. The allure of prices driving equilibrium was even embraced regarding nonrenewable resources, such as minerals. The American statistician Harold Hotelling published a "rule" in 1931, which posited that "that owners of non-renewable resources will only produce a supply of their basic commodity if it can yield more than available financial instruments, specifically interest-bearing securities." This rule led to a degree of complacence around pricing and investment return mechanisms that involved nonrenewable resource extraction. Economists such as Nobel Laureate Robert Solow and many of his intellectual progeny used this rule to develop dynamic equilibrium models, at the heart of which were key assumptions about substitutability of resources. The resource economists within the conventional annals of the field largely dismissed any ecological critiques which were offered of their models. Hotelling's Rule was subsequently linked to the question of how the profits from exhaustible resource extraction could be linked to creating a sustainable economic trajectory by Canadian economist John Hartwick in 1977. Much of Hartwick's work was focused on considering how resource-rich countries could use the wealth created from natural capital to develop a diversified economy which could reach sustainability. Among environmental studies scholars, the Hartwick-Solow approach to sustainability is referred to as "weak sustainability" because it ultimately assumes that resource depletion would transition from an extractive to a service-oriented economy (Table 5.1).

While "weak" implies a somewhat inferior form of sustainability, such a presumption is only applicable if one considers the primacy of natural capital's total asset base as somehow intrinsically more valuable than other forms of capital. However, if we inject technological adaptation and consider a qualitatively different global equilibrium for human well-being, then a weak sustainability outcome may be deemed preferable because higher levels of per capita consumption may be maintained as a result. The strong sustainability paradigm focuses on a replenishment of natural capital and maintaining its asset base at a certain level. However, such an equilibrium is more dependent on population stabilization as well as per unit consumption reduction. Strong sustainability is thus more normative and requires far more individual discipline on the part of the consumer and parameters around what forms of innovation and lifestyle adjustments can be accommodated within planetary ecological parameters. Finding an optimal economic order which can perhaps aspire toward strong

Table 5.1 Weak and strong sustainability

Weak sustainability	Strong sustainability
Natural, human, and reproducible capital can be substituted for each other.	Cannot always substitute for natural capital with reproducible or human capital.
Natural, human and, reproducible capital are an aggregate, homogeneous stock.	Cannot view natural, reproducible, and human capital as a homogeneous stock.
Natural capital should be used efficiently over time. As long as depleted natural capital is replaced with even more valuable reproducible and human capital, then the value of the aggregate stock will increase.	Certain environmental sinks, processes, and services are unique and essential, subject to irreversible loss, and there is uncertainty over their future value and importance.
Maintaining and enhancing the values of this aggregate capital stock is sufficient for sustainability.	Maintaining and enhancing the values of this aggregate capital stock is necessary but not sufficient. Sustainability also requires preserving unique and essential natural capital.

From Barbier, E., & Burgess, J. (2017). *Natural Resource Economics, Planetary Boundaries and Strong Sustainability.*

sustainability while realistically making weak sustainability efficient in the extension of natural capital is a core challenge for modern economies.

Novel fiscal ideas continue to be proposed to address such challenges, but one which gained considerable attention within the United States came from a most unlikely source: the founder of Godfather Pizza, the late Herman Cain, who died of COVID-19 in September 2020. Cain did not come across as an environmentalist by any stretch of the imagination. As the CEO of Godfather Pizza, he presided over an ecologically problematic food industry that clogged the arteries of millions. Cain also publicly derided the Environmental Protection Agency for even "regulating dust!" But, inadvertently, Mr. Cain proposed one of the most progressive environmental policy measures that could have had a transformative impact on American consumerism. Capitalizing on the populist allure of tax cuts and the order of sound bites, Cain's "999 tax proposal" was a hallmark talking point of the 2012 US presidential campaign. At the heart of the proposal was a flat tax of just over 9% on all new goods purchased and an exemption for all used goods; this would create incentives for moving from a disposable economy to a durable goods economy, with a greater emphasis on service of durable goods in creating jobs. Creating incentives to reduce disposable goods being consumed is the holy grail of environmentalism because it would reduce mining and energy usage. However, Mr. Cain was not motivated by environmental ideals in moving this forward. Rather his motivation came from the view that the tax system needs to be streamlined and made simpler. The plan would have regrettably raised taxes paid by poorer Americans and decreased the tax paid by wealthier Americans.

An increase of structural inequality is enough reason to oppose such plans, but environmentalist should acknowledge the possible green lining to "999." What was less clear to either Mr. Cain or environmentalists is how such a transition would impact livelihoods. Scant research exists on whether moving toward a durable goods economy through service-sector jobs can create livelihoods sufficient to support our population base. What is more likely is that a move toward a reuse and recycled manufacturing economy would provide both reduced impact and jobs. Thus, giving tax breaks to manufacturers and consumers who buy products made from recycled materials could provide a greater jobs base while mitigating environmental impacts (though not as much as the drastic shift of incentives toward used goods that 999 might provide). So Mr. Cain got himself embroiled in the conundrum of consumption: how to motivate changes in individual choice so as to have desirable outcomes in the aggregate.

Consumer Ecology and Varieties of Equilibria

Much of the order of consumption is defined dichotomously as "wants" and "needs." While no one resents the consumption of the bare necessities of survival, the ever-growing basket of "consumer goods" in modern societies is anathema to environmentalists. Activists agonize on how the collective good of the planet depends on the individual choices made by consumers. Yet the consumer's decision-making depends on signaling from society in terms of fashion, perception of privilege, and the quality of information about the products at hand. Such factors are not aligned with any particular natural order but rather with a set of often idiosyncratic preferences. To use the terminology of Nobel Laureate Thomas Schelling, individuals with self-serving "micro-motives" can control "macro-behavior" in ways that may not always be desirable. Schelling did not just simply challenge the wisdom of the invisible hand of markets but also considered the poverty of single equilibrium points in economic order. He posited that there was a very visible hand of individual human agency and that planning was less about control than about coordination of actions to achieve a desired outcome. While there was an allure to equilibrium in graphical terms, Schelling astutely reminded us that life had multiple equilibria, and the illusion of plucking out a single equilibrium must be eschewed. The tension between the "macro" and the "micro" realms of human perceptions is also reminiscent of what Richard Dawkins has called human bias of perceptions in the "middle world" between the microscopic and the macroscopic in which we reside. Thus, our brains have evolved to often tune out highly macroscopic phenomena at the level of physical space as well as highly microscopic phenomena at the quantum scale. Even though atoms are mostly made of empty space, that resolution of analysis is of no consequence to us since we deal with matter which is functionally palpable: "Just as monkeys live in an arboreal world and moles live in an underground world and water-striders live in a surface tension-dominated flatland, we live in a social world. We swim through a sea of people—a social version of Middle World."[5] Journalist Ziya Tong has referred to this phenomenon as the "Reality Bubble," where we are so influenced by our functional environment and ostensible order that we forget many other capricious realities. Calmly sitting on a peaceful porch, we forget that we are on a restless seething planet with molten mantle plumes or that 95% of all animal life is smaller than our fingernails—which are themselves a teaming microbial ecosystem.[6]

Ecological factors, like Dawkins exhortation to a Middle World, remained salient in Schelling's analysis even though he often irked many environmentalists by suggesting that with extreme trust deficits during the Cold War, the mutually assured destruction of nuclear weapons was the best deterrent. In one telling example from his book *Micromotives and Macrobehavior*, Schelling noted that economic equilibrium for whaling populations did not emanate from a stable population that was in harmony with the larger oceanic ecosystem. Rather the whaling population may reach economic equilibrium when the remaining stock of these mammals, prized for their meat and blubber, are so few that whalers can hardly catch enough to be profitable in their business. The few whalers who are able to invest in the high-stakes hunting are just enough to offset the new births in the small population. Another striking environmental example provided by Schelling pertained to the 20 to 30 million buffalo who had once roamed the plains west of the Mississippi at the end of US Civil War. Only their tongues and hides were marketable and carried a high price, but the "externality" of this market equilibrium was 20 billion pounds of rotting meat in the wild within 6 years. For every 5 pounds of buffalo meat left on the soil, a merchant got a penny for the hide. Had there been longer term thinking about when the transport infrastructure would have made the marketing of buffalo meat more conducive, the buffalo would have been worth more as live meat within 15 years. Yet the hunter's equilibrium was in that moment, and there was no mechanism to claim a property right to future live buffalo.[7] Equilibrium analysis thus oversimplifies outcomes by exaggerating points on a graph and neglecting processes of adjustment that depend on a broader range of parameters. Schelling was a master of metaphor, and his writings laid forth so elegantly as examples many of the complex concepts he analyzed through game theory.

> The point to make here is that there is nothing particularly attractive about an equilibrium.... Unless one is particularly interested in *how* dust settles, one can simplify analysis by concentrating on what happens after the dust has settled.... [T]he body of a hanged man is in equilibrium when it finally stops swinging, but nobody is going to insist that the man is all right.[8]

Not only are equilibria not inherently "right," but they may also be far more complicated in their temporal and spatial dimensions than we might otherwise realize. One of Schelling's key contributions to economic theory was the observation that equilibria may persist for a range of changing variables. Utility functions for consumers may encounter a rapid change only after certain thresholds are reached. The salience of the *threshold* had its roots in groundbreaking work by systems scientist Herbert Simon in his questioning of the traditional view of economists during the 1950s that consumers maximize their utility. Simon recognized that rational choice had its limits and was "bounded" by the information available to consumers and also their preferences for satiation. This insight added complexity and texture to our understanding of utility functions for consumers and ushered in the subfield of behavioral economics. Around the time that Simon was developing his views on bounded rationality, computer science was reaching maturity as well, and many of his ideas helped to nurture the field of artificial intelligence (AI), whose potential applications to earthly order we will discuss in the next chapter. Simon's greatest contribution to social science was his willingness to embrace nuance and always question the contrived elegance of equilibria. In this spirit he coined a new term "satisficing"—a portmanteau of satisfy and suffice—to describe human behavior in the context of bounded rationality. Unlike utility maximizing options, *satisficing options* were those which are suboptimal and dependent on a process of alternative searches until a threshold is met for decision. Simon's realization that mathematical optimization was often limited in human decisions because of "computational intractability" and a paucity of information transformed social science.

A year before Simon's death, in 2000, Carnegie Mellon University hosted a remarkable symposium titled "Earthware" to consider how the metaphors of "software" and "hardware," as the architecture of computational order, could be applied to the Earth. In introducing Simon as the keynote speaker, the President of Carnegie Mellon, Jared Cohon, noted that at the turn of the millennium the concept of *bounded rationality*, more than half a century after its coinage, still held dominant currency. Cohon had been one of my mentors during his time as Dean of the Yale School of the Environment when I pursued my master's degree there from 1994 to1996. Despite his hectic schedule as dean, Dr. Cohon taught a specialized course on decision analysis for the environment, and I was one of only five students who took that course in 1995. I got to know Cohon as a scholar who was always measured in his prose and not prone to exaggerate just for theatrical appeal. His accolade for Simon was, in that vein, not hyperbolic. The conceptual tools Simon had provided for economics, decision science, and computer engineering to consider suboptimality had profound implications for how we viewed social systems. The 1978 solo Nobel Prize citation noted his studies of "causal order" in complex social systems. A key aspect of his work was to elucidate the way directionality impacted order in natural and social systems. In his book, *Models of Discovery*, Simon used a specifically environmental example to demonstrate such directionality and its implications:[9]

poor growing weather → small wheat crops → increase in price of wheat

The price of wheat could not impact the weather and hence, in this causal order, "weather" is an "exogenous" variable while price of wheat is an "endogenous" variable. There is a certain level of drought tolerance that the wheat crop would have, and, following a threshold, the yield would deteriorate. All standard systems dynamic modeling software now utilize these insights.

Schelling built on this fundamental realization of thresholds for decisions. Figure 5.2 illustrates various models whereby multiple equilibria can evoke individual utility (or value) based on whether we are considering *utility maximizing assumptions* (functions 1) or *satisficing assumptions* (functions 2). Functions 3 and 4 show how the utility maximization and satisficing assumptions hold if the situations below the threshold also have a range of utility values rather than being binary. Some of the social real-world examples Schelling used to illustrate these multiple equilibria might make people uncomfortable now but are regrettably still salient. For example, he showed that as the number of multiple ethnicity residents in a white majority neighborhood increases, an equilibrium is maintained for a range of options after which there is "white flight," leading to the ghettoization of ethnic neighborhoods.

The threshold effect in the context of economic and social equilibria is also analogous to the notion of *punctuated equilibrium* in evolutionary biology. Stephen Jay Gould and Nils Eldridge observed that there were long temporal equilibria in evolutionary processes, followed by spurts of major growth triggered by some ecological disruption. Their analysis, analogous to Schelling's, was aimed at understanding how micro-scale phenomena (which in this case was evolution) are manifest in macro-scale outcomes. What they observed was that species evolution had a propensity to

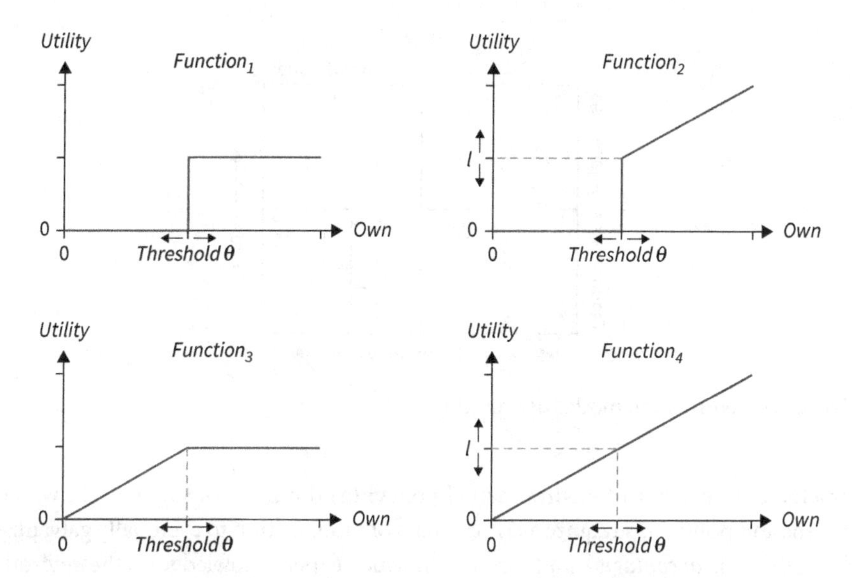

Figure 5.2 Schelling's models of how the size of one's own "type" of demographic (e.g., ethnicity or political persuasion) impacts the utility or value for the location to oneself depending on various scenarios.

undergo "stasis," whereby stability would occur in particular forms of species over long periods of time. Trilobites, starfish, and ginkgo leaves are examples of this phenomenon. Unfortunately, their highly evidenced theory was caricatured by Darwinian fundamentalists and misinterpreted as a validation of either "creationism" or "monster mutations."[10] Stasis could be errantly considered an absence of evolutionary agency or of "monster mutations" (since eventual rapid evolution could be confused with geneticist Richard Goldschmidt's notion of "hopeful monsters"). Such misunderstandings, as well as the metaphorical connections to social sciences, can be explained by considering issues of scale and the coexistence of various evolutionary mechanisms, as shown in Figure 5.3.

The diagram lays forth evolution in terms of two *modes* and *tempos*, terms which were coined by the paleontologist George Gaylord Simpson in 1944. *Cladogenesis* refers to the process by which evolution can occur through branching of multiple species, while *Anagenesis* refers to changes within the species that might lead to distinct interbreeding variations. The take-home message that Eldredge and Gould wanted to convey from this diagram was that all four pathways are possible and exist in nature. While punctuated equilibria might dominate in certain ways, all pathways are important and do not preclude each other. This fundamental message of coexistent pathways and varieties of equilibria in the natural world suggests even more prospects for heterogeneity in the socioeconomic world. Punctuated equilibrium holds a greater degree of sway in environmental policy settings as well. The resource economist Robert Repetto has observed that three key factors lead to the discontinuous (and

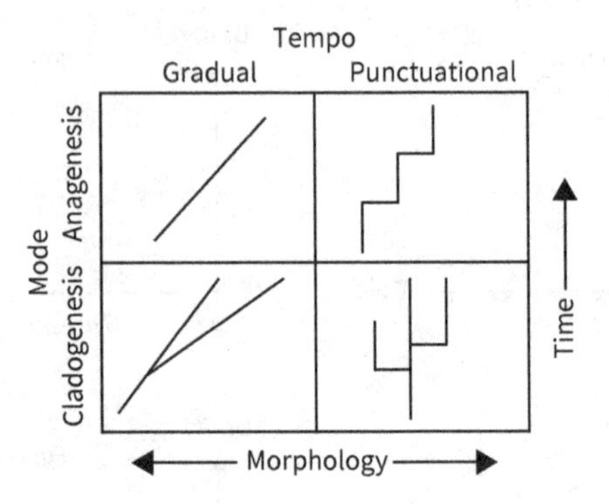

Figure 5.3 Tempos and modes of evolution.

punctuated) behavior of environmental policy: (a) the *Bandwagon effect*, in which citizens and politicians require a critical mass of mobilization to eventually galvanize change; (b) *social contagion and learning*, in which expert knowledge on the environment takes time to percolate to the level of social contagion; and (c) *media mimicry*, in which dominant news stories come in spurts, are copied by a range of communication media, and drive policy behavior sporadically rather than continuously.

The subfields of *evolutionary economics*, *bioeconomics*, and *neuroeconomics* have subsequently developed to consider these analogues from natural sciences to environmental economics and policy. These synthetic fields have further gained traction alongside ecological economics, which embraced the physical and chemical metaphors of Soddy and Georgescu-Roegen, as noted in the previous chapter. The concept of a "macro-state" emerging from entropic micro-states is also noteworthy in the context of comparing Schelling's insights about micro-motives and macro-behavior. Recall that in our discussion of physical entropy, individual particles at the micro-level were ultimately responsible for the macro-state (with properties such as temperature, pressure, volume); this is the fundamental premise of statistical mechanics. Ludwig Boltzmann, the progenitor of statistical mechanics, further noted that for a given set of large-scale observations, every possible configuration of micro-states that could give those properties is equally likely. Yet this is where divergence between natural systems and socioeconomic systems deserves clearer differentiation.

Although socioeconomic systems are of course constrained by key physical and biological limits of planetary sustainability, they have some fundamental differences in their range of prospective behavior. In his Nobel lecture, Herbert Simon noted with caution that the "social sciences have been accustomed to look for models in the most spectacular successes of the natural sciences. There is no harm in that provided that it is not done in a spirit of slavish imitation." He further elaborated that "if we wish to be guided by the natural science metaphor, I suggest one drawn from biology rather than physics." Part of this reasoning was that biology made greater allowance for variation and flexibility at the scale of macro-scale molecular complexity and was thus more akin to human behavior than was physics.

Toward an Optimal Economic Order for the Planet

Human behavior remains a capricious and confounding force in economic analysis as we try to ascribe monetary value to human values. Injecting morality into markets is part of the normative need in our times of rising planetary vulnerability, but it also remains elusive to economic order. A contemporary manifestation of the behavioral approach to economics that incorporates many of the fundamental insights of bounded rationality in a planetary context comes from the book *Humanomics*, coauthored by Vernon Smith and his associate Bart Wilson. Although the term "humanomics" has been appropriated by a variety of scholars, it has most robust relevance to our work from the insights of Smith and Wilson. They note that Adam Smith himself was influenced by Newtonian mechanics and rules-governed systems, but not by what Vernon Smith called "ecological rationality" in his Nobel lecture. Ecological rationality links human behavior to experiences with their environment that inevitably lead to specialization. While the specialization of economic markets might be considered analogous to speciation and ecological niches in natural systems, there is no inherent value in having more specialties of professions as there is in having more biodiverse species. Specialization in economic systems is value-neutral in and of itself, while speciation in biological systems is often deemed to have intrinsic worth. Yet specialization might also provide economic resilience in the same way that biodiversity has the potential for imparting ecological resilience, although the linkage is less direct than one may assume. For example, one study found that species richness increased temporal stability of the ecosystem but decreased resistance to a warming climate.[11] The virtue of specialization in professional terms harkens back to Ricardo's "comparative advantage" and is at some level inevitable with the diversity of human interests. There is also potential for greater bounding of information and knowledge through such specialized processes. Economic systems which are focused on the efficiency of specialization can gradually undermine the values of systems-based approaches needed for appreciating planetary sustainability. The quest then remains to move from "bounded rationality" to "comprehensive rationality" and find the appropriate bridging mechanism.

During the late 1960s, a freshly minted economist from MIT, George Akerlof, landed a junior professorship at the University of California at Berkeley and became intrigued by how defective cars or "lemons" were able to gain dominance in certain markets. As he pondered this question, he had an opportunity to spend a year at the Indian Statistical Institute to help find an optimal model for allocating the waters of the Bhakra-Nangal dam. Akerlof realized a strange connection between his inability to establish a timetable of water release to benefit peasants and the problem of car lemons. The core common problem of order in these cases was one of informational asymmetry. In the case of water regulation, it was indeterminate information on glacial melt and rainfall, while in the case of car lemons there was more discernible informational asymmetry between the buyer and the seller of the vehicle. Akerlof's subsequent paper on "The Market for Lemons" was to be the touchstone for his Nobel Memorial Prize citation more than 30 years later. His insights led us to realize that to achieve a societally optimal equilibrium, greater informational symmetry is needed through assurance mechanisms.

Information and its transmission were also a core aspect of understanding entropy in physical systems, and the same was also true of economic and social systems. The founder of information theory Claude Shannon's simple insights into the power of binary numbers as efficient transmitters of electronic information gave rise to the age of computing. While working at Bell Labs during the heyday of the telephone, Shannon postulated that any communication channel is also defined quite simply by two factors: bandwidth and noise. Given the metrics of these two factors, Shannon presented a seminal paper on the "mathematical theory of communication" and also stipulated the maximum limit or *channel capacity* by which information could be transmitted. In the case of the human brain, there is also a channel capacity for memory and information processing. The psychologist George Miller noted that seven units were often the cognitive limit of brain function in key attributes such as easily remembering telephone numbers, thus giving rise to his epic paper titled "The Magical Number Seven."[12] This observation should not be confused with other delineations of numerical order for matters like why there are 12 months in a year, 24 hours in a day, or 60 minutes in an hour. Many of those delineations are related to how easily divisible certain numbers are for record-keeping or lunar and solar cycles.

Shannon's additional key contribution was linking the concept of *entropy*, which was analogous to its thermodynamic connotation, in terms of the predictive attributes of a system. Information entropy increased with uncertainty about the state of a system. In thermodynamic terms, the location of particular particles in a system was also linked to informational attributes needed to ascertain their level of order. Entropy in this regard can also help us understand thresholds for "spontaneous" change in thermodynamic systems when we consider it in concert with heat and temperature change. Highly ordered structures rarely form spontaneously except when particular properties of the system favor the energetics. An example of spontaneous order arising in this context would be the formation of soap, which can trap dirt spontaneously due to the peculiar molecular structure of suds. Shannon's work resonated with a century-older thought experiment from physicist James Maxwell in which he had suggested that a box of mixed-speed molecules could be sorted by a "demon" who simply knew the speed of each molecule and allowed them to pass through a gated divider. Thus, greater order and lower entropy could be achieved not by investing direct energy but simply by having information about the entity itself.

Later research at IBM carried out by Rolf Landauer and associates suggested that there was a relationship between energy and information transfer in computational systems, which is manifest when we have irreversible computation. Ultimately, when information is deleted in a computational system to make room for new information, entropy increases and energy is employed—albeit in a very minute amount. Furthermore, with quantized energy, information also has discrete bounds in the binary realm of digits or "bits." In 2012, a team of European researchers were able to actually measure the energy erasure amount for each bit deletion, amounting to around 0.0175 electron volts at room temperature.[13] This seminal finding suggests a limit to how energy-efficient conventional computers can be made. Indeed, with our current advancements in computing technology, we expect to reach the "Landeur Limit" by 2048. Already the rapid rise in computational speed and the concomitant massive energy load of computation (estimated to be around 3% of global energy usage) is emblematic of this relationship between information ordering and the second law of thermodynamics. If we are to move to faster processing speeds of information transfer beyond this limit, we would need "reversible computing" similar to reversible chemical reactions. During the pandemic, the rise of telecommuting and high-quality information transfer in the form of video conferencing calls had a palpable energy footprint, even though we saved on transportation fuels for our own bodies. This is an area of immense research efforts globally as linkages to quantum computing infrastructure are developed that use photons rather than mass-based particles for computational transit.

In environmental systems as well, we are often dealing with a highly entropic informational nexus that can have peculiar social structure. This is why problems like climate change might be considered "wicked" problems, to use a term coined by sociologist C. West Churchman.[14] These problems have four characteristics: (a) incomplete or contradictory knowledge, (b) large number of people and opinions involved, (c) large economic burden, and (d) their interconnected nature with other problems. Further refinement of this concept was presented in the subsequent work of Kelly Levin and colleagues in what they termed "super-wicked problems,"[15] which are particularly relevant to issues of global environmental change. These problems include the following characteristics: (a) time is running out on action, (b) there is no central authority, (c) those seeking to solve the problem are also causing it, and (d) policies discount the future irrationally. Their solution to this challenge is to develop a *path-dependent policy*, one which has structural attributes that address these four challenges. In other words, they recognize that some ordered mechanism that is linked to underlying natural constraints is needed to address such challenges. In the past path-dependent policies have been criticized for inefficiencies and lack of adaptation to changing social norms, but if we consider them in the context of super-wicked problems, their necessity becomes more acute.

Returning to Akerlof's market for lemons, what was needed then was information assurance and signaling to create confidence in quality. The rise of *branding* as a mark of trust and the panoply of certification systems for quality goods from organic foods to conflict-free diamonds, to which we are now accustomed, are an outgrowth of this phenomenon. Trust is now also being operationalized through technologies such as *blockchain*, which allows for tamper-proof traceability through a highly devolved computational ledger assurance system. Yet the quality of the original input of data into a blockchain still needs to be assured for such "cryptogovernance" to be effective. The range and necessity of these systems depends on the kinds of goods and services we are considering. Broadly speaking we can delineate goods into the following three categories:

- *Search Goods*: Quality is obvious to a consumer merely by finding the good—generally in competitive markets
- *Experience Goods*: Quality is only observable after use by consumer and time-dependent
- *Credence Goods*: Quality comes from long-term display of effectiveness that is not directly observable by consumer

Figure 5.4 shows the how the demographics of the three types of goods and a range of services that we use may be considered on a scale pertaining to the ease of evaluation needed.

To address the challenge of information asymmetry, particularly if consumers are to be in the driving seat of instituting social change, we can consider the following approaches.

- Assurance from seller: a warranty based on trust guarantee
- External assurance against risk: insurance
- Crowd-sourced brand reputation of seller, provider, or manufacturer
- Prevalidation of product or service quality through a trusted party or certification

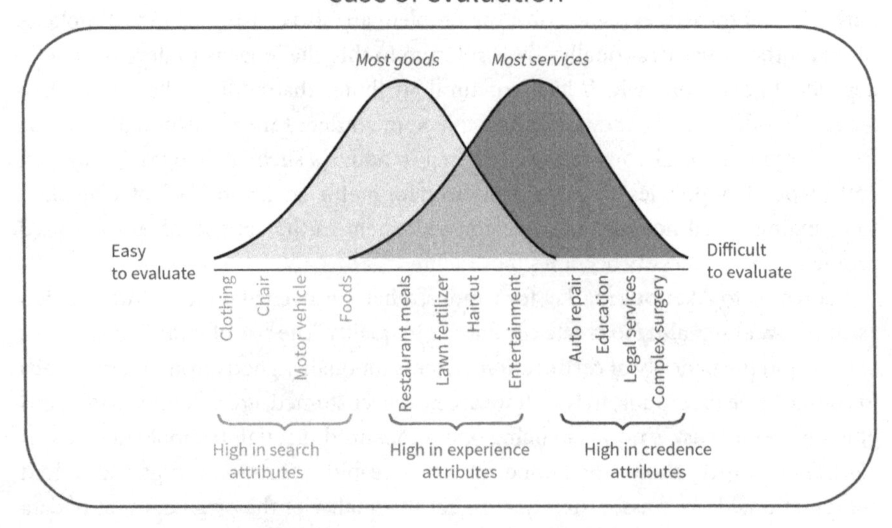

Figure 5.4 Informational evaluation of products and services
Adapted from figure by V. Zeithaml by D. Borman.

All these various mechanisms for providing consumers with more useful information to assist in decisions that are more societally efficient brings forth another important concept for our consideration in defining optimal economic order. The concept of *exergy*,—the effectively useful, available energy to undertake work—is particularly helpful in refining economic equilibrium analysis to better accommodate sustainability conditions. Although it was not explicitly mentioned at the time, the great ecologist Howard Odum presented energy and material flows through ecosystems as a means of efficient resource usage. At its core, exergy gives us the maximum theoretical work obtainable from an overall system considering that the system will come into equilibrium with its environment. When the system is in that equilibrium, it is referred to as the *dead state*, but in fact such a steady state may well be opportune ecologically under certain conditions (a form of quiescence or hibernation in times of crisis, for example).

How exergy ties in with our earlier discussion of information is that we are living in a world of data glut where useful knowledge about the economy or the environment may be also considered analogous to exergy in terms of effective decision-making. There is currently such immense noise in the range of statistics being presented that we feel paralyzed by an "infodemic." The level of data generation as a result of myriad sensors, electronic records of our activities, visual images, and satellite devices is astronomically large and growing at mind-boggling rates. In one remarkable statistic, from 2018 to 2020, we produced 90% of the data in all of recorded human history! Finding the exergetic value of these data while also considering the aforementioned notion of information entropy could be a very meaningful way for us to plan efficient outcomes for our planet. We are not just living in a world of information asymmetry but also of unprecedented informational redundancy whereby the temptation of analysis without insight haunts us. Furthermore, the information being produced is also far more networked. The connections between consumers and goods in the order of a network economy can grow much faster and have profound consequences for the impact and dominance of particular platforms. The information network economy has increasing returns to scale rather than the usual diminishing returns to investment. The value of a network increases exponentially with its users, which has in some ways even reversed the order of economic supply and demand curves. For internet applications, it makes sense to initially give away the product for free, even with high start-up costs. Through the phenomenon of *compound learning*, creating higher volumes of a product becomes easier to produce as we reach scale. At the same time, demand for a product is only going to accelerate once there is more volume—such as more users of a social media platform. This can also result in a particular clustering of natural monopolies but for a different reason than in conventional economies, where scale was linked to cost of production. In a network economy, scale is linked to the power of information sharing in terms of both the speed and quality of that information.

Information theorist Alex Wright, in his perceptive book *Glut*, has called this phenomenon an "escalating fugue"—the musical metaphor also noting how organizing the data can prevent dissonance and disarray.[16] He reveals through a fascinating exposition of human history how information management has defined economic order in human civilizations. Figure 5.5 is a simple representation of how we might consider high exergy output as a means of measuring economic and ecological efficiency. Exergy in a system with diverse attributes, as shown by the four different shapes in the diagram, is defined by the way we order and manage the information represented in each. By having a coherent and ordered nexus of information in any product or service, we can minimize material usage and waste while still achieving a functional livelihood and realizing the lifestyle values of consumerism.

The virtues of such analysis can also be considered at the level of consumer products that we would like to reuse, remanufacture, or recycle. A key feature of recycling and a circular economy is minimizing the system-wide impacts of consumer demand. In the exergetic hierarchy, we should try to reuse, then remanufacture, and then recycle. Preference for the recycle prospects would be given to "upcycling" whereby more durable products could be produced with less aggregate impact in the reprocessing of the materials. If we design products in modular form with exergetic principles in mind, the energy invested in these processes would perform the most "useful work." For example, cellular phones that can be deconstructed in modular form with replaceable or upgradable parts would be preferable to the full device being sent to a smelter for recycling and extraction of constituent metals. The macroeconomic system and its demand and supply for the product is not going to be able to internalize such factors unless there is either regulatory or consumer pressure to do so. The question then arises about how such pressure may be mobilized. How might we galvanize social order to ensure that economic systems are able to be in sync with natural order most effectively? The core of this debate harkens back to what social theorist Charles Lindblom epitomized in his landmark article *The Science of Muddling Through*. In this provocative piece, Lindblom critiqued centralized planning and "comprehensive rationality" for having the potential for autocracy. While exergetic efficiency at the level of cellular phone components may well be the innocuous task of a design engineer, the extrapolation of such design to social systems was deemed perilous. Instead, Lindblom argued for *incrementalism*, whereby social decisions and policies on

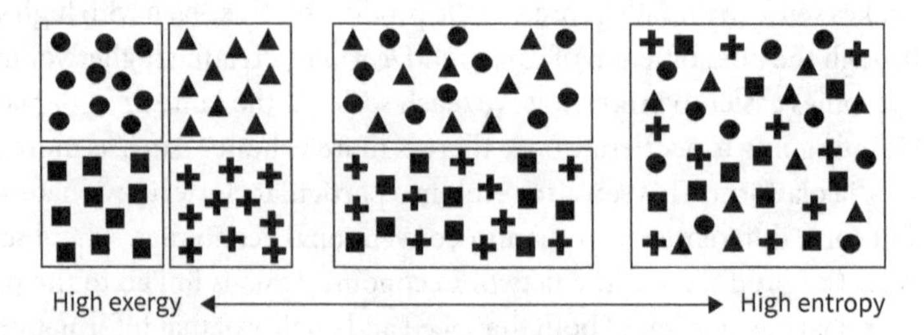

Figure 5.5 Relationship between exergy and entropy.

From Stremke, Sven, Andy van den Dobbelsteen, and Jusuck Koh. 2011. "Exergy Landscapes: Exploration of Second-law Thinking Towards Sustainable Landscape Design." International Journal of Exergy *8* (2): 148–174.

economic allocation, production, and consumption could be made through taking short steps and learning from them at each stage.

The metaphor used by Lindblom was of a rooted tree with many branches. Comprehensive rationality implied an effort to understand each and every systemic aspect of the tree's anatomy to arrive at a decision, whereas incrementalism focused on the branch on which we were poised to consider more limited options for immediate growth. Although the chances for error would be reduced if we had a more comprehensive understanding of the system, the time investment needed for such accuracy was disavowed by Lindblom. Along with his Yale colleague Robert Dahl, who coined the term "polyarchy" as means of economic and political power devolution, Lindblom was also concerned about the potential for abuse of any centralized decision-making power. The political implications of such an approach will be discussed in Part III of this book. Incrementalists argued that in a world where the future efficacy of particular central plans was largely indeterminate, such faltering flow forward was deemed efficient in the long term. How such trial-and-error approaches can be better calibrated to meet planetary goals of sustainability will require us to elucidate the order of errors.

Notes

1. For the ways in which markets can foster morality through the need for trust in trade, see Zak, P. J., and M. C. Jensen. 2007. *Moral Markets: The Critical Role of Values in the Economy*. Princeton University Press. For a critique of how morality has been superficially brought within the neoliberal paradigm, see Whyte, J. 2019. *The Morals of the Market: Human Rights and the Rise of Neoliberalism*. Verso.
2. Say, Jean Baptiste. 1834. *A Treatise on Political Economy: or the production, consumption, and distribution of wealth*. Grigg and Elliot (English translation).
3. President Dwight Eisenhower, Speech on the Military Industrial Complex. 1961. Accessible at The VAlon Project https://avalon.law.yale.edu/20th_century/eisenhower001.asp (accessed September 30, 2020).
4. Zencey, Eric. 2012. *The Other Road to Serfdom and the Path to a Sustainable Democracy*. University Press of New England.
5. Dawkins, Richard. TED Talk. "Why the Universe Seems So Strange?" https://www.ted.com/talks/richard_dawkins_why_the_universe_seems_so_strange#t-1198961.
6. Tong, Z. 2019. *The Reality Bubble: Blind Spots, Hidden Truths, and the Dangerous Illusions that Shape Our World*. Allen Lane.
7. Schelling, Thomas. "On the Ecology of Micromotives." In *The Corporate Society*, edited by Robert Morris.
8. Schelling, Thomas. 1978. *Micromotives and Macrobehavior*. W. W. Norton and Co., pp. 25–26,
9. Simon, Herbert. 1977. *Models of Discovery and Other Topics in the Methods of Science*. D. Reidel Publishing.

10. For an excellent review of the misunderstandings around punctuated equilibrium and lucid clarifications, see Stanley, Steven M. 1998. *Macroevolution: Pattern and Process*. Johns Hopkins University Press.
11. Pennekamp, F., et al. 2018. "Biodiversity Increases and Decreases Ecosystem Stability." *Nature* 563 (7729): 109–112. https://doi.org/10.1038/s41586-018-0627-8.
12. Miller, George A. 1956. "The Magical Number Seven." *Psychological Review* 63: 2.
13. Bérut, A., A. Arakelyan, A. Petrosyan, S. Ciliberto, R. Dillenschneider, and E. Lutz. 2012. "Experimental Verification of Landauer's Principle Linking Information and Thermodynamics." *Nature* 483 (7388): 187–189. https://doi.org/10.1038/nature10872.
14. Churchman, C. West. 1967. "Wicked Problems." *Management Science* 14 (4): B-141–B-146.
15. Levin, K., B. Cashore, S. Bernstein, and G. Auld. 2012. "Overcoming the Tragedy of Super Wicked Problems: Constraining Our Future Selves to Ameliorate Global Climate Change." *Policy Sciences* 45 (2): 123–152. https://doi.org/10.1007/s11077-012-9151-0.
16. Wright, Alex. 2008. *Glut: Mastering Information Through the Ages*. Cornell University Press.

6
Mindful Errors and Social Order

> An error does not make a mistake until you refuse to correct it.
>
> **—John F. Kennedy**

Following the tragic events of September 11, 2001, my land of ethnic origin, Pakistan, became the focus of intense scrutiny as a haven for terrorists. Initially, many Pakistanis complained about excessive screening at airports, but after a while we tried to find some wry humor in our predicament. A local clothing store in Islamabad started selling T-shirts labeled "Errorist," which our mind's eye would often see with a "T" at the start and raise alarm in passersby. A closer look would elicit puzzlement or perhaps a smirk. My nephew gave me one of these T-shirts and dared me to wear it back in the States. Anxiously amused, I reflected on the word itself and looked it up in the dictionary. "Errorist" as a word was only to be found in a few arcane literary manuscripts but a Google search finds it as a misspelling of the word "terrorist" thousands of times online. What was evident from this reflection was that our mind expects certain structures and anticipates errors. Yet this was merely one word which could play games with our mind. If we considered full sentences in English, there is clearly even more opportunity to define meaning when letters or even words are missing. Our mind is able to construct order out of incomplete or errant information as well. Languages, one may argue, have evolved to contain redundancy in syntax to create a more resilient means of ascertaining meaning, although we have now taken particular linguistic rules as marks of correctitude.

Insights about errors also tie in with the distributional order which is at the heart of statistics. As we count and enumerate vast amounts of data, we can generate graphs that can often take the form of proverbial "bell curves." Such a bell-shaped pattern is a manifestation of the *central limit theorem*, which is a key observation in statistics and suggests an underlying order to a variety of natural phenomena. Not all bell curves are *normal distributions*: these must follow a particular mathematical function for which a series of statistical techniques have been devised to test pattern adherence and errors. Furthermore, the allure of such distributions can lead us to make fundamental errors in judgment that we should also be on guard for. The mere exercise of defining mathematical patterns seeks to consider when anomalies and errors arise. As we set forth information in numerical series, they can converge, in the case of certain series of fractions, come closer to zero, or they can diverge either by increasing in size

Earthly Order. Saleem H. Ali, Oxford University Press. © Oxford University Press 2022.
DOI: 10.1093/oso/9780197640272.003.0007

toward positive or negative infinity, or they can oscillate between values. Yet even the notion of "infinity" is a mathematical abstraction that can lead to various paradoxes; it serves important heuristic functions but should not be construed as observable or even justifiable in its existence. There have even been arguments by scholars such as Max Tegmark to stop using infinity in physical calculations. Such confounding yet fundamental mathematical properties of numerical sets permeate natural and social science data in various ways.[1]

An extension of Claude Shannon's insights on information theory, which were discussed in the previous chapter, also reveal the importance of building redundancy into computing code to allow for such errors or *noise* to be addressed. Errors are a feature of both animate and inanimate existence. There are plenty of natural errors in the genetic code that can have deleterious manifestations in organisms. Our brain circuitry can make involuntary errors when we misspeak, or we can make deliberate errors of judgment. Computational machines can also make errors due to a variety of technical factors that can impede communication flows. Moving from economic to social order requires us to directly navigate these propensities for error and learn how to mitigate their eventuality. Managing such error transmission in human behavior is particularly important for planetary problem-solving where timely action is needed to prevent irreversible harm. In complex systems research there is also the concept of *renormalization*, whereby we can mitigate some resolution in data to get a usable and efficient picture—the metaphor of a low-pixel image that is still legible as you zoom out to view it would be an example of renormalization.

The complex systems scientist Scott Page has also highlighted the importance of diverse *collective intelligence* in mitigating social errors and facilitating more efficient order. He presents four *institutions* that can operationalize collective intelligence and that need to be considered: markets, hierarchies, democracies, and ecologies. We have already discussed the merits and limits of the collective intelligence of *market systems* in earlier chapters of this section. For *hierarchical systems*, collective intelligence arises from ordering data into useful information through its passage through an organizational management system, a process called *partitioning*. A partitioned task is essentially delegated to players best suited to address it, and this in turn leads to more efficient information analysis. In *democratic systems*, collective intelligence occurs through voting behavior, albeit often highly dependent on the quality of information on candidates and structural attributes of voting. Finally, in *ecological systems*, collective intelligence arises through ecosystem feedback loops. A key misunderstanding about complex systems stems from the notion that decentralized decision-making is incompatible with centralized planning. Indeed, the development of hierarchies that occurs in complex systems is what makes them most functional. What separates the scales is often the speed of change in those hierarchies and hence their endurance. Therefore, legal institutions and constitutions emerge out of necessity in complex political systems, as scale increases. Such institutions mitigate the uncertainty of outcomes that are important for the system's resilience. Yet the day-to-day behavior of citizens operates in a more devolved and discursive pattern that can eventually still lead to patterns of adaptive order. The key difference between the complexity approach to such political analysis and the classical approach is that the patterns are not necessarily imposed but rather evolve endogenously.

What is fascinating about the power of collective intelligence is that it is a compelling notion for either extreme of the political spectrum, but each side operationalizes its agency in different ways. Free marketers consider its virtues in the context of economics while ardent environmentalists consider its potency at the level of the ecosystem. Scott Page's research further shows that the more diverse the set of perspectives in the dataset, the more likely we would be in averting errors in the ultimate outcome. Diverse systems were more robust for a variety of mathematical reasons that have to do with the superior management of errors by crowds, particularly in the context of complex problems. Page lays this out in a simple equation:

Crowd Error = Average Error – Diversity of Prediction

The reason why such an equation holds true for complex problems like global environmental change is related to the same structural reasons for which our "muddling through" champion Charles Lindblom had expressed concerns about comprehensive rationality. Diversity and the ability to manage diverse output provides us an opportunity to harness far more information about a problem than would otherwise be possible, and the probability of recurring errors diminishes if diversity is higher. What is most surprising about the insights from Page and his associates' work is that in a given set of "high-caliber individuals," diversifying this pool via inclusion of lower average aptitude can still lead to better solution outcomes for complex problems. In other words, qualitative diversity trumps quantitative metrics of ability. However, the challenge with effectively incorporating diversity in social order is twofold: (a) varied perspectives which accentuate misinformation about established facts can muddle the value of diversity; and (b) in social systems without effective consensus-building processes, diversity can lead to destructive rather than constructive conflict. In a study conducted by the journal *Nature* of more than a million papers in 24 academic subfields, ethnic diversity correlates more strongly with citation counts than diversity in age, gender, or affiliation.[2] Thus, for us to capitalize on "the diversity dividend," we need to ensure that we have means of managing the rowdy phenotypically and perhaps genetically diverse crowd whose insights we wish to harness.

Lonely Crowds and the Greening of the Shared Economy

Finding analytical order in human crowds has been the preoccupation of sociologists. After World War II, the Baby Boom was characterized by a proliferation of critical reflection and scholarly work on contemporary civilization. Academic books rarely gain traction with the general public but in this momentous period of human history, one book, published in 1950, captivated the American imagination like no other: *The Lonely Crowd.* A troika of academics in New England at the time—David Riesman, Nathan Glazer, and Reuel Denney—crafted this scholarly work based on numerous in-depth interviews of American society. The lead author, Riesman, was a mild-mannered legal sociologist at Harvard who had no grand expectations for this volume. Yale University Press initially published only 3,000 copies. Yet the book captured the public's attention because of a simple typology of social order which it presented and whose prescience to this day persists as a staple of college course readings in sociology. By latest count, more than 1.5 million copies of this book have been sold, and it is by far the largest selling academic book in social science history.

The simple typology presented by Riesman, Glazer, and Denney divided society into three main cultural types: tradition-directed, inner-directed, and other-directed. The authors suggested that, initially, social systems favored *tradition-directed* people who adhered to norms that emerged out of survivalist tribal impulses and cultural norms. With the Industrial Revolution and the development of the markets came *inner-directed* persons who were more driven by individualistic impulses. Finally, the growth of mass communication led to the development of the *other-directed* person who is focused on how others behave and what they might think of each other in a network. The rise of the other-directed person created the conditions for mass consumerism to proliferate without necessarily considering the ecological implications of such a transition. Throughout life one human can pass through phases of being in any of these three categories of social behavior. While the authors of *The Lonely Crowd* appeared to yearn for the individual pragmatism of the inner-directed person, if we consider their insights of seven decades ago in a contemporary context, the other-directed person presents a new set of potential virtues.

Other-directed individuals present a formula for human co-dependence, which, if normatively chartered in the context of ecological literacy, can bring about social change far more efficiently and rapidly. Organic food and fair-trade product memes could be spread just as well as those for throwaway toys. Just as negative fashions and consumer fads can spread quickly, so, too, can positive attributes. Cultural "memes" (a term surprisingly coined by the geneticist Richard Dawkins) are able to spread in a networked society with alacrity. The key concern of societal leaders, whether at the level of a household, local councils, or national governments, should be how to improve the quality of such memes for societal benefit. Other-directed persons are not altruistic but rather seek to be "liked" in a way reminiscent of Generation Z and its quest for social media affirmations with thumbs up, hearts, and myriad other emojis. While such a yearning can lead to psychological pathologies, as demonstrated in the 2020 Netflix docudrama *Social Dilemma*, it can also be a means of harnessing collective intelligence. The physical reality of memes is manifest in neural networks showing how tightly communication systems are coupled between human brains and how memes are transmitted and received. The Collective Computation group at the Santa Fe Institute has recently conducted research on the reality of memes and found that the persistence of memory in human societies and the urge to share and communicate these memories is the key defining element of "memetics." Furthermore, as the work of adaptation ecologist John Holland showed in his classic work *Hidden Order*, the adaptive responses of humans are in and of themselves a key factor in creating complexity and accelerating the outcomes of evolution. The ability of humans to communicate through culture and organizational linkages further strengthens our ability to survive.

The advent of the "shared economy" is a direct positive outgrowth of such other-directed individuals. From adding crowd-sourced information to Wikipedia to reviews on Airbnb and Uber, we can all benefit from the collective intelligence of the crowd when modulated by effective technology. The social order of the shared economy also has potential for reducing ecological impact in various ways. In terms of material efficiency, shared mobility can lead to a reduction in average household vehicle ownership without commensurate increases in vehicle ownership in society as a whole. Thus, in such a context, material efficiency would be reflected in the reduction of materials used in automobiles and more precisely in the economy-wide quantity of materials used in passenger kilometers traveled. Shared mobility can also lead to a reduction in demand for parking space, with an associated reduction in materials consumed for that purpose. Research on car sharing in California and Sweden has shown potential for reducing private vehicle ownership, specifically for vehicles with infrequent use, such as seven-seater and larger passenger vehicles, all-wheel drive vehicles, long-range vehicles, or vehicles with a large trunk.[3]

Yet we must also be cautious of a rebound effect in the context of a shared economy situation, particularly when considering built infrastructure, such as shared apartments. While Airbnb claims that shared lodging leads to substantial greenhouse gas (GHG) benefits (equivalent to taking 33,000 cars off North American roads), the company does not provide access to the full study and methodology. In contrast, a recent study comparing the environmental impact of shared lodging with hotel stays got mixed results.[4] Growth in Peer to Peer (P2P) lodging can also increase hotel room vacancies, in which case there will be underutilization of hotel building stock that potentially offsets the gains from use of residential space.

Greening of the shared economy has much potential for aligning social order with natural order. However, the underlying networks' platforms that support this process require scrutiny. The sheer scale of data and the contagious speed with which computational speed has arisen can both be illuminating and confusing. On the one hand, computational speed presents a way out of the need to muddle through planning because we can digest comprehensively more information. On the other hand, the ability to communicate information to such a highly devolved set of stakeholders has the potential to create noisy confusion and anarchy. The polarization that has been observed as a result of the spread of conspiratorial thinking is an example. A disconcerting aspect of the misuse of the network power of a shared economic order is how information as well as misinformation are forced to compete with each other at an unprecedented level. Compounding this profusion of information is the monetization of human attention on social media networks. Psychological research is clear in showing that polarization and sensationalism pays in terms of keeping people tuned in to particular content. We are living in an age of what Shoshana Zuboff has referred to as "behavioral surplus" in her book the *Age of Surveillance Capitalism.* We have many "micro-behaviors" that can now be monitored and monetized by major social media companies. Many of these behaviors are not ecologically aligned, and the sheer volume of information overload has further removed us in many ways from the underlying natural processes that sustain us. In the poorly reviewed film *The Circle*, Tom Hanks's character, as the founder of a major technology firm, finds a way to gather all these data to create global order. A world where no crime would be possible because data could be used to identify criminals and prevent a range of malfeasant behaviors. Yet the human need for privacy is viscerally important as a mark of our own internal order, and complete transparency of the CEO's personal data dooms the film to an uneasy and incomplete end.

On the environmental front, polarization has also increased dramatically in this social media–based news environment. The environment used to be a unifying issue in American politics even up to the Gingrich era, when an otherwise polarizing Republican Speaker of the House wrote a book titled *Contract with the Earth,* with a foreword by leading ecologist E. O. Wilson.[5] However, social media has accentuated what environmental politics scholar Leah Stokes has called "the fog of enactment," wherein risk, uncertainty, and the perils of precaution have been lumped together in dysfunctional disorder. Unpacking these attributes is essential if we are to move to the next phase of our understanding of social determinants of earthly order.

Environmental Risk, Uncertainty, and Precautionary Disorder

For the ancient Greeks, the high seas were the earthly frontier of the future. Much of their mythology and their efforts at exploration revolved around the seas and the challenges of navigating them. There was always a chance that a boulder or a rooted reef would disrupt their marine travels. *Rhiza* or *rhizikon*—meaning root or stone cut way from the land—was the term they used as a metaphor for difficulties that could be encountered at sea. Our modern word "risk" comes from this Greek root, and its origins show that natural phenomena have always constituted dangers over which we feel least control. Yet the concept of risk in economic and social analysis implies that we have some idea of a range of possible futures and hence can offer probabilities. On the other hand, uncertainty implies that all possible futures may not be known, and the probability calculations of any particular occurrence are elusive. Most significantly, in the words of Chicago economist Frank Knight, "risk provides some basis for insurance whereas uncertainty cannot be insured against."[6]

Environmental variables and human activities which could impact natural systems have attributes of both risk and uncertainty. Further compounding this dynamic is the risk to an individual alongside a collective risk to a population. A decision to drink specific water from an unsafe location is an individual risk, but polluting a lake near your home from which everyone in a community may drink poses a collective risk. The COVID-19 pandemic laid bare the linkages between individual and collective risk in the context of mask wearing and social distancing, particularly for younger populations. The morbidity and mortality risks for youth were much less than the collective risk for the elderly. Social order and a need for a sort of intergenerational altruism was being demanded of youth who were often reluctant to oblige. The reason they were unwilling to do so was often steeped in perceived uncertainty about outcomes. Social networks have widened the spread of perceived uncertainty and legitimized greater risk-taking behavior. The rise of skepticism around climate change has been accentuated by pseudo-science and a vast network of lobbyists, whom the science historian Naomi Oreskes has called *merchants of doubt*. Psychological uncertainty is a highly disempowering process at one level, but it can also empower those who want to perpetuate the status quo. Social psychologist Gary Klein laid out a model for how uncertainty becomes a barrier to action (Figure 6.1).

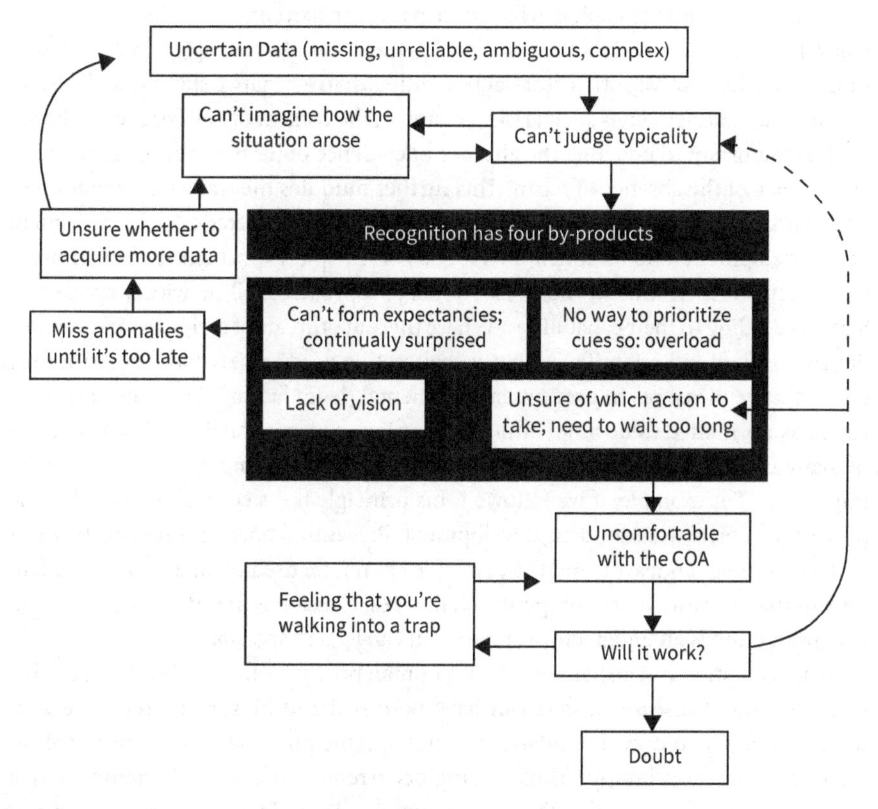

Figure 6.1 Klein model of uncertainty as a barrier to action. COA refers to "courses of action."

From Klein, Gary. 1998. Sources of Power: How People Make Decisions. Cambridge, MA; The MIT Press.

The discomfort felt when taking a course of action within a miasma of uncertainty has dire implications for imminent problems that have irreversible thresholds. Medical decisions on administering a certain experimental treatment fall into this category. Ecological concerns such as climate change might also fall into this category, though the diagnostics themselves are questioned. A patient with a fever or other symptoms is relatively easy to diagnose; a planet with short-term weather patterns that might occlude the long-term trend is a far more difficult diagnostic for the general public. Thus, uncertainty and doubt become a defining feature of policy paralysis and inaction. Such a prospect runs across the full spectrum of views around our ecological predicament. On the one hand, uncertainty is used by the fossil fuel industry to perpetuate the status quo around carbon emissions, while, on the other hand, many environmentalists have used uncertainty about the safety of nuclear technology to call for its phase-out. If we keep unpacking arguments for and against a particular technology, the conversation spirals into a battle of uncertainties. The general principle of risk aversion in the context of environmental decision-making is referred to as "the precautionary principle" and has become enshrined in many environmental regulations.

The origins of this principle can be found in the German concept of *Vorsorgeprinzip*, which by some measures is better translated as the "foresight principle." Such a translation suggests a positive anticipatory action rather than a negative status quo decision, but the essential element is social risk aversion in the context of environmental harm. Yet it is important to note that the absence of evidence of harm is not the same thing as evidence of the absence of harm. This further muddies the waters in operationalizing a principle that was put forth in the United Nations Conference on Environment and Development (Rio Summit), in 1992, as "Principle 15," which states: "In order to protect the environment the Precautionary Approach shall be widely applied by States according to their capabilities. Where there are threats of serious or irreversible damage, lack of full scientific certainty shall not be used as a reason for postponing cost-effective measures to prevent environmental degradation." Although this principle has normative value as an ethical aspiration, operationalizing it in the context of maintaining social order in a complex environment of competing goals is next to impossible. For example, if we followed this principle to its core, there would be no process of clinical trials for drug development. Precaution operates on a spectrum, as with any human endeavor, and the principle cannot be used as an excuse for indefinite inertia in a world with competing challenges. Caution is in order, but indeterminate precaution is an untenable postulate that can lead to societal paralysis.

Risk perception can also intersect with human proclivity for seeking patterns. The phenomenon of disease clusters can have both real and illusory currency because of our tendency to draw boundaries around specific observations. As risk analysts often say "chance is lumpy." Thus, having occurrences unevenly distributed across space or time is the rule rather than the exception. Similarly, exact averages are rarely observed, and far more common are clusters before and after the average. With cancer clusters, if the starting point of inquiry is a particular occurrence of cases, a researcher has to ask the question: Have we delineated an arbitrary boundary for such statistical occurrences? In their highly readable book, *How Much Risk*, Inge and Martin Goldstein illustrate a range of intervening factors and statistical anomalies that can mislead us into considering association when there is none.[7] They give the example of a Texas shooting range where a repeated cluster of shots going off target due to a shooter's physiological bias could simply be changed as on-the-mark by moving the target post. They analyze ostensible health risk clusters near nuclear power plants, exposure to pesticides, and a range of other human-induced pollution concerns. Their ultimate conclusion, which may well be unsatisfying but true, is that finding definitive patterns that provide causal explanations in a universe of lumpy chance is exceedingly difficult. Humans may try to set minimum standards for clean water or air, but we must be willing to show humility when anomalies occur and we must keep refining our causal apparatus.

Another way to think about the challenge of operationalizing risk analysis problems is to go back to a discussion of uncertainty in the context of statistical errors, of which there are essentially three types. A *type I (alpha)* error occurs when there is an errant conclusion about a phenomenon or association existing when in truth it does not (in the context of falsifiable scientific method, this happens when you reject the null hypothesis when it is really true). This is the classic "false-positive" outcome. This is the error noted in statistical terms as the "p" value (probability of such an error occurring is noted in calculations; it is usually set at 5% and deemed statistically significant). The *type II error* occurs when you fail to detect something that actually does exist. This is, conversely, the classic "false-negative" outcome and can be a bigger challenge in terms of ultimate consequence if there is serious environmental harm. In applying the precautionary principle, we have to find a way to figure out the tradeoff between type I and type II errors. Furthermore, *type III error*s, often less noticed, occur when one provides an accurate answer to the wrong problem. We may be evaluating a sample which is not adequately representative of a particular population or problem.[8] There are increasing concerns about the misuse of statistical significance in research, with one recent study noting that 51% of 291 articles in five highly ranked journals mistakenly assume that nonsignificance means no effect. In 2019, more than 800 researchers signed a letter calling for an end to the categorization of statistical significance as a threshold for cause–effect determination.[9]

In their landmark book, *Risk and Culture*, Mary Douglas and Aaron Wildavsky alerted us to the deceptive objectification of risk. Ultimately, risk in a complex world of competing and intersecting phenomena is a socially constructed phenomenon. Going back to the Bible, they note that the dietary laws of Leviticus may have stemmed from some degree of medical materialism but were ultimately about a cultural delineation of boundaries and the social construction of risk. Another curious insight from Douglas and Wildavsky's work is how "dirt" is a form of disruption to socially constructed order and, even when it is harmless—or perhaps even helpful in the case of nutrient transfer to arable land—it is deemed impure and repugnant. The authors of this classic collaboration between an anthropologist (Douglas) and a political scientist (Wildavsky) maintained that, strategically, keeping a balance between anticipating harm and trusting resilience is the essence of managing risk. The *law of diminishing returns* can be applied to this process, whereby each marginal risk prevention effort does not reap concomitant rewards. By their analysis, excessive safety targeted at a particular technology like nuclear power implementation could undermine overall systems' safety because alternatives can appear more attractive than they actually are when considering the full scale and scope of return on investment. The economic marginalization of nuclear power is an intriguing case in point. Massive safety upgrade requirements to existing nuclear power plants have rendered them uneconomical, thereby making the climate mitigation targets more challenging to reach in the short-term as other low carbon sources are up-scaled. Wildavsky was a firm believer in the iterative power of "trial and error" in generating a desired order for human safety. In an earlier book, *Searching for Safety,* he suggested the following:

> Trial and error is a device for courting small dangers in order to avoid or lessen the damage from big ones. Sequential trials by dispersed decision makers reduce the size of that unknown world to bite-sized, and hence manageable, chunks. An advantage of trial and error, therefore, is that it renders visible hitherto unforeseen errors. Because it is a discovery process that discloses latent errors so we can learn how to deal with them, trial and error also lowers risk by reducing the scope of unforeseen dangers. Trial and error samples the world of as yet unknown risks; by learning to cope with risks that become evident as the result of small-scale trial and error, we develop skills for dealing with whatever may come our way from the world of unknown risks.[10]

The role of modulated trial and error in risk management further relates to what the sociologist Charles Perrow called *normal accidents.* His book of this title was published in 1984 and presciently explained some key major accidents which took place soon thereafter, such as the Union Carbide chemical leak in Bhopal, India, in December of that year. Soon thereafter were the Challenger space shuttle disaster and the Chernobyl nuclear reactor meltdown, in 1986. The argument Perrow made was that accidents were a feature of complex interacting systems over which we could only have limited control in terms of causality. He did offer two points of guidance to mitigate the chance of accidents in such systems: (a) looser coupling between various parts of the system so that an initial fault would be less likely to lead to a cascade of catastrophic failures and (b) reducing the complexity of relationship between components. If either of these conditions is met, accidents become much less likely. For example, my employer—a university—is interactively complex but only loosely coupled. Thus, decisions may be influenced by factors that cannot be predicted and that might otherwise cause "normal accidents," but their effects are felt slowly. In contrast, modern factory production lines are often tightly coupled, with close and rapid transformations between various stages of a process. The relationships between these stages, however, is not *complex* (even though the machines may seem *complicated*).

Neither of these systems is prone to accidents, unlike a coal mine or a power plant.[11] Figure 6.2 presents one of the most significant charts from this book, which attempts to give some order to categorizing a range of phenomena within this framework.

So prescient was *Normal Accidents* that it even chillingly foresaw the Fukushima accident and suggested a preventative measure three decades before its occurrence. Perrow suggested that nuclear plants could be made marginally less complex if the spent storage pool were removed from the premises. Such pools often require constant cooling and attention, and a reactor accident could force a complete loss of power to the fuel cooling system, leading to a cascade of subsequent nonpreventable events, which is what happened in Fukushima. Yet prevention of accidents through a reduction of complexity or looser coupling could also have other implications for design. The quest should be for systems that are able to "harvest" complexity and use it to improve outcomes for society. How might we figuratively do a jujutsu move on disorder and use its ostensible opposition to our goals to instead benefit our outcomes?

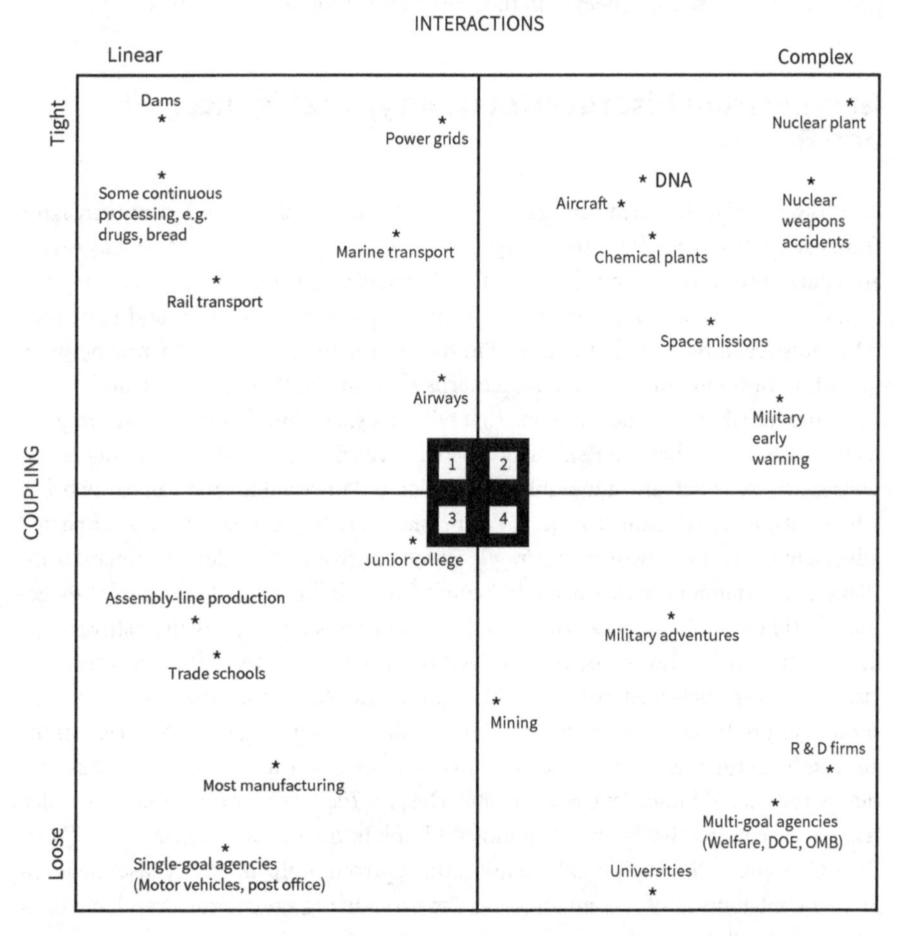

Figure 6.2 Perrow's typology of how "coupling" and "interactions" can be ordered.

The structure of particular social processes provides us with a mechanism to do so. These processes can also be imitated through synthetic decision instructions that we term "algorithms." Although often associated with computer science, algorithms are simply a set of instructions—and well-designed algorithms can help us make sense of seemingly disorganized data. However, an algorithm requires some organization of data if it is to be functional. There are even presumptions made by philosophers of science, such as Nick Bostrom, that the existence of key features of mathematical order in the universe signify that existence as we know it could be a massive simulation, one that might even be projected like a hologram. Elon Musk has embraced such a prospect, as have many science fiction authors such as Isaac Asimov and Daniel Galouye. Associating order with a stylized simulation is not new. The ancient Chinese manuscript *Zhuangzi* projected the view that existence was a "Butterfly Dream," and in ancient Indian philosophy the notion of *Maya* represents existence as a vast illusion. Careful algorithmic analysis of such a proposition in terms of observations from quantum physics suggests key constraints on the hypothesis, but there remains a lingering interest in probing deeper in the search for simulation algorithms.[12]

Gaining from Disorder: Immunity, Intelligence, and Religion

The power of algorithms in sifting through a vast amount of data and gaining insights from each misstep has given us the age of artificial intelligence (AI). Most AI systems are characterized by a complex web of such *machine learning* algorithms. Over the course of various faltering attempts at mimicking natural structures and networks, AI is now reaching a level of maturity in harnessing the power of dynamic network growth to develop durable learning patterns.[13] However, there are some underlying structural attributes of such systems that fall on a spectrum that runs from fragility, to robustness, to what the risk analyst Nassim Taleb calls *anti-fragility*. *Fragile systems* are those which are vulnerable to disorder and uncertainty and can easily break when outside their comfort zone. *Robust systems* are those which are not impacted adversely by disorder, while *anti-fragile systems* thrive in disorder and uncertainty. Planetary sustainability clearly could benefit from a better identification of management systems and intelligent processes that can harness uncertainty in positive ways. Taleb's "Incerto" series of books is derived from the Italian word for uncertainty or risk. The most celebrated primer of this volume was *The Black Swan*, which focused on rare unpredictable events (like the historically rare sighting of such a swan in the sixteenth century) and became a runaway best-seller because it was published just before the Global Financial Crisis in 2007. Despite *The Black Swan*'s popularity, Taleb has publicly noted that his most significant book in the series is *Anti-Fragile: Things That Gain from Disorder* (2012) He notes this upfront in the book because the intellectual contribution of recognizing ways of capitalizing on disorder can have huge social implications.

> With randomness, uncertainty, chaos; you want to use them not hide from them. You want to be the fire and wish for the wind.... The anti-fragile loves randomness and uncertainty, which also means—crucially—a love of errors, a certain class of errors. Anti-fragility, has the singular property of allowing us to deal with the unknown, to do things without understanding them—and do them well (p. 3).

Taleb's seminal insight on anti-fragility, as noted in the last sentence of this quotation, has resonance with what the Darwinian philosopher of cognition Daniel Dennett has called "competence without comprehension."[14] He also suggested that consciousness should itself be considered a "user interface" between biology and agency. I had the pleasure of being a student at Tufts University as an undergraduate where Dennett has taught for the past 40 or so years and attended many of his lectures. Although Dennett never used the term "anti-fragile," many of his explanations for the magic of consciousness (grounded in materiality, of course) were based on the brain's anti-fragile structure. A simple graphical presentation adapted from Taleb's insights in this regard is shown in Figure 6.3.

Another zoological metaphor that has gained traction around our understanding of risk is the *gray rhino*-popularized by journalist Michele Wucker.[15] Failures at risk management, Wucker argues, are not just about planning for unanticipated rare events but also about ignoring highly probable events which creep up on us and lead to calamitous outcomes. A charging rhino is likely to cause harm, but, like its pachyderm cousin, "the elephant in the room," may somehow seem invisible. This may be due to context and a numbing from either slow transitions or the distractions of short-term, spatially myopic vision. Our perception of climate change may fall into

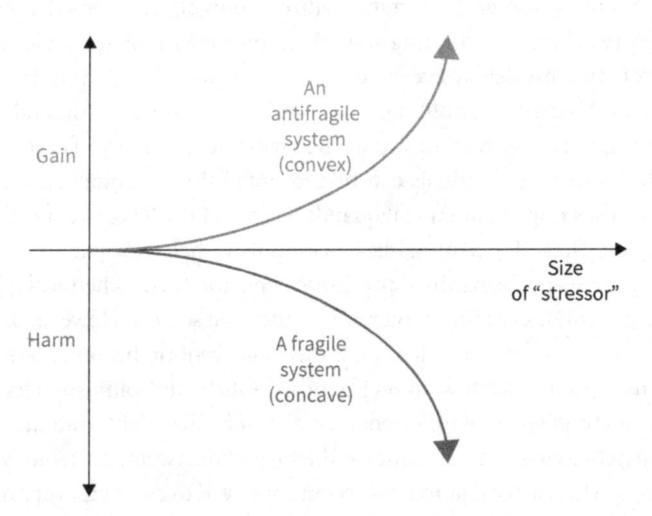

Figure 6.3 Adapted version of Taleb's representation of Fragile and anti-fragile systems.

such a structural category, in which our brains are not able to process impending harm due to lack of prior experience with a particular transition.

Neurons and synapses have evolved to gain from each disruptive experience, and, indeed, the very nature of "experience" is a testament to anti-fragility. Similarly, our muscles have the ability to gain strength from exercise and strain as the body attempts to learn from the impact of the stress and overcompensates in its restorative process by giving you more strength. The phenomenon of *hormesis* in toxicology is another example of such anti-fragile ordering in which small doses of an environmental agent can elicit a positive response while large doses can be poisonous. This "biphasic" dose-to-response relationship can, however, be misinterpreted if taken to an extreme, as it is in the context of alternative medical approaches like homeopathy. Ultimately, the dose–response relationship is still highly dependent on specific empirically determined criteria. However, there are now also some intriguing discoveries of how hunger or intermittent fasting within a nutritional range, as well as excessive short-term exercise, can play a similar therapeutic role in reducing cellular aging through hormesis. The exact processes are still being understood, but there appears to be a positive impact of such activities on *autophagy*—the process by which molecular detritus is removed from cells and positive regeneration can occur. Differentiating such empirical realities from metaphorically similar therapeutics remains an ongoing challenge in the struggle between conventional and alternative medicine, which in essence may still be tapping into the same structural processes. Even the remarkable strength of the *placebo effect* (the mere perception of a treatment without an actual dose) may be explained through the way the brain is able to modulate such biochemical responses through the power of perception. Conversely, a *nocebo effect* may also occur where a placebo makes you feel worse due to negative perceptions about a medicinal delivery process.

The ongoing debates on whether physical or at least biological processes determine human "free will" is also an important feature of how we can consider social order. The complexity of brain functioning and how human decisions are made continues to be a source of immense debate even without any theological underpinnings. Whether decisions are voluntary or involuntary has implications about credit and culpability, as well in terms of laws governing human behavior. Furthermore, if certain societally aberrant decisions by individuals can be prevented through biochemical intervention, then it raises important ethical quandaries about the level of control regulators should employ. Thus, if a criminal has a compulsive behavior pattern that is being dictated by genetically determined psychotic behavior, then a chemical intervention to mitigate that conduct becomes more incumbent on society. However, if we believe there is a greater level of free will at play, then intervention becomes less justifiable. Although the specific debates among neuroscientists and philosophers will likely rage on, our focus in terms of functional social order will benefit from the delineation offered by psychologist William James at the turn of the twentieth century. Although some features of his categorization may seem contrived to contemporary neuroscientists, the four categories he offered are useful in determining how we may respond to these issues in terms of social order.

1. *Instinctive selection*, for organisms with only genetically inherited behaviors
2. *Learned selection*, for organisms that remember their past to guide the future
3. *Predictive selection*, for animals with foresight who anticipate consequences
4. *Reflective and normative selection*, for humans who can think twice, then think again about the thought, evaluating it in the light of personal and societal values

The fourth category has most resonance with how the average human views "consciousness," regardless of its biological underpinnings. The potential for pseudo-science or para-science to creep into the conversation at this point is also particularly important around the elusive concept of consciousness. Libraries are replete with books that consider various structural orders and observations around the human mind and impute some intangible energy that instantiates life by resurrecting the theory of "vitalism." Although this theory in its eighteenth century form has been debunked there is massive energy transfer in living cells which is far greater per unit mass than for "non-living systems." Biophysicsts are trying to further understand the mechanisms by which such energy transfer is sustained in complex organisms, but this has also led to a genre of highly popular but scientifically questionable works from writers such as physicst Fritjof Capra and medical doctor Deepak Chopra. While humility should always be shown with highly complex fields of inquiry in this regard, there should be a willingness to differentiate those structural insights which are empirically determined and those which are not. This does not, however, diminish the functional value of these writings in giving a sense of transcendence to many in coping with adversity.

Whether it be biblical parables or the Dreamtime tales of Indigenous Australians, the power of stories, even if figurative rather than literal, should not be underestimated. It is important that we are not just enamored with scientism as a functional means of bringing meaning but that we should always be aware of physical reality versus emotional reality. Fields such as *panpsychism* are regaining currency as a third way out of the polarization between materialism and mysticism. Like materialism, panpsychism grants that the world is wholly physical. But like mysticism (or mind–body dualism), it views consciousness as being transcendent in terms of energy flows in the universe.[16] Certain characteristics of the energy flow itself can provide conscious order, just as material alignments can produce physical order. Some of the recent research using the BrainEx electrochemical apparatus at the Yale School of Medicine were able to reactivate some neural activity in decapitated pigs, though no sentience activity was observed. This research opens up new opportunities for understanding the mysterious order of consciousness.[17]

Another way to think about consciousness as a way of discerning order is to analyze what we consider to be "facts." Intuitively, we associate facts with truth, and hence trust in truth gives us greater chances of social order. One of the Oxford English Dictionary's definitions of a "fact" is surprisingly controversial among sociologists of knowledge: "a datum of experience, as distinguished from the conclusion that may be based upon it." Philosophers such as Bertrand Russell and Ludwig Wittgenstein agonized over the irreducibility of a fact, comparing it to the physical atom in the notion of *logical atomism*. Mary Poovey's masterful *History of the Modern Fact* (1998) expounds that, in scientific inquiry, it can be argued that a conscious observation of a physical phenomenon without its relationship to an explanation is not enough. She quotes from sociologist of science Thomas Kuhn who wrestled with the ways in which scientific paradigms change from existing observations of reality:

> Discovery commences with the awareness of anomaly i.e. with the recognition that nature has somehow violated the paradigm-induced expectations that govern normal science. It then continues with a more or less extended exploration of the area of anomaly. And it closes only when the paradigm theory has been adjusted so that the anomalous becomes the expected. Assimilating a new sort of "fact" demands a more than additive adjustment of the theory, and until that adjustment is completed—until the scientist has learned to see nature in a different way—*the new fact is not quite a scientific fact at all.*[18]

The anthropologist Gregory Bateson provoked further introspection on scientific inquiry as a means of understanding human social order. He questioned whether the two primary modes of research and reasoning—induction and deduction—were adequate to provide us with systems-level understandings of the human experience of existence. *Induction* focuses on drawing insights from particular observations and seeking to then infer broader conclusions; this is usually what is done with experiments or case analyses. *Deduction* considers broad observations and infers that particular instances would thus also be reflective of the larger pattern. Bateson was instead attracted to a third line of reasoning called *abduction*—a term coined by logician Charles Sanders Pierce—in which observations can be used to project possible patterns as a plausible explanation. Such an approach was more suited to Bateson's application of cybernetics to anthropology by differentiating between "Mind" and "Nature." The interconnections and patterns which exist beyond reductionist explanations is what he referred to as "Mind," and the physical reality that we observe and can reduce to specific forms he referred to as "Nature." Figure 6.4 provides a representation of these three forms of reasoning using Pierce's triangular framework. Induction can thus lead us to a specific rule-based outcome or a "law of nature" equation-based research proposition. Deduction is focused on the macro-level "result" analysis. Abduction is more case-specific and is not definitive in its ultimate conclusions, but it can still be meaningful in a particular context. Under certain conditions, particularly with modern tools of machine learning that can analyze linkages between cases, it can have explanatory power.

Inductive and deductive approaches might lead to an understanding of nature and consciousness, but Bateson also noted that Eastern approaches gave us an "ecology of mind," which required abductive reasoning to gain understanding. Related to this analysis is also a basic lesson in logic: false arguments can be logical, and you can arrive at the truth with an illogical argument. A lot depends on what assumptions

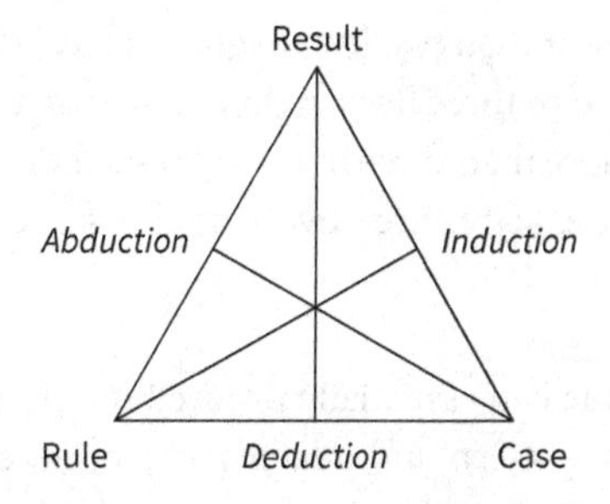

Figure 6.4 Relationship between three nodes of research premises and forms of reasoning to arrive at meaning.

or premises you are accepting. Thus, a set of the following logically correct statements leads us to a falsehood because premise B is false or providing incomplete information:

Premise A: Author Saleem Ali is a mammal.
Premise B: All mammals have four legs.
Argument: Therefore, Saleem has four legs.

We often begin an argument with an *axiom*, which may be simply defined as a "rule of the game" constructed by humans to make sense of a concept. When such a rule points us toward a specific structure, it is termed a *postulate*. Thus, the notion that any line can be extended to infinity is an axiom; the notion that two points in geometry can only be connected by one straight line may be termed a postulate. In essence, axioms are *assumptions* that are highly self-evident. We can also add *definitions* within mathematical structures to give us further meaning: for example, we are defining a straight line as a set of infinite points traveling in a constant direction. We can define numbers as odd, even, or prime. Inferring from an axiom, a specific argument can be derived and *proved* through a series of rigorous mathematical steps wherein the argument is called a *theorem*. If the statement is believed to be true due to indirect evidence but has not been proved, then it is called a *conjecture*. *Proofs* are examples of comprehensive deductive reasoning that ensconces a level of logical certainty. Arguments that rely on proofs are very different from empirical or nonexhaustive arguments which use inductive reasoning to establish a "reasonable expectation."

The interplay of mathematics and emergent uncertain systems led Jon von Neumann to develop *game theory* (the mathematical study of behavior under uncertainty of relations between individuals) and *cellular automata* (abstract algorithms that can replicate patterns like language development). An intriguing culmination of these vital abstractions was the "Game of Life" formulated by mathematician John Conway, which has particular relevance for understanding earthly order. The rules of the game for each "cell" are as follows:

1. Any live cell with fewer than two live neighbors dies, as if by underpopulation.
2. Any live cell with two or three live neighbors lives on to the next generation.
3. Any live cell with more than three live neighbors dies, as if by overpopulation.
4. Any dead cell with exactly three live neighbors becomes a live cell, as if by reproduction.

Once "seeded," the game can essentially proceed for eternity, but it is also "undecidable"—given an initial pattern and a later pattern, no algorithm exists that can tell whether the later pattern is ever going to appear. Various online representations of the game can be found to invigorate any intellectual conversation. The game also gives us insights into understanding Godel's incompleteness theorem, noted in this book's Introduction. Furthermore, it also provides insights that complex systems do not have to be consistent either, which has important ramifications for consciousness research. Daniel Dennett has also used the Game of Life to suggest the emergence of conscious systems from what may even be simple computational steps. He has also suggested that the notion of free will can be considered an evolutionary outcome of complex organisms such as humans. From events "happening," we move to "doing" work to make specific events happen. There may be a spectrum of possible events that could happen in terms of human agency, but they remain confined by boundaries of "decision landscapes." We can thus have determinism within a crucible of choice.

Computational systems operate on logic, so if incorrect premises are placed in a computing system, it will come to logical but false conclusions. Consciousness suggests the ability to consider multiple pathways *critically* and hence detect aberrations in premises which simple logic might not catch. Yet even if we increase the computational and knowledge base of AI systems, the next step toward "feeling" and "purpose" is more elusive. Figurative explanations of consciousness and resultant therapeutic avenues such as meditation are important for humans and can have positive value through the intermediate biochemical mediation of results. What we physically "feel" may well be reduced to a set of biochemical reactions and nerve pulses to our brain, but the stimulant effect and its perception is still "real" for us. The role of science in extricating these feelings has been suggested by historian Peter Dear to follow two paths: *science as natural philosophy* and *science as instrumentality*. Both roles are important because the former pertains to "knowing" while the latter focuses on "doing." One may also add ontologically the sense of "being" within the realm of natural philosophy.[19]

In the ecological arena, pantheism and abstract notions of natural spirituality, as well as concepts like *feng shui*, provide us with ways of finding meaning in ecological systems. Such concepts, alongside a range of what may be deemed religious or spiritual experiences, can also be highly "anti-fragile" if framed around the disaster–opportunity narrative. Indeed, the endurance of such concepts in human societies stems from how they are able to gain from disorder. The more poverty and adversity in a society, the more such approaches are able to provide solace and comfort, and they can even improve a sense of well-being. The power of absolutist ideologies in the context of political order will be discussed in Chapter 8, but suffice it to say for our purposes here that spiritual elements in social order can play an instrumental role in building resilience in societies.

The next question to ask, then, is whether the ability to develop such seemingly fictitious and empirically unreliable traits is a particular product of human "consciousness," or whether there are prospects for such concepts also making their way into AI. Norbert Wiener, the father of the field of cybernetics (which may be considered a constructivist precursor to AI), also wrote a monumental book in 1965 that won the National Book Award, titled *God and Golem, Inc.* Wiener draws on his Jewish heritage to use the metaphor of the *golem*: an embryonic Adam, amorphous and nascent, and hence deemed in Hebrew legend to be a monster. He then asks the philosophical question: God is to Golem as Man is to Machines? Several decades later, best-selling Israeli author Yuval Harari essentially suggests the same in his book *Homo Deus.* Metaphors aside, there is a very serious field of study around human–machine symbiosis prospects which has been termed "the singularity" by futurist Ray Kurzweil (not to be confused with the astronomical singularity of the universe's origin). I am more inclined to follow Stephen Jay Gould's advice to consider science and religion as "non-overlapping magisteria," each having its defining worldview for physical versus figurative aspects of human experience but best left epistemically separate. However, if such metaphorical flourishes add meaning to a reader's perceptions and elevate their interest in understanding the mysteries of the universe, that's fine with me!

The Conspiratorial Conundrum of Cause

Ultimately, humans yearn for a cause in our quest for changes in social order. Much of social science has emerged around attempts at establishing certainty of cause and differentiating such mechanisms from mere correlation. Yet the primacy of cause–effect reasoning in social order has not been favored by many philosophers of science even though it is such a fundamental part of how our common sense defines earthly order. One of the major reasons for this is that a unilateral search for cause can lead to conspiratorial thinking. If simple cause–effect relationships are not found, we yearn to find explanations that may not even be supported by evidence. This has been a particular hindrance to science-based problem-solving, particularly in issues around health and the environment. Furthermore, seeing patterns at an accentuated scale can be a gift in the case of prodigy chess players (as shown in the Netflix series *The Queen's Gambit*), but excessive pattern discernment where there are no patterns can be a symptom of *apophenia*, which in turn can lead to paranoia and conspiratorial thinking. There is clearly a spectrum between seeing intuitive patterns which could have meaning and consequence and imagining connections when there are in fact none.

Pathogens seem to provide fertile ground for conspiratorial thinking, as exemplified by the challenges of eradicating polio in my land of origin—Pakistan—due to conspiratorial rhetoric. It is easy to become exasperated by these conspiracy spinners, but a more considered and analytical response is in order. Conspiracy theories are a symptom of powerlessness. When people are unable to find causal answers or make sense of turmoil, they latch on to whatever fanciful explanation makes sense. Several brands of conspiracy theories exist in modern societies. Some are fueled by a suspicion of science and an inability to reconcile complexity of knowledge. For example, questioning the lunar landing has created an entire industry of books and websites in the United States where people question whether science could achieve such a feat. Skeptics couple a suspicion of science with a suspicion of government; suspicion of authority is central to conspiracy theories. Psychologists have also explored this topic as an example of *confirmation bias*, whereby people try to find an explanation for some deeper belief that they may espouse. There are theories that claim far more has been achieved in scientific knowledge than what the government is willing to reveal. This brand of conspiracy theorist is also very popular in the United States through a blend of science fiction pop culture and clandestine military activities in the southwestern part of the country. Contact with extraplanetary alien cultures is central to this group's narrative. The town of Roswell, New Mexico, has become ground zero for this counterculture. Hollywood has capitalized on this suspicion and perhaps even fueled it through popular TV series like *The X-Files*. I must confess to being a fan of this series, which ran for almost a decade. What fascinated me was how it took a grain of scientific fact or a true historic episode and wove a fictional web around it so deftly that even the most outlandish material could seem realistically appealing.

If there is any silver lining to conspiratorial thinking, it is a willingness to question what might seem obvious to the linear observer. As a scientist, I always consider such questioning to be positive. But when this curiosity becomes laced with predisposed dogma, it loses any charm. So, let us all feel comfortable in questioning the establishment but not be paralyzed by paranoia. International behavior changes just as much as human behavior, and we should always be willing to embrace positive change among nations. Countries such as the United States must confront conspiracy narratives head-on and show how they have clearly changed their modus operandi over the years. Foes of yesteryears can become friends today, and we should cautiously focus on such positive transformation rather than languishing in the past. Conspiracy theories reflect the human reluctance to process uncertainty and causal complexity. While the theories may seem complicated, they are usually very linear and deterministic in their causality.

Cause-and-effect relationships in a binary context are tempting because we are used to two-dimensional plotting on *x* and *y* axes and using techniques like regression in statistics. The nineteenth-century novella *Flatland* by theologian Edwin Abbott captured some of the ironies of a two-dimensional world in which the characters are all geometric figures rigidly governed by "laws of nature." The book was meant to be a social satire, but it highlighted the intellectual poverty of binary or linear relationships. Philosophers such as Bertrand Russell have even questioned the cognitive merit of cause–effect reasoning. In his essay, "On the Notion of Cause" (1913), Russel, who also won a Nobel Prize for Literature, stated, "The law of causality, I believe, like much that passes muster among philosophers, is a relic of a bygone age, surviving, like the monarchy, only because it is erroneously supposed to do no harm." Russel's concern as an avowed agnostic was that primacy of cause in our thinking can lead us down toward origin myths and religious order in ways similar to what we find in conspiracy theories. Yet there is no doubt that in the proximate context, cause and effect are real and have value. The salient question is more whether they constitute some broader social or ecological patterning of reality.

The physicist Richard Feynman suggested that an alternative worldview in the "Babylonian tradition" would indicate that physical and social reality comprises interconnected structures with no unique, context-independent starting point for our observations. Given such structures, Feynman conceded, "I am never quite sure of where I am supposed to begin or where I am supposed to end."[20] Such an approach can also be upscaled to the ultimate question of the origin of the universe through the theory of "Conformal Cyclic Cosmology" (CCC) suggested by Roger Penrose in his 2010 book *Cycles of Time.*

Attempts at finding a means of coexistence between science and religion in social perceptions of order have also followed such an approach. The triumph of asymmetry and noncausal complexity in creating order has been noted by Marcelo Gleiser[21] as a means of considering the coexistence of seemingly contradictory human intellectual experiences such as science and religion. Gleiser was awarded the $1.4 million Templeton Prize for "celebrating scientific and spiritual curiosity" in 2019. As a professor of physics at Dartmouth College for much of his career, he has followed orthodox scientific methods in his research but also notes that "science, philosophy, and religion are the three pillars that humans use to try to make sense of the world. They are different, but they all try to respond to existential anxiety." The fundamental difference between science and religion, he adds, is that "with science, eventually we get data about the universe which helps us distinguish among the possibilities."[22]

The differentiation between possibilities of action depends on the units of agency in any social system. Each individual human being has consciousness and agency that accelerates and accentuates the complexity of cause and the perplexing challenges of belief. Evaluating the various aptitudes of individuals and how they can deliver societal good within the bounds of ethical norms defined by society is yet another key challenge. Understanding the mechanisms by which human populations have procreated and nurtured their progeny and how they have, in turn, professionalized their contributions to human civilization is an essential next point of inquiry.

Notes

1. Georg Cantor's diagonal argument, which showed that there can be countable and uncountable infinities, is a famous example of such. Tegmark's essay in the following volume

covers the arguments against use of infinity in physics. Brockman, J. 2015. *This Idea Must Die: Scientific Theories That Are Blocking Progress*. Harper Perennial.

2. Powell, K. 2018. "The Power of Diversity." *Nature* 558:19–22.
3. Sprei, F., and D. Ginnebaugh. 2018. "Unbundling Cars to Daily Use and Infrequent Use Vehicles: The Potential Role of Car Sharing." *Energy Efficiency* 11 (6): 1433–1447. https://doi.org/10.1007/s12053-018-9636-6.
4. Zervas, G., D. Proserpio, and J. W. Byers. 2017. "The Rise of the Sharing Economy: Estimating the Impact of Airbnb on the Hotel Industry." *Journal of Marketing Research*. https://doi.org/10.1509/jmr.15.0204.
5. Gingrich, N. 2007. *A Contract with the Earth*. Johns Hopkins University Press.
6. Knight, F. H. 1921. *Risk, Uncertainty and Profit*. Houghton Mifflin.
7. Goldstein, I. F., and M. Goldstein. 2002. *How Much Risk?: A Guide to Understanding Environmental Health Hazards*. Oxford University Press.
8. This paragraph draws heavily on insights in Kriebel, D., et al. 2001. "The Precautionary Principle in Environmental Science." *Environmental Health Perspectives* 109 (9): 871–876.
9. Amerhein, V., et al. 2019. "Retire Statistical Significance." *Nature* 567: 305–307.
10. Wildavsky, A. 1988. *Searching for Safety* (1st edition). Transaction Books, 37.
11. Pidgeon, N. 2011. "In Retrospect: Normal Accidents." *Nature* 477 (7365): 404–405.
12. Beane, S. R., Z. Davoudi, and J. M. Savage. 2014. "Constraints on the Universe as a Numerical Simulation." *European Physical Journal A* 50 (9): 148. https://doi.org/10.1140/epja/i2014-14148-0. A series of experiments were also proposed to test such a hypothesis; see International Journal of Quantum Foundations. (n.d.). "On Testing the Simulation Theory." Retrieved March 24, 2021. https://ijqf.org/archives/4105.
13. Refer to Hiesinger, P. R. 2021. *The Self-Assembling Brain: How Neural Networks Grow Smarter*. Princeton University Press.
14. Dennett, Daniel. Lecture to the Royal Institution. April 6, 2017.
15. Wucker, M. 2016. *The Gray Rhino: How to Recognize and Act on the Obvious Dangers We Ignore*. St. Martin's Press.
16. Goff, P. 2020. *Galileo's Error: Foundations for a New Science of Consciousness*. Vintage.
17. Vrselja, Z., et al. 2019. "Restoration of Brain Circulation and Cellular Functions Hours Post-Mortem." *Nature* 568 (7752): 336–343. https://doi.org/10.1038/s41586-019-1099-1.
18. Quoted from Kuhn's classic work *Structure of Scientific Revolutions* (1962) in Poovey, M. 1998. *A History of the Modern Fact*. University of Chicago Press
19. Dear, P. 2008. *The Intelligibility of Nature: How Science Makes Sense of the World* (reprint edition). University of Chicago Press.
20. Frisch, M. 2017. *Causal Reasoning in Physics* (reprint edition). Cambridge University Press.
21. Gleiser, M. 2010. *A Tear at the Edge of Creation: A Radical New Vision for Life in an Imperfect Universe*. Free Press.
22. Ibid.

7
Sex, Population, and Sustainability

> So long as you are alive you are just the moment, perhaps, but when you are dead then you are all your life from the first moment to the last.
>
> —**H. G. Wells, *The World Set Free***

Procreation of any living form is a prerequisite for survival of a species and hence has essential implications for earthly order. Yet the urge to engage in mechanisms that would lead to procreation is complex and has generated the most intricate set of rituals and taboos in both humans and in many other advanced organisms. The mating dances of the bird of paradise, the monogamous winged embraces of albatross couples, and the shy abstinence of endangered pachyderms have graced media screens for decades. Within human societies, sexual reproduction has even been linked to an enduring mythology of horoscopes and the astrological order of stars from thousands of years back. To this day, even highly educated societies like Taiwan see major changes in birth patterns linked to particular Chinese-designated animal years. Even though the original 12 zodiac signs have actually shifted in their celestial order over the past millennia, astrologers hold sway, and most local newspapers worldwide still carry horoscopes. Since 1970, they have, at times, even included the thirteenth constellation of Ophiuchus to account for astronomical changes in stellar patterns. Birth and marriage—the effect and ordered instigators of sex—remain the portents of astrological pseudoscience.

On the scientific front, researchers have been working to develop a biological "Tree of Sex" similar to the taxonomy for a "Tree of Life," which is more familiar to students of biology—albeit contentiously.[1] The project published a series of notable papers which covered the range of organismal sexual habits from the physical fusion of parasitic tapeworm mates to the promiscuity of bonobos. It was a remarkable effort to consider whether order could be found in the context of eukaryotic organisms and their sexual reproductive evolution. Yet humans and other advanced animals have evolved complex forms of signaling in reproductive behavior that have major consequences for how we impact planetary order. Until the advent of the Neolithic revolution around 12,000 years ago, anthropologists believe humans were promiscuous. The dominant order of monogamy that arose around the time we learned to settle in agrarian societies to grow our own food appears to be closely tied in time. Monogamy provided an advantage for social

Earthly Order. Saleem H. Ali, Oxford University Press. © Oxford University Press 2022.
DOI: 10.1093/oso/9780197640272.003.0008

order around property rights, although it may well have also had positive value for the nurturing of progeny. Property rights, incidentally, were also a way to avoid the "tragedy of the commons" that population biologist Garret Hardin warned against in his most celebrated paper, a staple reading in most environmental studies college curricula.

The primacy of private property in maintaining ordered governance was also challenged by economics Nobel Laureate Elinor Ostrom in her work on common-property resources in many nomadic societies. Ostrom showed through fieldwork and game theoretic models that the commons could be governed without private property or public interventions when there were specific norms in place within cultures and under key constraints of scale.[2] Even when there are divergent interests, individuals can develop norms of orderly behavior through what may be termed "dilemmas of common aversion." At a crossing, two parties may have different directional interests, but they form some mechanism of maintaining order to avoid a crash, such as a hand signal or eye contact that gives one party a signal to pass. As the number of parties in the crossing grows, we may need some enforcing mechanism like a stop sign or even a traffic light. Still self-enforcing order can emerge in many social systems.

Unlike Ostrom, who embraced a wide range of cultural worldviews on property in her scholarship, Hardin was an ardent Neo-Malthusian, Eugenicist, and, by some accounts, a White Supremacist, who was prompted to write this article because of an obsession among academics in the 1960s around population growth as an imminent crisis. The term "Neo-Malthusian" that Hardin wore with pride stems from the work of eighteenth-century economist Thomas Malthus who had brought population within the ambit of economic inquiry and had also inspired some of Darwin's writings on "survival of the fittest." Although he had four children himself, Hardin advocated population control to the extreme level of even proposing "lifeboat ethics," using the metaphor of a limited-weight vessel on water which could only float if some people had to be thrown overboard or not allowed to board. A master of extreme metaphors, Hardin's last book was titled *The Ostrich Fact*or, referring to humanity's inability to perceive population growth as a challenge, and he called for coercive constraints on "unqualified reproductive rights." He and his wife were also members of the erstwhile Hemlock Society (now called End-of-Life Choices), which believed in the right for individuals to choose their time to die. On their sixty-second wedding anniversary, in 2003, the couple committed suicide together.

Hardin exemplified one of the great perils of trying to contrive social order under the guise of resource scarcity or Darwinian evolution. In trying to find the "fittest" humans, the scientific mind of some scholars was lured into a categorization of abilities. Intellectual and physical abilities that were below some preordained metric were branded a burden in the quest for efficient social order. Darwin's half-cousin Sir Francis Galton was a polymath who became one of the key proponents of these approaches that paved the way for the eugenics movement at the end of the nineteenth century and was later appropriated by Nazi ideology. Galton wrote a science fiction novel to get his points across, titled *The Eugenic College of Kantsaywhere.*

Less than a century ago, the intellectually disabled were put into the repugnant categories of "morons," "imbeciles," and "idiots." Even the US Supreme Court, in the infamous *Buck v. Bell* decision in 1927, legalized forced sterilization of such individuals deemed to be in particular categories of intellectual performance. Justice Oliver Wendell Homes wrote a chilling opinion that stated that sterilization programs were "in order to prevent our being swamped with incompetence." The poignant irony of such impositions of stylized perfection at the behest of Darwinism is that they defy a core principle of natural selection: variation is essential for evolution to proceed. By homogenizing a population, we mitigate the potential for evolution. Genes and our environment interact in complex ways, and traits which may seem ostensibly limiting on the surface may provide resilience in other ways. The classic example is sickle cell anemia, which is a debilitating disease at one level but protects its victims from malaria. This led to its dominance in certain malaria-infested regions of the world. While finding cures for such ailments should be within the ambit of scientific discovery through gene editing technologies, their use must recognize that the correction of any aberration may have unforeseen costs. We may be willing to accept such contingencies for the greater good, but we should always approach interventions for order with some degree of humility.

Social Darwinism also provides us with another twisted lesson about being cautious with arguments and their ultimate ends. Many eugenicists were ardent peace activists who campaigned against war. The ichthyologist David Starr Jordan, who became the first president of Stanford University, was an evangelical eugenicist who was also president of the World Peace Foundation. However, his argument against war had a dark Darwinian twist. Since soldiers had to meet certain basic physical standards of performance, he agonized over the danger to the fittest of society in exposing them to warfare. Jordan's call for peace was thus to perpetuate the dominance of the fittest rather than to safeguard the sanctity of human life. Instead, Jordan lamented how the village of Aosta in northern Italy was a sanctuary for the disabled to find jobs and procreate, calling it a "veritable chamber of horrors." Jordan also had an errant view of natural order whereby organisms that were "underperformers" would degenerate into lesser beings. Thus, he believed that sea cucumbers had once had the disposition of fish but lapsed into lesser forms due to acquiring resources through parasitism.[3] Just as arguments for world peace can have both benevolent and malevolent intent, population control and resource security arguments need to be unpacked to consider their motives. Garrett Hardin's advocacy for the right to contraception or resource conservation was also motivated by similarly problematic rationales, albeit both are worthy objectives if considered within the context of a more deliberative order of human choice.

From Tragedy to Comedy of the Commons

Repugnant as Hardin's personal views on humanity and the ethics of population may seem to us now, his one article on the dilemmas of common property resources spawned a vast genre of literature in social sciences on this topic. *The Tragedy of the Commons* is one of the most widely cited papers in environmental science because it laid bare the sharp fracture between natural constraints and social choice with regard to human procreation. By doing so it connected demography with ecology and governance, raising questions about the role of government in regulating property and human behavior. It alerted us also to how human population growth itself was closely tied to the land. Much of the highest growth in global population had occurred in places with fertile agrarian societies such as South and East Asia, areas that still house more than half of the world's population. The 1960s, when Hardin wrote his paper, were an inflection point in demographic order as the Baby Boom years were giving way to the advent of contraceptive medications and technologies. The anxiety that ecologists like Garrett Hardin and Paul Ehrlich were experiencing stemmed from the rapid population growth heralded by the Industrial Revolution. It had taken 1,600 years for Earth's population to go from a quarter of a billion to half a billion, but it took only an additional 200 years to double our population to a billion by the year 1800.

The population growth during this period was more than exponential and unprecedented and was only possible due to a momentous confluence of food access and human mobility. European navigation to the Americas led to what demographers call "The Columbian Exchange," a movement of high starch foods like corn and potatoes to Europe and a reciprocal movement of people to the Americas. Europeans thus gained greater reproductive and survival capacity. The native population of the Americas was sadly decimated by diseases carried by European settlers during this time, to which they did not have immunity, but overall global populations continued to rise. Our population "doubling time" began to shrink dramatically: we went from 1 billion to 2 billion population in around 130 years, and from 2 billion to 4 billion in only 44 years (1930 to 1974). This doubling period occurred when ecological anxieties over population were at their highest, but when, behind the scenes, a quiet reversal of trends also began to occur. Demographers and environmentalists also became conscious of the enormous inequality in consumption patterns across the planet and became more guarded in their recriminations against population growth. Instead, they shifted their attention to employment and livelihood potential for citizens. There was a realization that humanity was shifting from subsistence lifestyles to a market economy, and this meant there was a desperate need for jobs.

In terms of global employment profiles, the changes that have occurred in the past 200 years have created an irrevocable shift in which the service sector will now always dominate the order of livelihoods. This is largely due to efficiency improvements and automation in food production and industrial manufacturing. Figure 7.1 shows this

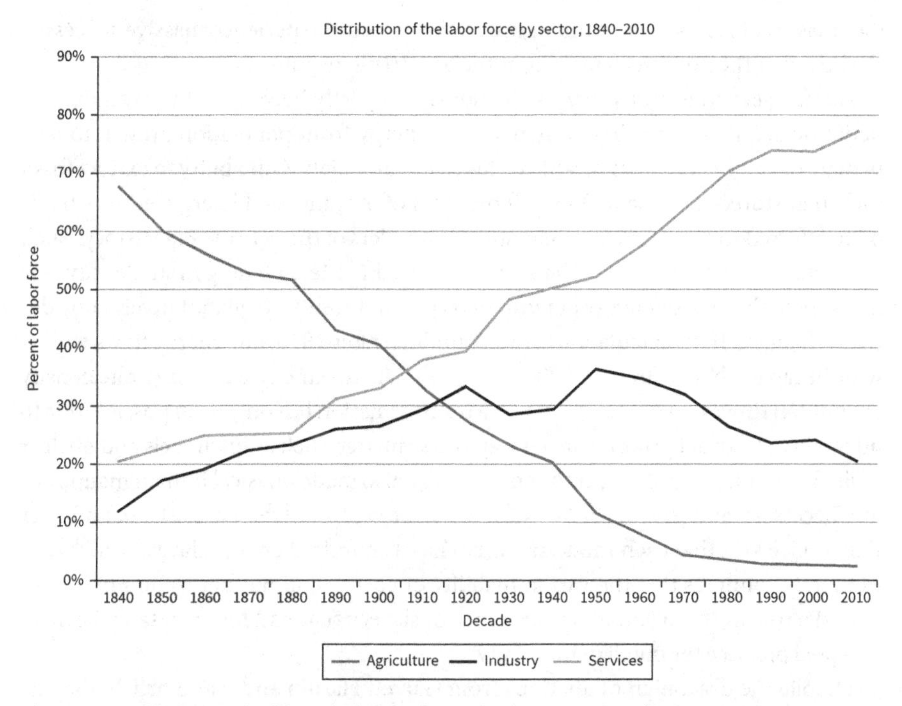

Figure 7.1 The Great Livelihood Shift in the United States over the past 200 years. From Bureau of Economic Statistics.

trend, which is likely to widen even further, whereby as much as 90% of the workforce at the turn of the twenty-second century may be employed in the service sector. The question is whether we will have enough need for such a range of service sector jobs to add that additional value to the marketplace. How many programmers, teachers, bankers, writers, and sundry consultants or bureaucrats does the world really need? The World Economic Forum estimated in October 2020 that, as a result of automation and also accounting for disruptions such as the COVID-19 pandemic, an estimated 85 million jobs would be lost worldwide by 2025, but an additional 97 million jobs would be created—thus netting the world an additional 12 million jobs.[4] Yet the absolute population of the world will still be adding more people than jobs.

Another way to interpret Figure 7.1 is by going further back in time to consider how humanity has seen a rise in global energy flows as we became more efficient at cultivating natural systems for food delivery. Following the advent of agriculture around 10,000 years ago, we as a species had the ability to have "excess energy" at our disposal. This energy was in turn used, first of all, to grow the human population, which in turn led to a stratification of society into professions, which in turn led to an elite class that used a substantive portion of those energy flows until the Industrial Revolution. Like the ordered bricks of a pyramid, the ascendant apex was supported by the heavy lifting of the baser bricks. The proverbial "pyramid schemes" that often loot masses of funds have a similar order: the tiny elite experience massive success at the behest of the minions scurrying at the base to fill inventories.

Human servitude and slavery sadly flourished partly because of the ravenous appetite we acquired for using human kinetic energy from population growth to perpetuate the highly unequal, stratified status quo of society. Our ability to extract fossil fuels that stored more than 300 million years of accumulated energy led to what is often referred to as the "great acceleration." Founders of the field of "Big History," such as David Christian and Eric Chaisson, have used the term "energy rate density"—a measure of the flow of energy per unit mass per unit time—to quantitatively consider these changes. Their calculations are astounding: after 1850, our energy flows worldwide began to change dramatically to the point where our current energy rate density is a million times that of the sun![5] Such a transformation has only been possible due to our ability to extract power from energy-dense matter, such as fossil fuels and nuclear fuels. The availability of vast amounts of energy also made physical human manpower less necessary and gave each of us "excess energy" to utilize. Futurist Buckminster Fuller once said that each modern human has a "hundred energy slaves," which also indirectly connotes that the vast availability of energy and mechanization may have played a role in the eventual obsolescence of slavery that had hitherto sadly been an accepted practice for much of humanity.

Despite the draconian predictions from Garrett Hardin and Paul Ehrlich, the rate of population growth began to slow largely owing to greater gender empowerment and reduced infant mortality. People had been having more kids out of risk management behavior, given their child's chances of survival. With more confidence in survival rates, and access to birth control as well, the demographic order was completely disrupted. Until 1950, the average number of children for women at a global level was around 5; now it is just over 2. Yet the overall population of the planet is still growing, and, in certain regions, growth is particularly concerning because there are fewer options for mobility in ecologies that are confined by political borders (which will be discussed in more detail in Chapter 9). Furthermore, the agonizing reality is that those areas where population growth is highest are often the poorest. The world is beset by a structural inequality whose underlying causes are complex. One explanation is that population growth necessitated more food via agrarian production systems, which also led to surplus and the need for managers and specialized roles. This in turn led to corresponding social classes and hierarchies. Meanwhile, we began to use up natural resources and needed to venture farther afield to seek them out. This expansion bred conflict and conquest, with the conquered becoming the underclass. Thus, if egalitarianism is to subvert this unequal and hierarchical human order, a key variable to consider is "social mobility": the opportunities available for those with few resources to gain property and income.[6]

Population increase in the commons does not always have to be a tragedy, as astutely noted by Yale Law School Professor Carol Rose in her famous essay *Comedy of the Commons*. Rose suggests that public property can also exhibit "increasing returns to scale" where the good in question is "non-rivalrous"—like knowledge or group recreational activities. Wikipedia is a fantastic example of how increasing returns to scale are manifest through sharing the commons with more people who are invested in improving its quality. Thus, population growth in particular contexts can deliver advantages if the population is invested in the system. In the context of livelihoods, this is known as the *demographic dividend*—a young workforce that is invested in and capable of making their livelihoods more productive for all of society. China was able to harness the demographic dividend from its twentieth-century population growth because it had a large working-age population. This happened despite the terrible famines and setbacks of Mao's forced and utterly misguided agrarian "reforms"—known as The Great Leap Forward—that left as many as 45 million people dead due to famines between 1958 and 1962. This was followed by the controversial "one-child policy," in 1979, which also attempted to deal with ecological ravages during Mao's era.[7] The power of the demographic dividend is best reaped when both the death rate and birth rate decline. Figure 7.2 shows how the change in birth and death rates presents opportunities for economic growth and the usual trajectories such growth may take in four stages of human development.

What Stage 5 holds will depend a lot on a range of factors with which futurists will have a field day in terms of possibilities, particularly regarding the potential reduction in our death rate. Finding a mechanism to biochemically mitigate the aging process remains a major wildcard in population projects, even though much of the consensus remains that world population should stabilize by 2050 at around 11 billion, and we would have a steady-state demographic profile. However, even slight changes in life

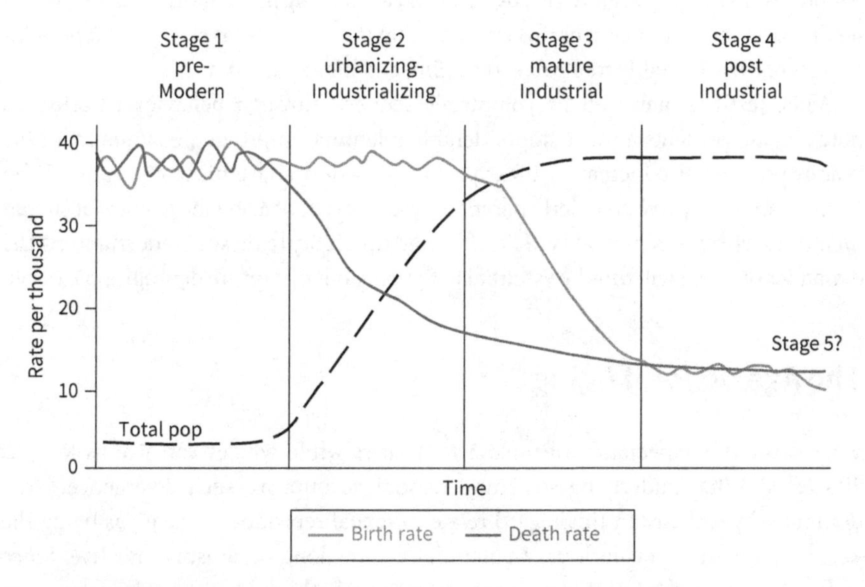

Figure 7.2 Harnessing the demographic dividend toward a steady-state development path.

Adapted from an online presentation by Marissa Pine Yeakey, Population Reference Bureau.

expectancy and assumed reductions in family size could lead to astronomically different outcomes: such are the structural aspects of population growth or decline at local or regional levels. There is a huge variation possible between what are termed *low-* and *high-fertility traps*.

Fertility results from a combination of the aspirations of either gender to be parents linked to expected income, although the rights of women in expressing their choice has unfortunately been severely constrained for much of human history. In places like China or Italy, a rapid decline in the fertility rate and a rise in cultural individualism could lead to a "low-fertility trap" in which fewer potential mothers in the future will result in fewer births. In fact, ideal family size for younger cohorts is declining as a consequence of the lower actual fertility seen in previous cohorts. Similarly, a high-fertility trap could occur in other contexts. For example, if resistance to family size reduction for entrenched cultural reasons or lack of access to contraceptives continues in some parts of sub-Saharan Africa but life expectancy increases, then this could lead to a situation where rapid population growth inhibits the conditions for development needed to bring the fertility rate down. In their long-term projection scenarios for population up to the year 2300, researchers at the International Institute for Applied Systems Analysis in Austria concluded that if these fertility traps come to fruition, wildly different population outcomes could result. A reduction of fertility rates in some Chinese urban centers, which are now at 0.75 (2.1 is approximately replacement value), would lead to a population size decline to a maximum of 5.6 billion people by 2200. On the other hand, if the fertility rates in sub-Saharan Africa remain at around 5 (which they are currently in several countries), we could have 12 billion people just in sub-Saharan Africa by 2100 and a stunning 355 billion by 2200.[8] While neither of these extreme scenarios is likely, what the calculations show is that population scenarios are highly malleable to fertility and life expectancy.

While fertility may well be voluntarily reduced through behavioral norms, as noted, aging presents a more imponderable dilemma. Improving our quality of life year by year and also increasing life expectancy is a desirable ethical imperative. Who would want to deprive an elderly parent or grandparent of a few more years of shared memories with a decent quality of life if we had the ability to do so? Understanding the dynamics of aging will thus be essential to figuring out our future demographic order.

The Age Beyond Aging

Why is the life expectancy of mice 3 to 4 years while whales can live as long as 200 years? What underlying structural reasons account for such divergences? The dominant hypotheses in this regard relate to sexual reproductive years as being the key determinant in evolutionary outcomes of how long organisms may live. Since rodents can reproduce within a few months of birth, the evolutionary investment in perpetuating longer life expectancy is not warranted. Human beings reach sexual maturity after 12–15 years, and females are able to reproduce for a few decades; hence our life expectancy has emerged accordingly. An additional factor which plays into life expectancy is how long a species has been around to overcome some aspects of "evolutionary neglect" and still increase life expectancy. Humans have only been around for a few hundred thousand years, while whales have been around for at least 30 million years. Hence, despite similar reproductive maturity timelines, whales currently live much longer. If humans continue to evolve over millennia, our life expectancy may well evolutionarily increase as well. Yet the issue at hand is how, through our intellect, we are able to intervene in the biochemistry of aging and thereby disrupt the overall social order of human habitation on the planet.

Genetics, microbiology, and biochemistry have given us extraordinary tools in recent years to disrupt demographic order. For nonhuman species, we are already making major interventions at the ecological level. The advent of CRISPR technology and the "evolutionary engineering" of reproductive cells has led to trials in which genes are introduced that would lead to key traits being passed on in certain populations to control particular diseases. The challenge of using such interventions in reproductive cells is the long-term intergenerational impact that may be possible, unlike *gene therapy*, which is conducted on nonreproductive cells. However, researchers at MIT's Media Lab have found a way to limit the population carryover of this impact as well. A fascinating example of this is the "daisy gene drive" system which is being trialed on Chappaquiddick Island in Massachusetts to eradicate tick transmission in mice.[9]

Similar gene drive systems are under way that make mosquitoes sterile and thus quell the spread of insidious diseases such as dengue, malaria, and zika virus. The potential for this technology to have a long-lasting impact on species populations led to major international deliberations through the Convention on Biological Diversity. Activists against genetic modification of organisms, who adhere to a primordial order wherein human beings are considered apart from nature, resisted vigorously. The issue was debated at length among the 196 parties (countries and other self-governing entities) to this global treaty. The US National Academy of Sciences reports on the topic were also considered in terms of ethical safeguards needed for such an enterprise. A final decision was made to proceed with careful prototyping of this technology and a rejection of activist calls for a moratorium. Members of the Convention recognized that humans have been making genetic interventions in populations through plant and animal breeding for millennia. From broccoli to bull mastiffs, selective breeding is a macro-physical form of genetic manipulation of the earthly order that we have embraced; hence, carefully monitored manipulation of pest populations was deemed very much in order. This momentous decision was a reckoning for the "precautionary principle" discussed in Chapter 6.

Returning to how such genetic manipulations may impact human aging, we need to first consider the underlying structural causes of aging and whether there is opportunity for their reversal. In 2005, MIT's *Technology Review*, in cooperation with the Methuselah Foundation, announced a $20,000 prize for any molecular biologist who could demonstrate that aging reversal was *not* possible. This competition was prompted by claims being made by anti-aging researchers such as Aubrey de Grey at Cambridge University who worked on Strategies for Engineered Negligible Senescence (SENS)—essentially a fancy term for a range of regenerative medical therapies to reverse aging. There are also examples in the natural world of organisms such as hydra (micro-animals related to jellyfish) and lobsters that do not show overt cellular signs of aging. Five entries were received by the judges who hailed from a variety of disciplines: Rodney Brooks, director of MIT's Computer Science and Artificial Intelligence Laboratory and the chief technology officer of iRobot; Anita Goel, founder and chief executive of the biotechnology firm Nanobiosym; Vikram Kumar, cofounder and chief executive of the data science company Dimagi and a pathologist at Brigham and Women's Hospital in Boston; Nathan Myhrvold, cofounder and chief executive of Intellectual Ventures and the former chief technology officer of Microsoft; and J. Craig Venter, the founder and president of the Venter Institute, whose computational methods hastened the completion of the Human Genome Project.

The distinguished panel did not find any of the competitors convincing enough to award the full prize to any submission, although they found one submission elegant enough in its argumentation to receive half the prize. We will come back to this group a bit later in this section, but first let us consider how the multidisciplinary and multi-professional panel presented their reasoning.

> The scientific process requires evidence through independent experimentation or observation in order to accord credibility to a hypothesis. SENS is a collection of hypotheses that have mostly not been subjected to that process and thus cannot rise to the level of being scientifically verified. However, by the same token, the ideas of SENS have not been conclusively disproved. SENS exists in a middle ground of yet-to-be-tested ideas that some people may find intriguing but which others are free to doubt. . . . We need to remember that *all* hypotheses go through a stage where one or a small number of investigators believe something and others raise doubts. The conventional wisdom is usually correct. But while most radical ideas are in fact wrong, it is a hallmark of the scientific process that it is fair about considering new propositions; every now and then, radical ideas turn out to be true. Indeed, these exceptions are often the most momentous discoveries in science.[10]

The judges were prescient in their reasoning since, a decade later, some convincing evidence on aging reversal continues to emerge and is entering the mainstream. Among the leading lights in the field is Australian-American health scientist David Sinclair who has even devised a mechanism to estimate the molecular age of a person by examining certain chemical changes in their DNA. Sinclair self-medicates with a variety of supplements that are purported to mitigate aging biochemistry. In his book, *Lifespan: Why We Age and Why We Don't Have To*, he has listed nine possible molecular biological pathways by which aging occurs and corresponding potential pathways for mitigating each of these. He is particularly keen to explain the "nature–nurture" dichotomy in aging by applying concepts from information theory—similar to Claude Shannon's work. Aging in essence is a process by which the body loses information. DNA and genetics presents the body with a sort of "digital" form of information; the body must read that information at the molecular level through an analogue process of *transcription*, which Sinclair refers to as the *epigenome*. The prefix "epi" refers to environmental factors that can lead to particular genes being activated or not, for better or for worse: hence, nature–nurture complementarity is perennially relevant (approximately one-third nature and two-thirds nurture dependence). Sinclair, like de Grey, continues to be controversial in his research, but there are key discoveries in biology which suggest that some of his approaches may well be workable. Most notably, the Nobel Prize-winning discoveries of John Gurdon and Shinya Yamanaka found that mature cells could be reprogrammed to become *pluripotent*, whereby they could essentially go back to an earlier stem state. This revolutionary insight indicated that we could turn back the cellular clock in what was previously believed to be an irreversible process.

Despite their optimism for the success of anti-aging therapies, Sinclair and DeGray are not worried about overpopulation at the aggregate level because they believe fertility rate decline will be able to maintain a balance. Yet this raises challenges for science in terms of whether we want the same old people or more new people to populate the planet through birth and death cycles.

At this juncture let us return to that partial winning entry in the aforementioned *Technology Review* challenge to disprove longevity research. Preston W. Estep and his team put forward a detailed critique of the SENS approach to aging reversal. (Estep is now leading the Rapid Deployment Vaccine Collaborative (RadVac) established during the COVID-19 pandemic.) Estep's critique of aging reversal research likened the work to the Russian political campaign of "Lysenkoism" (Russian: *Lysenkovshchina*). This was a movement led by agronomist Trofim Lysenko in the 1930s and 1940s, which dismissed modern genetics and brought back the ideas of French biologist Jean Baptiste Lamarck about environmentally acquired traits being passed on to subsequent progeny of an organism. Such ideas were attractive to Stalin in the context of his communal farming austerity measures. Lysenko's physical methods of crop manipulation provided wider opportunities for year-round work in agriculture. I find such a critique of longevity research harsh, but the social and political implications of aging reversal cannot be neglected, and Estep astutely pointed to these in his critique, which went beyond biology.

Though critical of the SENS approach to aging, Estep has suggested other ways in which immortality may be in the offing for all practical purposes. Human beings, in essence, "live" through their memories and their personalities. If those neural attributes of what we call our "mind" can be "uploaded" into a digital form that could then be manifest in either a physical or virtual persona, we could have mortality without the same degree of resource constraints that a biological form would entail. This is also referred to as *whole brain emulation* (WBE). This concept has long been championed by futurist Ray Kurzweil, currently head of engineering at Google, as well as by Israeli-Swiss neuroscientist Henry Markram who started the Blue Brain Project in 2005. This project aims to develop detailed digital reconstructions and simulations of the mouse brain, ultimately leading to similar options for human brains. Using data to generate neural order of this kind would require enormous amounts of energy. Futurists in Russia, as part of the country's Transhumanist Movement, have proposed using a massive orbital solar energy harvesting device known as a *Dyson sphere* to accomplish such a feat with the help of artificial intelligence mechanisms.

While we consider ways to create a data-harvested immortal conscious order, efforts also exist to mitigate genetic "disorder." Ongoing research is gaining commercial traction and may herald the age of "designer babies" who have fewer possibilities for genetic disease, to the point where geneticist Stephen Hsu, cofounder of the firm Genomic Prediction says, "sex will be only for recreation and labs will be for reproduction." Thus sex, birth, and aging have a complex structural story that is at the forefront of how the future of human social order will be recrafted in coming years. Medical technologies, human reproductive behavior, and what form of immortality we aspire to will determine the ultimate social order.

Gender, Culture, and Reconciling Anomalies

While aging is a determining variable for population growth, there are peculiar anomalies in how human biological systems have developed to differentiate fertility between males and females. Men are able to produce sperm throughout their lives, but women have only a few decades of fertility in which to procreate. Although in vitro fertilization has largely disrupted this ostensible natural order, the number of children being born to women has been declining rapidly as countries continue to develop. Such a move seems to some conservatives to work against our notions of animal instincts and the Darwinian imperative for organisms to pass along their genetic material to posterity. Social transformation appears to have trumped underlying biological proclivities. The evolutionary theorist Ernst Mayr also identified another key feature of Darwinian worldviews that led to "population thinking" approaches in understanding natural order. A key feature of the Darwinists view is to consider the primacy of each individual within a species as being unique. Categorization in the "typological" sense had been a feature of Plato's notion of *eidos* (ideas) as the only original features in the universe. This additional manifestation of the nature–nurture debate is informed by some intriguing structural science that goes back to the work of Scottish biologist D'Arcy Thompson, whose influence is felt from anthropology to zoology.

Thompson questioned the Darwinian orthodoxy through careful scientific observations. While embracing the fundamental tenets of natural selection, he considered other structural forms and templates in nature. His book *On Growth and Form* (published in 1917) is a massive tome that builds on his training in classical philosophy but links the insights of the Greek masters to physics and biology. He suggests that the underlying patterns observed in natural systems are a result of physical constraints. Many of Thompson's ideas inspired the complex systems researchers noted in the Introduction to this book. However, for me, his most enduring legacy is the way he brought structural research into the social sciences. His work profoundly influenced French anthropologist Claude Levi-Strauss, who wrote a momentous book titled *Structural Anthropology*, in 1937. Just as Thompson searched for patterns in the physicality of biological forms, Levi-Strauss considered the role of patterns and structures in cultural forms. For him, notions such as language, gender identity, sexual morals, and, more broadly, any culture were products of the mind, and since human beings fundamentally had the same brain biology, there were likely underlying structures that permeated these more abstruse manifestations of human behavior as well. Therefore, a seminal contribution that emanated from his work was the assertion that no cultural ascendance should be claimed by what were previously categorized as "civilized" cultures over "savage cultures." He also analyzed genders across cultures and found that although the dualism of binary opposites was dominant, there were always some "mediators" created by the mind to find complementarity between nature and culture.

Among the most persistent ways in which social order has been maintained in human societies is through notions of gender roles that often manifest themselves in patriarchy or matriarchy. Why did institutions such as marriage arise, and why they are now obsolescent? What evidence from research on natural systems of sexual behavior needs to be considered? Are there key advantages to dispensing with order and norms when it comes to sexual behavior? What might be the limitations of such a disavowal of order in terms of procreative outcomes or social cohesion as seen through historical processes and contemporary empirical research? What is the significance of stereotypes in the construction of a pseudo-order for gender behavior and beyond, and how must we transcend such tendencies? Modern society has disavowed many of the presumed structures of gender and culture and found an increasing number of "mediators"—as the term is used by Levi-Strauss. For example, the categorization of psychological "disorders" within the *Diagnostic and Statistical Manual of Mental Disorders* (DSM) has evolved over time, starting with the history of how sexual orientation and preferences were deleted as a "disorder" in 1973. The manual is still highly controversial among some psychiatrists who see many of the categories as medicalizing a spectrum of diverse human normality. Thus, attempts at finding patterns in social order can also sow the seeds of prejudice if we are not careful.

Allow me to share a somewhat protracted case example from my own ethnic homeland of Pakistan to consider how even cultures that we perceive to be absolutist can show social order malleability even in these areas of personal identity. On December 22, 2009, the erstwhile chief justice of the Pakistani Supreme Court, Iftikhar Muhammad Chaudhary, ordered the government to create a "third gender" category on all national identity cards and to make the police directly accountable for any discrimination against transgendered individuals. Such individuals are also given a legal right to inherit property that was previously divided based on gender under the Muslim Family Laws Ordinance. For a highly conservative Islamic society where transgendered individuals have at best been pitied and marginalized, this was a revolutionary ruling that augurs well for a country that is so painfully wrestling with modernity.

Gender politics are central to reformation efforts in any society because sexual norms are the most personal and persistent traits where discrimination can elude pluralism. The Muslim states of the Gulf, particularly Saudi Arabia, manifest this asymmetry with their ostensibly modern infrastructure but an inertial stance on gender politics and an atavistic adherence to arcane structuralism. Indonesia, Pakistan, and Bangladesh, which now rank as the world's three largest Muslim countries, have exhibited a perplexing ambivalence on issues of women's rights that appear to be stratified by class. On the one hand, all three countries have boasted women heads of state in their recent history and have leading women's rights activists. On the other hand, the status of women in rural communities in all these states remains highly constrained by conservative norms.

Beyond the status of women, the rights of transgendered individuals represent an even more difficult issue for Muslim countries to consider. The challenge has been the strong taboo against homosexuality in Islam, similar to fundamentalist traditions in most other religions as well. Eunuchs and transgendered individuals were immune from much of this taboo because there was no blame ascribed to their sexual preferences. Religious scholars could also give these individuals a pass because it could be argued that they were simply "victims" of biological abnormalities or of extenuating social circumstances that may have led them to be castrated at birth. This enshrinement of tradition or superstition under the guise of "scientific order" is sometimes termed "sciosophy." The Pakistani Supreme Court ruling in this regard was perhaps a modest step in considering the complex nature of sexuality but without addressing the underlying issue of voluntary sexual preferences. Indeed, the court could not possibly address those more contentious questions, such as gay marriage, because the Pakistani penal code still has clear injunctions against homosexuality and mandates prison sentences against such "unnatural acts" of up to 10 years. Sharia courts can have more stringent corporal punishments in this context as well. Yet, despite this public stance against alternative sexual lifestyles, there is tacit acceptance of homosexuality in even the most rural and conservative parts of Pakistan and in many other Muslim societies. While data on such issues are largely anecdotal, my own research on Islamic schools in Pakistan suggests that there is general consensus around a "don't ask, don't tell" principle. However, it is also important to note that affection expressed between same-gendered individuals in Muslim societies can also have nonsexual manifestations that are often misinterpreted by Western commentators because of their own cultural biases. For example, embracing and kissing each other on the cheeks, which is common among Arab males, or holding hands, which is common among South Asian males, are practices that do not have any sexual underpinnings.

Structural explanations have potency but are in some ways similar to stereotypes in any cultural observation, whether considering gender or any of the myriad other forms of human expression. While giving credit to the currency of such descriptors, the Nigerian-American writer Chimamanda Adichie, in her essay "The Danger of a Single Story," wisely states that "the problem with stereotypes is not that they are wrong but that they are incomplete." Our continuing search for structure in social order reveals some irrefutable human insights, but also a realization that definitive descriptors in this arena are constantly adaptive and elusive. The way to handle these multiple adaptive stories of human social construction has remarkable implications for how humanity will reconcile nature and nurture debates in the future. Our mechanism to reach such reconciliation lies next in the realm of political order.

Notes

1. Ashman, T.-L., D. Bachtrog, H. Blackmon, E. E. Goldberg, M. W. Hahn, M. Kirkpatrick, J. Kitano, J. E. Mank, I. Mayrose, R. Ming, S. P. Otto, C. L. Peichel, M. W. Pennell, N. Perrin, L. Ross, N. Valenzuela, J. C. Vamosi, and the Tree of Sex Consortium. 2014. "Tree of Sex: A Database of Sexual Systems." *Scientific Data* 1 (1): 140015. https://doi.org/10.1038/sdata.2014.15.
2. Ostrom, Elinor. 2015. *Governing the Commons: The Evolution of Institutions for Collective Action* (reissue edition). Cambridge University Press.
3. A fantastic account of Jordan's life and its lessons for finding social order can be found in Miller, Lulu. 2020. *Why Fish Don't Exist: A Story of Loss, Love, and the Hidden Order of Life*. Simon & Schuster.
4. World Economic Forum. 2020. "The Future of Jobs Report." https://www.weforum.org/reports/the-future-of-jobs-report-2020.
5. Chaisson, E. 2011. "Energy Rate Density as a Complexity Metric and Evolutionary Driver." *Complexity* 16: 27–40. https://doi.org/10.1002/cplx.20323.
6. Rogers, D. 2012. "Inequality: Why Egalitarian Societies Died Out." Https://www.newscientist.com/article/dn22071-inequality-why-egalitarian-societies-died-out/#ixzz6dYOUfc6b.
7. Shapiro, J. 2001. *Mao's War Against Nature: Politics and the Environment in Revolutionary China* (illustrated edition). Cambridge University Press.
8. Gietel-Basten, S., W. Lutz, and S. Scherbov. 2013. "Very Long Range Global Population Scenarios to 2300 and the Implications of Sustained Low Fertility." *Demographic Research* 28 (39): 1145–1166. https://doi.org/10.4054/DemRes.2013.28.39.
9. Buchthal, J., S. W. Evans, J. Lunshof, S. R. Telford, and K. M. Esvelt. 2019. "Mice Against Ticks: An Experimental Community-Guided Effort to Prevent Tick-Borne Disease by Altering the Shared Environment." *Philosophical Transactions of the Royal Society B: Biological Sciences* 374 (1772). Also refer to Noble, C., J. Min, J. Olejarz, J. Buchthal, A.

Chavez, A. L. Smidler, E. A. DeBenedictis, G. M. Church, M. A. Nowak, and K. M. Esvelt. 2019. "Daisy-Chain Gene Drives for the Alteration of Local Populations." *Proceedings of the National Academy of Sciences* 116 (17): 8275–8282. https://doi.org/10.1073/pnas.1716358116.

10. Pontin, J. 2006. "Is Defeating Aging Only a Dream?" *Technology Review*. https://www2.technologyreview.com/sens/.

PART III

POLITICAL ORDER

A nation should be just as full of conflict as it can contain.

—Robert Frost

[T]here is no reason to doubt our present ability to destroy all organic life on earth. The question is only whether we wish to use our new scientific and technical knowledge in this direction, and this question cannot be decided by scientific means; it is a political question of the first order and therefore can hardly be left to the decision of professional scientists or professional politicians.

—Hannah Arendt, *The Human Condition*, 1958

8
Empires and Edens

> Nothing is more dangerous than an idea when it is the only one you have.
>
> —**Émile Chartier**

Among the many arguments against aging reversal research to change demographic order on the planet is the dystopian fear that if we had such elixirs of perennial youth, dictators might live forever! Despite many concerns about despotism, human societies have evolved mechanisms of governance which in essence necessitate some form of control—albeit not the kind of absolute power that many dictators may appropriate. Such control may be necessary for mitigating human conflict, but it may also be necessary for ecological sustainability if we are concerned about immediate and definitive action. Political order through such control mechanisms has its origins in both selfish and communal goals. The etymological root for politics comes from the *polises* (meaning "citizens") in ancient Greek "city-states." Citizens have a reciprocal relationship with the state that they inhabit: they offer their allegiance, and the state recognizes that they have certain fundamental rights therein. The concept of the state can anthropologically be deconstructed also to the level of a tribe or a clan, though in modern states the concept is tied to a nation that may have multiple tiers of identities under one umbrella. Ultimately, states that tend to have aspirational goals and the narratives of a "promised land" are replete in statecraft. Such a search for Eden is an important motivating force for states, but the vision of such an Eden might not be in sync with ecological order, regardless of the ostensibly green biblical metaphor.

Francis Fukuyama, in his detailed history of political order, suggests a trinity of institutions that are essential for mature political order, although he misses out on natural constraints to any process of political ordering. First, regardless of what we may call the entity, some form of *statehood* is a core component of political order. A state may be constituted around the concept of a *nation*, which is a self-defined community linked by a common set of identities, or even a tribe, which may have more specific cultural underpinnings. Fukuyama favors sociological doyen Max Weber's definition of the state, with which I concur: paraphrased as *an entity which has legitimate monopoly of force over a defined territory.* A state may be defined in terms of nations or tribes, but that is not essential. Second, institutions which can enforce the rule of law are essential to ensure that citizens and the state abide by certain tenets and norms that apply to all without prejudice. Third, there must be accountability of institutions,

Earthly Order. Saleem H. Ali, Oxford University Press. © Oxford University Press 2022.
DOI: 10.1093/oso/9780197640272.003.0009

which one can assume is possible through a democratic voting system, but there may also be other technocratic means of accountability. It is particularly the third point which becomes relevant in discussions around how to reconcile urgent ecological order with political inclusion.

Economics Nobel Laureate Kenneth Arrow had alerted us to the limitations of democratic voting systems in delivering the ranked preferences of individuals into a society-wide ranking of social preferences. Arrow's "voters' paradox" suggests that democratic systems, while projecting individual preferences, may end up leading to outcomes which will not satisfy all constituents. The rules of aggregation will determine the outcome, which will not be able to meet each individual preference. This becomes even more challenging when the problems at hand are global and the consequential voting systems may be defined by states. We have no global government per se, and the United Nations is predicated on a notion of the primacy of state sovereignty. The next question to consider then is how states themselves arise and whether there are ways to operationalize their functioning so that they are more aligned to earthly order. There are three broad theories which have been presented by anthropologists as ways in which "pristine states" arose.

1. *State formation through social contracts* between populations around particular services that may be provided more easily in scale, such as environmental and health services and public infrastructure. Although suggested by classic theorists like Hobbes, the actual process of how such a contract might be negotiated is hard to conceive, although once a state is formed from extant factors such a social contract may materialize.
2. *Environmental determinism based on riparian access*, labeled "the hydraulic theory" of state formation in locations such as the Yellow River, the Nile, or the Tigris and Euphrates.[1] While widely contested for being too simplistic, riparian geography has undoubtedly played a part in human settlements.
3. *Population density as the key determinant of state formation*, with origins in urban centers that may in turn be linked to particular natural resource or commercial centers.[2] This could potentially also explain how island societies like Tonga were able to establish states while isolated mountain tribal societies like those in Papua New Guinea found it more difficult to do so.

All three of these theories are not particularly satisfying in terms of explaining the wide range of state formation experiences over history. Further political order in the context of state formation and entrenchment can also be explained by the notion of "competitive states," an idea which has more resonance with the work of Samuel Huntington, who was also Fukuyama's doctoral advisor at Harvard. Competitive states arise out of resource necessities or the erosion of trust. Much revered and reviled, Huntington was interested in macro-historical questions with dominant variables. One of his other protégés, Fareed Zakaria, commented that Huntington felt it the obligation of political scientists to find "big independent variables" (cause) and "big dependent" variables (effect) to explain phenomena. Furthermore, he had opined that "if you tell people the world is complicated, you are not doing your job as a social scientist—they already know it is—your job is to distill the powerful causes." Zakaria astutely summed it up with anatomical metaphors: "in social science your instinct is to go for the capillary but Sam went for the jugular!"[3] With such a stance of sweeping grand theories, Huntington had vehement critics who even led a campaign to deny him membership in the National Academy of Sciences. However, while as social scientists we may disagree with this forcing of dominant variables, there is little denying the fact that human beings generally see the world in stark macro-variables when it comes to political discourse. Even if particular events cannot be explained by grand theories, the individuals involved can often perceive them as such—regardless of actual complexity.

Huntington, who is famous for his controversial "clash of civilizations" proposition regarding Islam and the West, also wrote a provocative book on *Political Order in Changing Societies* (1968) which is still very much in currency today. His main argument was against the conventional view among political scientists of the time that economic development and democratization will consequently lead to political order. Instead, he argued that modernization does not necessarily lead to political development but can also lead to political decay. Thus, in Huntington's view, it was important to ensure political order if we were concerned about particular societal outcomes: this could be through democratic pathways but not necessarily so. Although environmental factors were not at the forefront of his mind, Huntington did note urbanization as one of the factors that ensue from economic growth which could lead to political decay rather than development. In his worldview for societal good, political order itself was an important goal of developing societies, regardless of the pathways by which it was achieved. Draconian as this may seem now to our predilection for inclusive democracy, such a view has important alignments with ecological order. Fortunately, two of the world's largest countries provide us with a natural comparative experiment in this regard.

The Dragon and the Wild Goose

In the twilight days of the Cold War, senior US Foreign Service officer Jay Taylor considered the world's two largest countries (by population) from the vantage point of a diplomat. China's and India's trajectories as two neighboring nation-states with vast demographic power are instructive in considering the relative pros and cons of democratic versus authoritarian order. Jay Taylor had spent 11 years in China soon after America opened its diplomatic relations with the communist country in the 1970s. He had observed how the Chinese had transitioned from the follies of Mao's cultural revolution to the pragmatic economic opening of Deng Xiaoping. Underlying this transition was also a more subtle philosophical shift back to ancient Chinese Confucian traditions of technocracy and meritocracy which Mao had disavowed in favor of a whimsical personality cult. Throughout history, there has been a tendency to centralize authority in key alpha personalities, perhaps harkening back to our primate roots. As a civilization, we have evolved beyond this "emperor impulse" but remain captivated by it as exemplified by the rise and popularity of strong authoritarian leaders in contemporary times. What are some of the key attributes of these captivating leaders that have reinvigorated populism toward them? Rather than being judgmental about these individuals or their proponents, might there be ways of addressing such a yearning through a more "entropic" political system? Such a redemption of autocracy through a hybrid form of democratic governance has been deemed a threat to national unity by the Chinese Communist Party. The Chinese had gone through a major shift in their outward engagement with the world which was coupled with a quiet atavistic return to what has been referred to as a "civilizational state."

Jay Taylor observed that this shift in China was deeply rooted in cultural symbolism and identity, and so when he decided to write a book about the country, he looked for the most apt cultural metaphor: a dragon. These mythical fire-breathing serpentine beasts were considered redeeming forces by the Chinese and were omens of progress and power. The dragon symbolized China as a nation since the advent of the Qing dynasty, when flags with dragon imagery were first used. It is thus quite fitting that contemporary narratives of China's progress frequently use the imagery of a "racing dragon" to capture the country's alarming rate of economic growth.[4] Meanwhile, to the south, across the mighty Himalayas, was Taylor's other country of literary interest. India was arguably more heterogeneous than China, and Taylor was enthralled by how complicated mythology was still such a pivotal part of Indian rituals. The stratified caste system created an artificial order which was in some ways more constraining to its development. The metaphor he used for the country was that of a wild goose, the mount of Hinduism's creative force, Brahma. As he notes in his opening epigraph, the goose swims on the surface of the water but is not bound by it and can fly or dive with ease. Hence came the title of his book comparing the two countries *The Dragon and the Wild Goose.*

Since its publication in 1987, myriad articles and books have been written on a similar topic, but the core arguments have largely rested across a spectrum of value-based views on the ascendance of Western democratic systems versus the virtues of Oriental traditionalism. India was able to form a secular democratic system following British decolonization but has struggled to be decisive about development. Yet the deliberative processes followed by the system arguably may be more resilient than an autocratic system that festers dissent, as may be the case in China. Defenders of the Chinese model argue that efficiency through meritocracy lays the foundations for building trust in government, thereby mitigating dissent. Zhang Weiwei, one of the most prominent defenders of contemporary Chinese political order, has suggested that the West misrepresents the contemporary Chinese system, which he describes as "selection plus election." Although there is one dominant political party, there is a "consultative cooperation" mechanism with various local parties, supplemented by a process of opinion-gathering and local-level elections. In comparison with the Western system described by Churchill as "the least bad option" (or *Xiaxiace* in Chinese), the Chinese system aspires for *Shang Shang Ce* (best of the best option) based on merit, as with any other job.

India's history also provides an interesting corollary to the Chinese system, one going back to the reign of the emperor Ashoka (268–232 BCE). In his youth, Ashoka was a highly autocratic and violent ruler, but after a bloody campaign in Kalinga he is believed to have reflected on the teachings of the Buddha and converted to Buddhism soon thereafter. There was a tectonic shift in his worldview from autocracy to meritocracy. He became instrumental in establishing centers of learning and arboreta to conserve trees and wildlife. Hunting of wild animals was abolished as a violation of natural order. In terms of his ecological ethos, Ashoka may well be deemed "the Green Emperor." He was also the main force in allowing Buddhism to gain evangelical momentum and spread from India to the north across the Himalayas to China, where it also found fertile congruence with the Chinese philosophies of Lao Tzu, the founder of Taoism. Both philosophical traditions place primacy in natural order, with the word "Tao" often translated as "The Way of Nature." In contrast with Lao Tzu, the founder of Confucianism, Kong Fuzi (transliterated as Confucius), had a more codified view of the world, one in which moral edicts and rituals reinforce technocracy rather than the pantheistic flair for natural order that Lao Tzu espoused. Confucius and Lao Tzu were contemporaries, and legend has it that the two also met each other a few times at a library but did not see eye to eye. Yet contemporary China has reconciled many of these contending ancient philosophies in syncretic form, and all three philosophical traditions are now equally valued.

India's democratic traditions have necessitated a more short-term view of political order in comparison with China. Politicians are kept on their toes to deliver and may be more accountable on one hand, but, on the other hand, the electoral cycle can impede the tangible progress of longer-term planning. Managing democratic order at the scale of a country like India has required elections to be carried out over several weeks, and a vast array of special interest groups play a vigorous role in the electoral process. In democratic systems, these groups can hold more sway and there is also greater opportunity for the average citizen to gain power. India has prided itself on having given an erstwhile tea-seller like Narendra Modi the opportunity to become the country's Prime Minister. At the same time, his rise to power has been at the behest of a variety of special interest groups, such as the coal and steel industries, which can exert undue influence in a transactional way as well. Mancur Olson also explored the scourge of interest group politics from an economic perspective in his classic work, *The Logic of Collective Action*, in which he laid out a theory of how concentrated benefits can trump diffuse cost factors.[5] Thus lobbying groups that arise in democracies can lead to an undermining of the collective good because the costs borne by society are not within the calculation of the benefits accrued by each group. Climate change linked to lobbying by the fossil fuel industry versus conservationists is a classic case of such an outcome in countries like Australia and the United

States, a situation that often leads to national policy paralysis and effective action only being possible at the local level. This remains one of the challenges to democratic order gaining effectiveness in large polities such as India. However, the rise of hyper-communication technologies creates opportunities for harnessing popular insights, as does what is often referred to as "citizen science." Such efforts are only possible if there is adequate literacy and trust in reliable sources of information. Unfortunately, the converse has happened in many large democracies with regard to access to misinformation and the resultant influence of darker forces that are reducing trust in technical knowledge. Fukuyama alerts us to this reality in the context of the COVID-19 pandemic, which hit India particularly hard, but the same observations apply to many other democratic countries.

> The link between technocratic expertise and public policy is weaker today than in the past when elites held more power. The democratization of authority spurred by the digital revolution has flattened cognitive hierarchies along with other hierarchies, and political decision-making is now driven by often weaponized babble. That is hardly an ideal environment for constructive, collective self-examination, and some polities may remain irrational longer than they can remain solvent.[6]

India's deep history has been shaped by fusion, more so than China's, with multiple ethnicities and religions crossing paths. The diversity was partially held together by a very long-range view of universal order from Hindu tradition in which reincarnation and karma can allow for recompense where immediate order might not be forthcoming. Caste stratification also entrenched a negative order that hindered social mobility but also led to quiescence. Intraspecies diversity can be just as important for resilience as interspecies diversity (which we discussed in Part I of this book). While there may be initial struggles and attrition in diverse human societies such as India, the long-term outlook for such societies may well be more robust if they can translate demographic diversity into pluralism of expression.[7] The public is also increasingly able to discern various layers of diversity, which are especially notable in populations from the Indian subcontinent. The findings of a 2020 study of diversity in the United Kingdom, which has one of the longest records of Asian diaspora populations residing in a Western country, are instructive. For South Asians, intolerance levels were not the same if you differentiated ethnicity versus religious affiliation data. For ethnicity, there was far greater willingness to embrace diversity, but far less so for religion.[8] Until more inclusive forms of religious doctrines are adopted, the potential for exclusion and prejudice in this context is likely to continue both within Asia and in relations with diaspora communities abroad. This in turn can have serious implications for migration flows and managing demographic order from the world's most populous countries, which could supply labor and tax base to countries that need both. Indeed, if migration was linked to resource allocation, it could be argued that human mobility is a means of harmonizing natural order with social and economic order. As a member of the United Nations International Resource Panel, I have coordinated a global assessment on the resource implications of migration. While in the past human migration was driven by resource location along riparian zones or fertile soils, contemporary migration flows rarely align to such "forces of nature," to use the title of Barry Vann's book on this topic.[9]

India's and China's development paths are going to be particularly consequential for planetary order in terms of their sheer demographic and geographic size, but also because they are important experiments in divergent political systems. Former US Secretary of State Henry Kissinger, who won a controversial Nobel Peace Prize for normalizing US–China relations, writes in his magnum opus *World Order* (2014) that "to strike a balance between the two aspects of order—power and legitimacy—is the essence of statesmanship." Political power can be articulated in military or economic terms, but legitimacy remains more elusive in a globalized world. From indigenous primacy, to settler statecraft, to retributive justice, there are many paths to claiming legitimacy. The key question will be whether legitimacy can be articulated through alignment with natural order and science rather than entrenched identity politics. India and China present us a test for nationalism in the years to come if this question is to be answered most consequentially for ecological sustainability. As natural resources become more scarce, the forces of nationalism may want a more parochial political order.

Resource Nationalism

In the dying days of the Soviet Union, there was an unsettling feeling that the crumbling of an empire would lead to disordered states and anarchy. A young Stanford graduate student, Ian Bremmer, who had learned Russian and was keenly familiar with the geopolitics of the time, capitalized astutely on the moment. He formed a consulting company called the Eurasia Group, in 1998, to advise private and public sector entities in navigating this new landscape of nation-states that had emerged. Bremmer was a political scientist who had written his doctoral thesis on Russian ethnic minorities in Ukraine, an issue which, a few decades later, would lead to the Crimean conflict. Bremmer's consulting firm grew rapidly within a few years, opening offices from Sao Paulo to Singapore, with eager clients interested in post-Soviet national dividends. The 15 new countries that suddenly arose after the collapse of the Soviet Union took highly divergent paths, from democratic socialism in Estonia to personality cult despotism in Turkmenistan.

Considering this diversity of outcomes, Bremmer began to develop a theory of how state stability related to political openness for dissent. He posited that the relationship was shaped like a "J" in which very low openness can lead to a "cold peace," as seen in Turkmenistan or North Korea. And, as societies open up, stability diminishes as more anger against repression finds venting space. Beyond a certain inflexion point, however, openness leads to more stability as a social contract between the state and citizens emerges and trust is developed in government institutions. The *J-curve theory* also considers the potential for backsliding into the trough of instability and potentially even reversing course, as was the case during the Arab Spring in Egypt. These factors were further studied by researchers at Carleton University in Canada, a decade after Bremmer's original work by plotting various states using quantitative metrics of stability and political openness. The results are shown in Figure 8.1. A note of caution about the use of two variables (bivariate analysis) to suggest causality: we can only consider initial patterns as suggestive, and each case needs further elucidation to establish any definitive insights. This illustration further codes the countries in terms of their economic and functional status and notes those considered at high risk of failure. The three states in the circled region, China, Saudi Arabia, and Russia, suggest anomalous outcomes unless one considers a nested J curve from their current state of equilibrium.

A key feature of these three anomalous countries is their dominance of natural resource capital which can lead to a distortion of political order. On the one hand, large natural resources such as minerals or forests can lead to a desire to covet and consolidate nationalism around the resource. This happened with Russian oil and gas, as

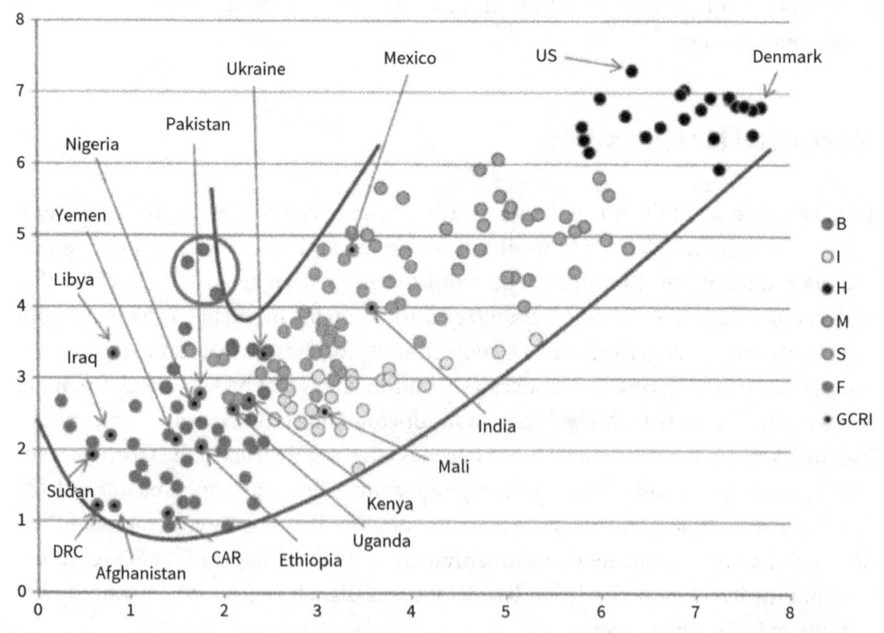

Note: colour-coded by cluster type (B = brittle, I = improverished, H = highly functional, M = moderately functional, S = struggling functional, and F = fragile).

Figure 8.1 Scatter plot of stability versus openness indices of 172 countries. States within the circle, clockwise from the top, are Saudi Arabia, Russia, and China. The 16 states labeled "GCRI" (Global Conflict Risk Index) are among the top 20 predicted at risk of violent internal conflict.

Adapted from Carment, D. et al. 2019. "Backsliding and Reversal: The J Curve Revisited." *Democracy and Security* 15 (1): 1–24.

well as with resources from some of the post-Soviet republics like Kazakhstan and Turkmenistan. On the other hand, it can lead to a *rentier state*, whereby authoritarian regimes can essentially "rent" the consent of the public through hand-outs, which is what happened in many of the Gulf monarchies. What incentive do you have to call for democracy when you are getting a tax-free minimum income that is more than sufficient to meet your needs? Either way, the broader ecological order gets missed, and this trajectory can also lead to highly inefficient development projects. The extravagant capital of Kazakhstan, which was built in the middle of a largely uninhabited, bitterly cold steppe region in the center of the country, is one such example. Artificial luxury islands and sky-licking towers built in many resource-rich territories with immense ecological impact are also a result of such distortions. Occasionally one gets boutique projects of "sustainability," such as the Masdar carbon neutral city in Abu Dhabi, but they are often a distraction from the dominant paradigm.

While countries with relatively small populations or very large resource bases can endure such largesse and maintain tenuous stability over long periods of time, impoverished countries with sudden resource windfalls may encounter a very different prospect for order. Instructive in this context is the case of Africa's most populous country. Nigeria is among the world's most ethnically diverse countries, with at least 376 tribal groups, many of whom retain their own ceremonial kings and chiefs. The country also straddles the 10th parallel, which is, for a variety of reasons of colonial expansion and ecology, defined as a theological transition zone. Continuing east, the same divide is also seen in the Sudan, and it ultimately led to the eventual separation of the larger country into two parts, partly sparked by resource nationalism and partly by theological differences. Nigeria remains one country, though it endured a bloody civil war soon after independence in the 1960s. This war created the secessionist state of Biafra (1967–1970) in the oil-rich southeastern Niger Delta region of the country, but also led to more than a million deaths from violence and starvation. Cold War ideological allegiances were surprisingly not as evident in this particular struggle–France supported the Biafrans while the United Kingdom and Soviet Union supported the Nigerian government in a divisive African "Great Game." Only Cote d'Ivoire, Gabon, Haiti, Tanzania, and Zambia formally recognized Biafra.

Eventually, the Nigerian State prevailed, and it has endured as a country despite persistent challenges, the most recent of which has been the rise of Boko Haram in the impoverished north of the country. The year 2014 marked the centenary of "nationhood" for Nigeria, when the Northern and Southern parts of the territory colonized by the British were administratively united. Although the country did not gain independence from the United Kingdom until 1960, the current Nigerian government heralded 2014 as the "Nigerian Centenary" with celebratory billboards emblazoned along the streets of the country. There is some degree of synthetic resolution to any country's nationhood, but Africa's borders accentuate the scars of colonialism most acutely. Nevertheless, for Nigeria to recognize its national identity going back to a colonial period also reflects a level of maturity in its attempts to pragmatically reconcile with its history. Nigerians have a rich and proud history of civilizational accomplishments by their various tribal ancestors. Benin City in the South as well as Kano in the North were centers of trade and commerce with impressive walled cities over a thousand years ago. The ramparts of Benin city, which was once the world's second largest man-made structure after China's Great Wall, are still visible today.

Resource nationalism and strife continued in the Niger Delta region long after the Biafran war for the next three decades. This nationalistic order was buoyed by the federal government's heavy-handed military tactics against tribal members, most notably the execution of the Ogoni environmentalist Ken Saro-Wiwa by the Abacha dictatorship in 1995. Lawsuits and violence against oil companies, which were often considered culpable for neglect in the absence of government services, continued for years. However, within the past decade or so, there has been a palpable albeit chaotic peace in the Niger Delta. Most of the lawsuits brought overseas were either settled out of court or the jurisdiction of statute was highly constrained by the courts with reference to recovery of any torts claims. After violence and crime forced American oil giant Chevron to close operations in 2003, there was a concerted effort to establish a "Global Memorandum of Understanding" to improve the impact of community development programs. Despite continuing criticism over the slow speed of development, much has changed for the better in Southeastern Nigeria. Terrorist movements such as the Movement for the Emancipation of the Niger Delta (MEND), which were involved in the 2010 Abuja bombings, have lost much popular support as economic conditions improve. Might Nigeria have move up the J curve in due course? A lot will depend on how it can negotiate the potential for economic order from its demographic dividend with the constraints of its natural resource-based order.

Poverty levels in Nigeria are hard to measure, but by some metrics the states in the Niger Delta now have the lowest poverty levels in the country, whereas only a few years ago states like Bayelsa were ranked among the poorest. Inequality remains a challenge for the entire country, but the overall inequality levels in the Niger Delta are also below the national average. The much-neglected North, where terrorist groups like Boko Haram are still active, also has core competencies in terms of its natural resource base as well as established institutions that could be further cultivated to provide a firm economic footing for the region. Such investment would make Boko Haram's vengeful nihilism less attractive to troubled youth and allow for the voices of moderate Nigerian Muslims to prevail. Interestingly enough, the American University of Nigeria is actually based in Yola, the capital of the currently restive Northeastern state of Adamawa. The university, which is part of the greater American University network, is one of the finer institutions of higher learning in the country, and there is much potential to use its presence in developing a knowledge economy cluster in this region. The Chad Basin National Park and the adjoining Waza National Park in Cameroon boast some of the finest large mammal populations in Central West Africa; the area is also a UNESCO Biosphere reserve. With proper investment and security, there is much potential for a tourist economy here as well.

Yet such a confluence of natural, economic, and social order still requires us to raise the ecological imperative more directly within social consciousness. To use

Shakespeare's metaphor of the world being a stage, many of the influential characters remain embroiled in the "high politics" of security and survival. Natural systems are either a backdrop on the set or mere props to be moved around to maintain an appealing vista on which a plot can be enacted. How might such a paradigm be changed so political entities like Nigeria might develop their resources beyond just a vacuous call for nationalism?

Great Powers Concerts and Radical Salvations

In political science there is a striking use of language to define objective reality. Those who espouse a stark view of the world in which competition for power is paramount among states and where, therefore, conflict is the norm and peace is an aberration are termed "realists." In contrast, those who consider cooperation to be the norm and conflict as an aberration are termed "liberals." The mere use of language entrenches assumptions about political order on a spectrum between "realism," as the world "is," and other notions deemed somewhat idealistic and normative about how the world "should be." Within the pantheon of political order theorists, Hedley Bull was among the first to consider a much broader comparative set of variables around the functionality of order. He identified Henry Kissinger as one of the major cheerleaders of the realist school of thinkers that orchestrated a "great powers concert" view of order. In this view those with the greatest amount of power, such as perhaps the G7 countries or maybe even the G20 countries, form an orchestra of competitive and cooperative tunes with each other and create a "structure of peace." Yet the peace upheld in this structure is between these powers more so than the peace of the world at large. Even in cases where such cooperation between great powers can have global benefits, such gains are incidental rather than deliberate.

From the point of view of finding confluence between natural order and political order, the work of contentious Princeton legal scholar Richard Falk, in his 1972 book *The Endangered Planet*, is instructive. He suggested what may be called a "radical salvationist" view of political order, one motivated by the urgency of the moment with reference to ecological decline. Falk, who spent much of his career on issues of human rights law and policy, was among the first to also link the environment to the high politics of war, peace, and security. In one chapter of *The Endangered Planet*, titled "Designing a New World Order System," Falk sets out the goals for his system in somewhat Malthusian terms, with ceilings on population increase, resource use, and waste disposal, as well as minimum environmental quality standards. To promote these goals, he suggests a structure whose core components are a strengthening of the current United Nations with a reconfiguration of the Security Council among the five most populous states (rather than the ones who won World War II). He also outlines a program of "world order activism" or "consciousness-raising" through "declarations of ecological emergency," "survival universities," "peacekeepers academies," and a "world political party." Falk further suggests we invest in an "ark of renewal" that would be humanity's insurance policy in case of an ecological cataclysm. The Global Seed Vault in Svalbard, Norway, already extant, is perhaps a starting point for such a concept. However, the absence of any hard global governance structure at present makes the prospect of a greater "ark" infeasible.

The closest any politician came to making such a clear connection between national security and the environment was US President Franklin Delano Roosevelt in establishing the Soil Conservation Service within his "New Deal." Yet this action was prompted by an epic crisis. The appalling extent of land degradation due to deep ploughing of topsoil by farmers in the American heartland during the 1930s led to massive erosion, threatened food security for the country, and left more than half a million people homeless. The tillage practices, which are estimated to have eroded more than 70% of the topsoil in Northern Texas, Oklahoma, Colorado, and Kansas, were accentuated by droughts. The massive dust storms and bleak horizons of this period are captured with literary flair by John Steinbeck in *The Grapes of Wrath*. The Dust Bowl led to the single largest internal displacement in American history, with more than 3 million people relocating from the center of the country to the coasts. Clearly, quality of life in this case was linked to environmental performance.

In revisiting the J curve, we may want to also consider other indices and relationships when trying to reconcile natural order with social, economic, and political order. Figure 8.2 shows the relationship between one measure of quality of life (the World Happiness Score, calculated annually by the United Nations) and the Environmental

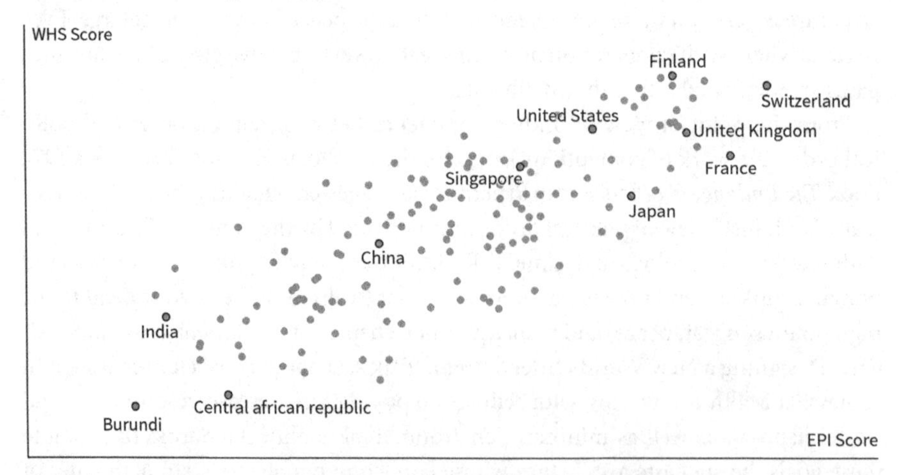

The Environmental Performance Index scores countries on 24 performance indicators across ten issue categories covering environmental health and ecosystem vitality. The World Happiness Survey measures quality of life as assessed through a variety of subjective well-being measures.

Figure 8.2 Relationship between Environmental Performance Index and World Happiness Survey Score.

Source: EPI 2018. World Happiness Report 2018.

Performance Index. The clear relationship between quality of life and environmental performance suggests that social order has positive synergies with ecological order. Yet bringing this to the realization of political elite who may still be focused on the grand concert theory of world order remains a challenge. Although using a country as a unit of analysis is somewhat problematic due to wide subnational variations, such index comparisons build on underlying governance factors that make such country-level presentations plausible.

This graph also begs the question: Does a "happy society" have more political order, or, conversely, does a happy populace confer order in society? What is clear from even a cursory observation of humanity is that there are many paths to perceived happiness. America's founding fathers enshrined the pursuit of happiness alongside "life and liberty" as our ultimate quest. But let's ponder further on this blissful term. Philosopher Bertrand Russell grappled with this question in his seminal book *The Conquest of Happiness* in terms of the intrinsic worth of happiness at the societal level. To quote Russell: "The good life, as I conceive it, is a happy life. I do not mean that if you are good, you will be happy; I mean if you are happy you will be good." Let's challenge Russel's premise a bit further and ask: Can we also be happy by being good? There are indeed limits to our quest for happiness. Most psychologists who have studied happiness agree that about 50% of our day-to-day happiness is determined by temperamental factors that are beyond our direct control. An additional 10% is determined by social circumstances, and perhaps around 40% is within our control in terms of behavioral choices.[10] It is essentially this last component that the "happiness industry" has been catering to with remedies ranging from shopping therapy to yoga.

Usually, the most potent question in this regard is: Can money "buy happiness?" One might wonder what level of wealth is minimal for happiness? Can lessons from a remote kingdom be applied to the United States? Economic psychologist Daniel Kahneman has estimated that for our country's average family size and annual expenditure, an income of around $60,000 is needed to attain the monetary component of happiness. Further income is unlikely to contribute to our feeling of "happiness." Around 65% of the American population is still under this income bracket. So what do we do? Research suggests that while meeting material needs can contribute to well-being, such factors cannot sustain happiness. Surprisingly, the mere act of helping others contributes most dependably to human happiness. For those of us who have surplus wealth, we can go a long way in helping ourselves and others in attaining collective bliss by showing some generosity. If money is too hard to part with, then we can use our time and volunteer in ways that provide services to those in need who might otherwise have to pay for them. In this way, we can move toward a more communal view of happiness that is likely to improve our society as a whole rather than trying to find individual happiness in our own very limited lives.

An important insight from this research on happiness and its connection to political order is that grassroots efforts at creating a community of giving are just as important in promoting happiness as top-down social welfare efforts. The World Happiness Survey, which is used to develop the indicators represented on the y-axis of Figure

8.2, is now conducted by the United Nations regularly to further understand the linkages between well-being and governance as well. The study weights the ranking based on seven components in most effectively explaining the aggregate score: (a) gross domestic product (GDP) per capita as a corollary for overall economic size of country and opportunity, (b) social support programs, (c) healthy life expectancy, (d) freedom to make life choices, (e) generosity, (f) perceptions of corruptions, and (g) Dystopia + residual. This final category has a peculiar name that needs to be better understood in the context of our reflections on political order because it is the largest explanatory factor for the top 20 countries in the ranking. The methods section of the report states that "Dystopia is an imaginary country that has the world's least-happy people. The purpose in establishing Dystopia is to have a benchmark against which all countries can be favorably compared (no country performs more poorly than Dystopia) in terms of each of the six key variables, thus allowing each sub-bar to be of positive (or zero, in six instances) width. The lowest scores observed for the six key variables, therefore, characterize Dystopia."

The need for Dystopia as a category suggests the importance of benchmarking in any efforts at developing ranked orders in any categorization. We aspire for Eden (Utopia) in our lives but have to calibrate our well-being with a fictional nadir of well-being (Dystopia). Even though the top 20 countries ranked in this index are all liberal democracies, this large dystopian explanatory factor still remains a puzzle. Among authoritarian states, the United Arab Emirates (UAE) had the highest ranking in the 2020 survey, coming in at number 21 and ranked above some European democracies such as France and Italy. The UAE presents an intriguing case of how a country has used its resources and wealth to create a fairly diversified economy and also a more cosmopolitan society than many of its neighboring states. While concerns about the human rights of the migrant peoples that comprises more than 75% of its population continue, the overall happiness metrics (that cover not just citizens but residents) are impressive. Will the UAE be able to climb further up the ladder of the index, and at what ecological cost would this occur? The country has recently begun to take ecological factors more seriously in developing its social order, but much of its infrastructure has been developed with profligacy. Through building lavish land reclamation projects from the sea and indoor skiing resorts in the hot desert, the Emirates have tried to defy rather than embrace natural systems. Yet it is the only country so far that has a "Ministry of the Future," which suggests a longer-term planning vision that accepts that ecological constraints are beginning to dawn on the rulers. Will the efficiency of the empire take this desert wonderland into a pathway toward Eden while still remaining autocratic?

The UAE also highlights another aspect of national order: federalism. Seven small tribal entities in the region came together in the twilight days of British colonialism to form the UAE, just the same as so many other nations, too. What is the right scale for establishing a state and then gaining legitimacy as a country that can join the international order? Seeing all the dots on scatter plots illustrated in this chapter reflects that countries are the unit of analysis for determining international order and leads us to also consider another key feature of the dynamics of governance. Ultimately, we have to demarcate territories with borders within which different tiers of governance operate. The next step in understanding the ramifications of politics for earthly order requires us to analyze those boundaries and understand how they are determined and what they may in turn determine about the fate of human societies.

Notes

1. A good review article that considers the recent literature on this theory, building on the classic and controversial work by Karl Wittfogel, is Scarborough, V. L. 2017. "The Hydraulic Lift of Early States Societies." *Proceedings of the National Academy of Sciences of the United States of America* 114 (52): 13600–13601. https://doi.org/10.1073/pnas.1719536115.
2. Much of this line of theory can be traced to the Danish demographer Ester Boserup, who was very influential in the United Nations as well. A retrospective of her contributions can be found at "Ester Boserup: An Interdisciplinary Visionary Relevant for Sustainability." *PNAS*. (n.d.). Retrieved November 19, 2020, https://www.pnas.org/content/107/51/21963.
3. Seminar, *The Legacy of Samuel Huntington*, Harvard University Kennedy School of Government. Available online at Harvard Kennedy School Institute of Politics Youtube channel January 10, 2017. https://www.youtube.com/watch?v=3M-vwHWCT1g.
4. Roy Bates. 2002. *Chinese Dragons*. Oxford University Press.
5. Olson, M. 1965. *Logic of Collective Action*. Harvard University Press.
6. Fukuyama, F. 2020. "The Pandemic and Political Order." *Foreign Affairs* 99 (4) (July/August): 32.
7. Abascal, M., and D. Baldassarri. 2015). "Love Thy Neighbor? Ethnoracial Diversity and Trust Reexamined." *American Journal of Sociology* 121 (3): 722–782. https://doi.org/10.1086/683144.
8. Woolf Institute. 2020. "How We Get Along." https://www.woolf.cam.ac.uk/research/projects/diversity.
9. Vann, B. A. 2012. *Forces of Nature: Our Quest to Conquer the Planet* (illustrated edition). Prometheus.
10. Layard, Richard, and George Ward. 2020. *Can We Be Happier?: Evidence and Ethics* (illustrated edition). Pelican.

9
Borders and Functional Political Order

> Whether the borders which divide us are picket fences or national boundaries, we are all neighbors in a global community.
>
> —**Jimmy Carter**

At the confluence of the Sava and Danube Rivers lies the great city of Belgrade, which has been center stage to arguably the most acrimonious conflict Europe has endured since World War II. I first visited Belgrade in 1988, as a teenager on a short tourist excursion with my mother, enroute from Pakistan to the United States. It was then then the capital of Yugoslavia, the center of power for an experiment in synthetic nationalism commandeered by Marshal Tito. Six disparate republics with a complex history of tensions based on religion and ethnicity were brought together under one banner of nonaligned socialism. Nationalism in this context was actually an antidote to tribalism, whereby similar physical or linguistic attributes had brought people together in antiquity (termed "homophily" by political psychologists). Nationalism raised identity to a higher scale than the family or an ethnic grouping. Nations like Yugoslavia were larger tribes with aspirations for transcendence beyond the primeval draws of tribalism. Nationalism was noticeably also different from fascism, whose etymological roots come from the Roman word for a bundle of rods, implying an absolutist order. Nationalism, in the context of states like Yugoslavia, was not absolutist, even though it was not democratic. In a polarized world between communism and capitalism, Yugoslavia appeared as a beacon of hope during the Cold War. Warring states such as Pakistan and India found an ideological refuge in Belgrade at the summit of Non-Aligned States. Yugoslavs traveled widely and could traverse borders with ease, even in a time of tough passport regulations.

More than 25 years later, I visited Belgrade again, now the capital of Serbia. My maverick elderly driver in Belgrade, Rocky, informed me how, in the 1970s, he would do truck trips from Munich to Kabul for an Afghan trucking company. They preferred hiring Yugoslav drivers who could cross borders with ease and move across the troubled terrains of Eastern Turkey and Iran without visas. Much of the industrial might which had been generated by Yugoslav nationalism has declined. No longer is there the famed Yugo car—it has been absorbed by the Italian automaker Fiat—but the skill sets of those who worked in those factories has been transferred through the universities and technical centers that still endure. Belgrade University is abuzz with

Earthly Order. Saleem H. Ali, Oxford University Press. © Oxford University Press 2022.
DOI: 10.1093/oso/9780197640272.003.0010

activity, and, while Serbia's most celebrated scientific son, Nikola Tesla, never studied in Belgrade, the city's airport bears his name with pride. Yet the question still haunts many residents of the seven separate countries that arose from the disintegration of Yugoslavia: Would we have been better off as one borderless nation? It is impossible to answer such historical counterfactuals, but there is little doubt that physical borders between nation states remain highly consequential to forms of political order.

The end of Yugoslavia, at one level, brought forth a sense of despair for those who believed in the transcendence of ethno-nationalism. It showed that tribalism is still rife, even in industrialized and developed societies. During its heyday, Yugoslavia was an industrial powerhouse producing cars and planes and boasting a highly skilled workforce. No doubt the Yugoslav wars undermined the development path of the country, but those fractures have started to heal, partly because the prize of greater European unity is at stake. A new bridge is rising across the Sava River with a spire that my Serbian driver pointed out was reminiscent of a towering minaret. But this semblance to a largely bygone Islamic identity no longer troubles the residents of the city who are instead looking toward building figurative bridges to other faiths as well. No doubt there are still ethnic tensions in many parts of the country, particularly in the Southern region, bordering Kosovo. Yet the divisive forces that split apart the country are largely in abeyance. Sometimes it makes sense for fractures to emerge in nations that have not yet matured to the point of transcendent governance and then to allow them to organically cohere with time over those issues which are of most consequence: economic development, health, environmental protection, and education. Belgrade's transformation within 25 years from the inviolable capital of a multiethnic federation to a war-torn despot's den and then to a vibrant post-conflict metropolis suggests that we should never underestimate urban resilience or the human capacity to heal. Geopolitical borders form the most enduring form of order in international relations but are just as salient at the local level.

Through the Schengen Treaty process, the Europeans began to dissolve barriers to access along their political borders as a remarkable testament to peacebuilding between erstwhile adversaries. Yet the refugee crisis following the Iraqi, Syrian, and Afghan wars after 2010 have led to a reemergence of "hard" borders. On November 14, 2015, I was arriving at the airport in Zagreb, Croatia, on a brief visit to observe the impact of the Refugee Crisis on border communities in the Balkans. It was the day of the dreadful Paris bombings conducted by radicalized Jihadist terrorists. There was a sobering sense of connectivity between the news flashing on my mobile about the Paris attacks and the refugee predicament. Crossing the border from Croatia to Slovenia then was far more complicated than it would have been a year earlier when cars could zoom through without any passport checks. For the Balkan states that are at the frontlines of transitioning the refugees toward the promised land of Germany with its generous welfare support programs, the reemergence of borders has a particular emotional sting. There is a sense of loss for many older citizens, remembering how the former Yugoslavia had fallen apart and how so many of their own citizens had been relegated to being refugees. Borders had arisen then and, after much effort, had begun to dissipate as economic expediency triumphed over ethno-religious tribalism.

The gradual accession of some former Yugoslav republics into the European Union and the Schengen System again led to some border restrictions emerging. Slovenia had largely escaped the Yugoslav wars of the 1990s since its secession as a mostly land-locked state of 2 million had proceeded largely unopposed by Slobodan Milošević. Therefore, Slovenia had been the former Yugoslav republic to most easily become part of the European Union and indeed even of the Euro Zone. The Croatian–Slovenian border thus has particular salience in the history of the Balkan conflicts because it was perceived to be the frontier where wars of the East ended and the peace of the modern West began. Sadly, that peace is now illusory for East and West in what Pope Francis has ominously called a "piece-meal Third World War." Razor wire is again being laid across the Croatian–Slovenian border and military vehicles can be seen patrolling the small border villages that speckle the gentle landscape.

The modest hills in this region have been sculpted over millennia by the Sava River, which starts in the mountains of Slovenia, flows by Zagreb, forms the Croatian border with Bosnia, and eventually reaches its confluence with the Danube near the Serbian capital of Belgrade. The ecological connectivity of this region has always fascinated me, and I have also co-taught an experiential learning course with environmental field educator Todd Walters, focusing on various conservation science and policy aspects of developing a peace park in the postwar border areas. I was thus keen to see how the fallout from the current conflicts was impacting this sensitive region. There was palpable concern in the streets of Zagreb about the renewed "securitization" of the region. Protesters in the city center held caricatured flags of the European Union, replacing the circle of stars with a circle of barbed wire. Political order was trumping ecological order or even social order. There were structural lessons to be drawn in this context. Natural systems can also have borders: the question is, when are these borders most functionally useful for achieving goals of human well-being?

The Ambivalence of Ecological Borders

Political borders have become a fraught area for ecological engagement as global inequality leads to greater human migration.[1] Calls for harder physical barriers across frontiers are gaining traction worldwide and particularly within the United States. Even those borders that had until recently been dissolving, such as those within the European Union, are now being more acutely demarcated, as manifest by Brexit as well as by calls for changes to the Schengen Treaty that provides visa-free borders in many parts of Europe. The European Union was formed on the political order of *subsidiarity*, a doctrine which has its roots in the Catholic notion that central governance should have a subsidiary function, performing only those tasks which cannot be performed at a more local level. Environmental factors were noted within the European Union's mandate as having cross-border relevance and hence were managed more centrally through a series of directives. Yet the assertion of parochial sovereignty has challenged these norms. Such developments have major implications not only for ecosystem fragmentation but also for environmentally efficient trade flows.

Growing food, mining minerals, or manufacturing products with both ecological and economic efficiency is only possible through green-smart trade policy. At the same time, some border restrictions on immigration are also important as we consider how best to globally harness the *demographic dividend* to the economy that comes from a large working-age population. Political borders also serve as an important safeguard to contain the spread of pathogens, as they did during the Ebola epidemic of 2014. This chapter considers how the pragmatic consideration of political borders might be advanced by using smart technologies and migration flow analysis based on ecological and social carrying capacity. The integrated social and ecological approach to border management proposed in this chapter considers the dynamic nature of carrying capacity and how to best develop a policy for making decisions on migration and border protection accordingly.

Russian researcher Anton Kireev,[2] in his important work on borderland studies, developed a typology of how one might consider borders in social science as zones of regulation (a modified version of this is shown in Table 9.1). This framework offers an important starting point for our analysis because regulatory jurisdictions are often the most palpable and consequential aspects of borders. The original typology omitted ecological regulation, which I have added, because it is where key innovations are most necessary to ensure positive environmental outcomes.

The North American conservation movement has recognized the importance of ecosystem connectivity across political borders ever since the first transboundary international park was established between Montana and Alberta in 1932. Through grassroots efforts by Rotary International chapters on both sides of the border, Waterton-Glacier International Peace Park was established by an Act of Congress in the United States and a parallel Act of Parliament in Canada. To this day park service officials on both sides of the border manage the area jointly. The park is now also a designated United Nations Educational, Scientific, and Cultural Organization (UNESCO) World Heritage Site (despite the United States' withdrawal from UNESCO in January 2019). The border within the park has an intriguing landscape. Instead of a wall or fence, the US Department of Homeland Security has defoliated a stretch of woodland to be able to more visibly monitor any illegal border crossing activity within the park. Yet even in the highly securitized political climate after 9/11, someone in Canada without a valid US visa can still get on a boat, cross Waterton Lake, touch US soil on the Montanan shore, walk around under the watchful eye of rangers, and then return to Canada.

This flexible approach highlights how biological and physical porosity across such frontiers can be managed. The International Boundaries Commission, which manages the US–Canadian border was established in 1907 and has more than a hundred years of experience in collective management of the world's longest continuous border (more than 5,500 miles). No doubt the soft enforcement along this border

Table 9.1 Border functionalism following the Kireev typology

Function of the border	Objects of regulation	Examples of regulation
Political regulation	Relations of political powers, their influence on their participants, means and resources	Fighting international terrorism or conducting intelligence activities
Economic regulation	Movements of material goods, factors of production, objects of exchange and consumption, actors, means and resources	Customs taxation of goods, quotas for the import of foreign labor, national sanitary and technical standards
Social regulation	Transborder processes of production and reproduction of social capital, their participants, means and resources	Rules for obtaining residence, marriage to foreigners, measures to encourage educational migration
Cultural regulation	Ethnic consciousness, information, knowledge, values, behavioral patterns, their actors, means and resources	Censorship of imported foreign literature, registration of foreign media, cultural exchange and assimilation programs
Ecological regulation	Flow of natural resources in the form of water, wildlife, and biotic resources	Water impoundments/dams; fencing/walls, quarantine mechanisms

From Ali, Saleem H. "Extracting at the Borders: Negotiating Political and Ecological Geographies of Movement in Mineral Frontiers." *Sustainable Development* 26 (5): 481–90, 2018.

owes to the greater economic congruence between the United States and Canada in comparison with the US–Mexico border. There are, however, opportunities to ensure ecological connectivity while having a harder border in areas where conservation zones are less salient. For example, a similar transboundary conservation area has been proposed near Big Bend National Park in Texas. This designation would allow for better transboundary management but the idea has stalled even though no border wall exists in that region and the Trump administration's latest proposals don't call for one. A formidable mountain range forms enough of a deterrent for any itinerant illegal migrants in this region. Indeed, ecological boundaries such as mountains or rivers have traditionally formed political borders as well and should compose the primary means of demarcating borders where possible. Policy makers should manage border zones through more technically advanced mapping of ecosystem functioning in terms of water flows and wildlife migration corridors.

Nations also shouldn't consider political borders completely indelible on maps. History is replete with examples of borders being renegotiated based on changing circumstances, while hybrid border governance mechanisms evolve based on the needs of the time. Just as borders such as the Iron Curtain disintegrated while new borders in the Balkans formed, countries can indeed renegotiate borders where necessary to consider ecological norms. The International Union for Conservation of Nature has a long-standing taskforce on transboundary conservation that deserves to be further engaged in such conversations. Ultimately, border management should consider ecosystem viability and the sustainability of important natural resources for jurisdictions on either side of the boundary.

In addition to physical borders, nationalism has also led to a wide swath of economic regulations that could have serious ecological implications. Although there has been a rich literature on the greening of the World Trade Organization,[3] it has mostly emphasized how environmental regulations can fit better within the mandate of fair competition.[4] Less attention has been paid to ensuring countries' *comparative advantage*—their ability to perform certain economic functions more efficiently—for certain products based on their geography. Such an approach would favor mining in areas where deposits are more accessible and less environmentally harmful to extract, engaging in agriculture and forestry where water is more accessible and soil is more conducive to growth without intervention, or siting manufacturing infrastructure where renewable energy supply is easily harnessed. Yet the world can only mobilize such comparative advantage if countries feel secure about the supply of important goods through international accords. Without such surety, security narratives that focus on economic self-reliance will dominate no matter how ecologically inefficient that might be.

In addition to the production cycle, waste management and a move toward a *circular economy* that recycles its own byproducts also require us to consider borders quite differently. Excessive risk aversion around waste materials can prevent a transition to a circular economy if countries are not willing to accept wastes for reprocessing across borders. International agreements such as the Basel Convention on Transboundary Movement of Hazardous Wastes and their Disposal (1989) need reform to better align with opportunities for reprocessing while containing pollution exposure. Where reprocessing is not possible, countries need to also choose disposal sites through a confluence of technical and social criteria involving the UN's principles of free, prior, and informed consent.

To follow such a wide-ranging approach to the true "greening of trade," nations must better coordinate comparative advantage through evaluating and valuing their natural resources and the goods and services that they provide to us all. The United States, Canada, and Mexico have a long history of partnership through the Commission on Environmental Cooperation established under the North American Free Trade Agreement (NAFTA) and partially continued under the renegotiated 2019 US-Mexico-Canada Agreement (USMCA). The role of this commission should be expanded to consider such "green comparative advantage" and take a more long-term view of sustainable development for the region. Other trade agreements can advance similar arrangements, as can the eventual incorporation of metrics for ecological efficiency into the World Trade Organization.

Comparative advantage does not have to be confined to natural resources alone. It could also reflect research and knowledge clusters, including traditional ecological knowledge. The global system of intellectual property rights can also better allow for more universal sharing of green innovation. Considering sustainability targets within the constraints of earthly order, we may consider what economist Marianna Mazzucato has termed "mission-oriented innovation." In 2008, the World Business Council on Sustainable Development established the first "Eco-Patent Commons," where major member companies pledged to make the green technology patents that they develop downloadable and available for use in manufacturing without any charge, anywhere in the world. The World Intellectual Property Organization (WIPO) has also recognized the importance of broader dissemination of green technologies and launched WIPO GREEN in 2013, an online marketplace for green technologies that allows donors to assist with licensing arrangements. Thus, a broad set of mechanisms can allow for natural comparative advantage to emerge and the more ecologically efficient sharing of resources across borders.

Political borders may seem a hindrance to trade and migration in a globalized world, but we should not neglect the very significant safeguards that they provide in containing the spread of pathogens, invasive species, and other noxious agents. Even in cases where borders have been an unfortunate result of colonialism, as in Africa, they have now acquired a functionality which we cannot discount in times of crisis. For example, hard frontiers between Guinea and its neighbors Mali, Senegal, and Côte d'Ivoire prevented the Ebola epidemic of 2013–2014 from becoming a far greater tragedy across Africa. The case raises the challenging question of how to balance positive movement toward economic deregulation of borders with the very important social regulation that borders provide during crises. Côte d'Ivoire had no Ebola outbreak in part because of its ability to close its border more effectively due to a better security environment, coupled with better preparedness of its medical staff.[5] Indeed, the failure of governments to enforce hard border restrictions could be partially blamed for the initial wide spread of the COVID-19 pandemic, though a tipping point was reached with the Omicron variant in early 2022 when such restrictions became less effective.

Quarantine procedures have been commonplace at borders to prevent the spread of invasive species, but until the most recent pandemic they were well off the radar of the general public. Countries with high levels of indigenous wildlife species—such as Australia and New Zealand—often strictly enforce these laws with major fines and potential for imprisonment. Countries even practice internal border control for such biotic agents. For example, domestic flights to and from Hawaii to the US mainland have additional screening for biological materials. "Living modified organisms" that have been altered genetically or through other biotechnological mechanisms have also come under scrutiny within the international Convention on Biological Diversity through the Cartagena Protocol on Biosafety. While the United States is the only nonparty country to the convention, it has regular contact with the signatories and could consider ways of incorporating some of its key insights on how to better regulate such agents at its borders as well.

The real challenge lies in balancing the ease of human travel for commerce and cross-cultural exchange versus the exercise of caution to prevent maladies from crossing frontiers. The only plausible win-win solution in this regard is more refined technology to detect unwanted biotic agents while allowing for ease of human movement. Fortunately, optical recognition technology coupled with artificial intelligence algorithms are now advancing to the point where airports can rapidly identify biological security threats with high levels of accuracy. Australia will spend more than US$250 million between 2019 and 2024 on biological security, a substantial portion of which will go toward smart technologies that could provide some greater win-win strategies in this regard for cargo and luggage screening. Ultimately, countries need to coordinate the adoption of such technologies in the least intrusive ways that maintain border functionality for biological security without compromising our human need for physical connection.

The Order of Environmental Peacebuilding

The ecological transition from the high Andes mountains to the Amazon rainforest is dramatic and constitutes an *ecotone* of rugged forested hills that also serve as a political border region between two states. Ecuador and Peru share this high biodiversity region despite a territorial dispute going back to the Spanish colonial period in the nineteenth century, when Peru and Ecuador gained independence. In 1995, following several failed attempts at conflict resolution, an armed conflict broke out between the two countries that lasted for about 3 weeks. A peace agreement signed in February of 1995 committed both countries to the withdrawal of forces "far" from the disputed zone. This plan was overseen by four guarantor countries: Argentina, Brazil, Chile, and the United States. In compliance with the plan, both nations organized the withdrawal of 5,000 troops from the Cenepa Valley and supervised the demobilization of 140,000 troops on both sides.

Due to the high ecological value of this region, conservation groups became very active in lobbying for a peace park that would strengthen conservation prospects and also help to solidify peace. It should be noted that Conservation International was actively involved in biodiversity fieldwork even before the resolution of the conflict; it had worked closely with the military when fieldwork on documenting the biodiversity of the region was conducted in 1993. Therefore, they were gradually able to influence more "hawkish" army officers about the collective importance of conservation and its instrumental use for conflict resolution. In November 1997, the two nations agreed in the Declaration of Brasilia to address four areas: (1) a commerce and navigation treaty, (2) a border integration agreement that would stimulate much-needed development in both countries, (3) a mutual security agreement designed to prevent future conflicts, and (4) a completion of demarcation of land borders. By February 1998, they were able to agree on the first three, but that left the most important one: the demarcation of land borders. Tensions arose again in August 1998, as

300 Ecuadorian soldiers spread out along an 11-kilometer line, 3 kilometers inside Peru and 20 kilometers from the demilitarized zone.

To prevent further escalation, and with pressure from conservation groups, the presidents of Ecuador and Peru both met with President Clinton on October 9, 1998, and asked that the guarantor nations make a proposal to mark the border for them.

With US satellite mapping they were able to arrive at an agreeable border demarcation. The terms of the peace agreement had some innovative features. The disputed stretch of border would be demarcated according to the Rio Protocol's line of division,[6] going back to a 1948 map, which was a major concession from the Ecuadorians. In return, Ecuador was given a square kilometer of private—but not sovereign—property across the Peruvian side of the border, extending to Tiwintza. Both countries would establish ecological parks on either side of the border, where unimpeded transit would be guaranteed and no military forces would be allowed. Ecuador was also granted nonsovereign navigation access to the Amazon and its tributaries in Peru and also allowed to establish two trading centers along the river. Initially both countries declared national parks on their respective sides of the border. However, in 2000, Conservation International and the International Tropical Timber Organization (ITTO) partnered with local conservation groups in Ecuador and Peru and with the Indigenous communities (particularly the Shuar of Ecuador) to establish a bioregional management regime. The ITTO is an unusual intergovernmental organization that was set up in 1986 to promote sustainable forest management. The organization has its origins in the International Tropical Timber Trade Agreement of the United Nations Conference on Trade and Development. Currently, the membership of ITTO comprises 35 timber-producing countries and 38 timber-consuming countries. The organization is headquartered in Yokohama, Japan, and has been one of the leading supporters of transboundary conservation projects in forested regions.[7]

While the overall armed conflict has stopped, the implementation of various features of the peace agreement remains unfulfilled. The structural aspects of the peace treaty have also prevented the formation of a functional "peace park" where access to both sides of the border would be guaranteed. Instead of creating a shared zone, the peace treaty demarcated borders and established conservation areas as buffer zones. The operations of the Cordillera del Condor Transboundary Protected area (TBPA) could be vastly improved if these commitments were met. In addition there is scant evidence of implementation of the Regulation of the Ecuadorian Peruvian Border Committees; the Programs of the Binational Plan for border areas and development; and the Comprehensive Peruvian Ecuadorian Agreement on Border Integration, Development and Neighborhood that includes as annexes the Rules of the Peruvian-Ecuadorian Neighborhood Commission.[8] The Cordillera del Condor was heralded as a true instrumental peace park when the establishment of the conservation zone helped to end a conflict between Ecuador and Peru in 1998. It was also a sterling example of a public–private partnership between Conservation International, the governments of the two countries, and the ITTO. Yet the diminution of funding to sustain conservation efforts, as well as a lack of governance planning in the post-conflict period, has led to a situation wherein environmental conflicts with Indigenous communities have taken root. Lack of environmental enforcement remains a concern for both legal and illegal industrial activities in the area.

During the military conflicts of yesteryear, external extractive interests had limited access to the region and there was a level of default conservation as a result. This was in some ways reminiscent to the Demilitarized Zone between North and South Korea, which has conserved a forested region despite the land being infested with explosive mines. Perhaps the peace dividends of the Condor Corridor rapprochement deserve to be considered at a broader level than just the park zone itself. The cessation of hostilities between Ecuador and Peru in this area has led to a concerted and comprehensive Binational Plan for development and security. This binational planning process has also included the establishment of a joint UNESCO biosphere reserve. The current focus of this program has been on the Bosque Seco dryland forest biome, which is much closer to population centers and could also be a greater tourism draw. Since the peace agreement, the Condor Corridor itself has lost priority for the government due to its remoteness, but the process of cooperation begun with the peace agreements has no doubt reaped many positive dividends in other parts of the two countries.

Thus we need to view demilitarization and the advent of peace along borders from a more regional perspective rather than only focusing on the border zone of conflict itself. Military presence can have a huge ecological impact as well, and it would be naïve to consider the mere presence of military forces as an antidote to destructive development. Nevertheless, what we need is more effective planning for demilitarization and post-conflict development. This can perhaps be achieved in transboundary areas with greater monitoring and enforcement of international agreements as well as by using universal metrics of natural capital such as ecosystem services. At the end of the day, the peace dividends in cases such as the Cordillera del Condor are a direct function of ensuring that there is no governance deficit following the end of hostilities.

The Condor Corridor case reveals how borders are precarious zones, where "uncertainty has found its most exact recognition," to quote the French political theorist of frontiers Michael Agier.[9] The writer Salman Rushdie, who agonizes over the aftermath of national ruptures through border creation in books like *Midnight's Children*, says in one of his essays that the "freest of free societies is unfree at the edge . . . where only the 'right' things and people must go in and out."[10] To use a technical term, borders have *liminality* in their order; they are places "in-between"—not just in space but also in time. A peace treaty can be a border in time, one that divides the memory of past discord with hope toward a future of concord. In an essay written earlier about this conflict, I called the Cordillera del Condor case "a casualty of peace" since the ecological order that had been envisaged was not realized even though political order was cemented.

Identity, Borders, and Order

While the Condor Corridor case provides an ambivalent outcome for the potential of aligning ecological order with political order, it was still relatively easy to resolve because the populations on both sides of the border had great ethnic, linguistic, and religious commonality. Add to this mix the challenge of "identity," and borders become a far more challenging place. Inject the notion of "civilization" into our demarcation of identity, as was done during colonial times, and we have what Michael Agier has called "a hyper-border." Such a border transcends ecology as well as social norms and delineation. This notion can hold tremendous sway over tribal impulses in human societies. A group of border studies researchers got together around the turn of the millennium in Las Cruces, New Mexico—not far from restive American border with Mexico—to wrestle with the notion of identity around borders. They came up with a simple triad which can help us further understand the political parameters of order. Figure 9.1 shows their simple diagram that links identity, border, and order in what they refer to as the *IBO framework* that permeates 11 chapters of a major anthology that was subsequently compiled. The framework recognizes that often conversations are only being conducted along an edge of the triangle and yet a stable outcome is only possible by considering the system as a whole, with also potential for more nodes

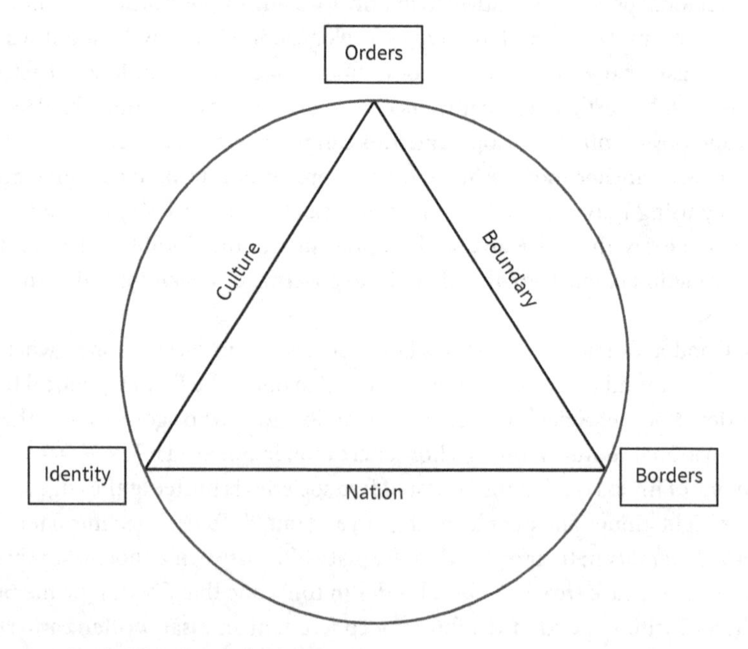

Figure 9.1 Identity-Border-Order triad.

Adapted from the Las Cruces Group. Albert, M. et al., eds. (2001). Identities, Borders, Orders: Rethinking International Relations Theory: Rethinking International Relations Theory. Univ of Minnesota Press.

of interaction along the circle of broader natural and social orders that are superimposed on the proximate triangle.

This triad appealed to me because I had grown up with terrible tales of my grandparents enduring the convulsion of the post-colonial Partition of the Indian subcontinent into Pakistan and India, which led to more than a million deaths. I often wondered if the creation of Pakistan was a good idea or not, and, as turmoil continued in neighboring Afghanistan, I decided to test an idea for reordering the political geography of the region. In February 2011, at the height of the Afghan war, I wrote a piece for *Foreign Policy* magazine titled "The Islamic Republic of Talibanistan," which suggested a radical idea based on an analysis of identity and order in a region with which I have close familiarity. I was partly inspired to write this from my chemistry background: like the *immiscibility* of certain liquids, such as oil and water, there can be fundamental structural aspects that prevent some societies from coexisting. At that point, forcing a union is a recipe for either conflict or dominance of one absolutist ideology over the other. A decade of war had elapsed at that point and a military solution was simply not in sight. In this vein I suggested that a small enclave republic for the Taliban be created, with containment so those who wanted to live under *shariah* could do so there in exchange for complete peace and noninterference with the rest of the country. Just as the creation of Pakistan involved a migration, or *hijrah*, the radical elements in both countries who yearn for an Islamic emirate could be allowed to migrate to this hinterland and help build their new political order.

Within Pakistan's conservative establishment, there is a persistent folklore of Taliban justice, claiming that the Islamists reduced crime and brought a pristine sense of order to the frontier during their heyday of power in the 1990s. The same is true for conservative Afghans who recall the incorruptibility of the Taliban *mullahs* despite their draconian punishments. Giving the Islamists an autonomous region would put those memories to the test. If the West allowed the Taliban to shoulder responsibility for a self-proclaimed "sinless state," the Islamists could no longer blame their destructive indiscretions on the vicissitudes of war. They could no longer earn money through the drug trade. In 2011, the Taliban encouraged opium cultivation as an instrument of war, earning an estimated $400 million per year even though one of the claims to piety during their heyday was a ban on opium. And when they are responsible for their own economy, the Taliban would perhaps realize the need for a broader education system than their meager *madrasas*—religious institutions that, in their current form, cannot produce doctors or any other professionals needed for a functioning contemporary society. With the Taliban takeover of Afghanistan in August 2021, I began to get messages from people who had read my article from a decade ago noting that the hybrid order of semi-autonomy I had suggested might have averted conflict. A negotiated hybrid form of governance might have also prevented the nihilistic order that has now taken root with the Taliban's complete dominance over the country.

Of course, what such a solution misses in terms of our broader goals of earthly order is any sense of bioregionalism. The aforementioned border "solution" was purely a pragmatic "high politics" proposal as a means of mitigating conflict and allowing for an experiment in governance to be considered. An ecological interface for demarcating such a new state could take two paths: (a) choosing a region which had some level of domestic resources within its borders to be self-sustaining or (b) deliberately choosing a demarcation which would necessitate trade between the two state entities to promote interdependence. Border delineation for political ends is no fool's errand. It is highly consequential and, without foresight about all aspects of order, can be a recipe for disaster. Consider the Nigerian government's efforts to deal with internal fragmentation through the rebordering of states from 3 to 4, to 12 to 20, and now 36! Yet the country is still beset by civil strife.

There are more than 300 land borders between 196 nation states within the United Nations system. These are still the most consequential kind of borders for human engagement. They can be zones of divergence but also zones of convergence. These political borders derive from historical events that have largely ignored planetary processes.[11] They are nevertheless unavoidable and, in some cases, have the potential for protecting ecosystems from harm. As human migration flows cause greater anxiety in a world of structural economic inequality, we need a more reasoned approach to border functions that balances the needs of the environment and society. Such an approach follows the six-step framework researchers have developed at the University of California, Irvine.[12]

1. Identify a phenomenon as a social problem.
2. View the problem from multiple levels and methods of analysis.
3. Apply diverse theoretical perspectives.
4. Recognize human–environment interactions as dynamic and active processes.
5. Consider the social, historical, cultural, and institutional contexts of human–environment relations.
6. Understand people's lives in an everyday sense.

This six-step process may seem fairly straightforward and intuitively reasonable, but decisions are often made on borders with an urgency that neglects such a structured, multifaceted approach. Countries often manage borders through commissions: the United States has the International Boundary Commission on the Canadian side and more topically specific commissions on the Mexican side, such as the United States–Mexico Border Health Commission or the International Boundary and Water Commission. These commissions have the potential to operationalize an integrated social and ecological approach to border policy through a confluence of science and social metrics. Although win-win opportunities might not exist in all cases of border delineation and enforcement, reason is more likely to prevail over rhetoric through such an incremental and considered path. The German legal theorist Carl Schmitt, in his book *Nomos of the Earth* (*Nomos* being the ancient Greek word for "body of law"), considers border demarcation as the basis for all legal and political spatial order. Yet human longing for connection continues. If you look at the US–Mexican border, despite the securitization, border cities like El Paso and Ciudad Juarez have grown to geographically embrace one other. Without such an embrace and with more fortifications and divisions through walls, there is documented evidence of serious psychological ailments in border communities. The German psychiatrist Dietfried Muller-Hegemann coined the term "wall disease" to describe the pathologies confronted by communities in such disrupted landscapes.[13]

Returning to European borders, which have etched most profoundly the consequences of world wars in our memory, there are remarkable examples of healing, convergence, and renewal. Consider Viadrina University on the German side of the border with Poland, on the banks of the Oder River; as a mark of healing, it has required that at least a third of its students come from Poland and it has strong cooperative ties with Adam Mickiewicz University on the other side of the border. The European Greenbelt project and the Iron Curtain Trail are initiatives aimed at creating cycling and hiking tracks along the entire length of the old points of political divergence and making them zones of convergence. A fascinating example of the political vicissitudes of borders in this region is the Venn Bahn railway route between present-day Belgium and Germany. Through a series of conflicts and border delineations, the former railway route now has the world's weirdest border, which zigzags between two countries. There is a sliver of Belgium between German territory, and a bridge across the sliver to connect the two sides is itself in Belgium. Crossing the street can take you from Germany to Belgium, but then visiting a house on the other side may mean you are back in Germany! Since the Schengen Treaty such borders are not consequential, and the region is a biking and hiking track. These transformations have come about because people were willing to consider order beyond parochial notions of tribal identity and consider more humanistic, ecological, and even planetary identities.

As an avid *Star Trek* fan, one of the distinctive features of that fictional universe was an absence of countries. Within the mythology created by Gene Roddenberry, after ravages of a nuclear World War III, humanity was on the brink of collapse but was saved by an unexpected alien encounter thanks to the discovery of the "warp drive." On April 5, 2063, there is a dramatic shift in earth governance when humanity is visited by extraterrestrial life—Vulcans—as depicted in the film *Star Trek: First Contact* (1996). In the film, humanity's ultimate enemy, the cybernetic hybrid life forms known as "the Borg," are going back in time to prevent the occurrence of this emergence of planetary governance. The Borg themselves embody a collective governance whereby there is no individual entity, but rather a collective "queen" to the hive who says in the film that her purpose is to "bring order to chaos." The *Enterprise*'s crew is able to prevent this temporal change without disobeying the "Prime Directive" of not messing with the order of time. The Star Trek universe has no countries and no money because political conflict and scarcity have been rendered obsolete. Yet the management culture of the United Federation of Planets is still highly ordered and hierarchical, with strict protocols of obedience to authority. The utility of such authority goes back to the Socratic view that every ship needs a captain for functional order during times of indecision and potential anarchy. The tension between rationality and emotion reigns supreme in arguments between humans and Vulcans, as summed up in Spock's pithy but profound statement in *Star Trek VI: The Undiscovered Count*ry: "logic is the beginning of wisdom—not the end."

There is a powerful "othering" effect on humanity at First Contact with other planetary civilizations, thereby leading to an earthly order in which borders disintegrate and unity takes hold through a planetary government. Humanity unites when we see a much bigger "other" than the differences among ourselves. Might such a future eventuate, regardless of alien visitations, because we can recognize the salience of cooperation from international to global order? Futurists have suggested we need an Earth Constitution, building on documents like the Earth Charter, with a clear set of governance parameters.[14] As Paul Raskin, founder of the Tellus Institute has noted, the planet's globalized economy already functions as a "state" in practical terms, but it is a failed state. Yet, as we have learned from examples in this chapter, borders can have important functional value, too, depending on the context. The remaining question is how we move from international to global order while recognizing that boundaries and barriers are also natural processes in many ways. Just as a cell needs membranes, and there is always an edge to any organ, we need to find a mechanism by which political borders align with a global order that recognizes such modalities.

Notes

1. Kaplan, Robert D. 2012. *The Revenge of Geography: What the Map Tells Us About Coming Conflicts and the Battle against Fate*. Random House.
2. Kireev, Anton A. 2013. "The Historical Typology of Boundaries and Some Peculiarities of Russian Limogenesis." In *Borders and Transborder Processes in Eurasia*, edited by Sergei V. Sevastianov, Paul Richardson, and Anton A. Kireev, 45–67. Dalnauka.
3. Esty, Daniel C. 1994. *Greening the GATT: Trade, Environment, and the Future*. Columbia University Press; World Trade Organization. "Harnessing Trade for Sustainable Development and a Green Economy." https://www.wto.org/english/res_e/publications_e/brochure_rio_20_e.pdf; Hufbauer, Gary Clyde, and Jisun Kim. September 9, 2009. "The World Trade Organization and Climate Change: Challenges and Options." Peterson Institute Working Paper. https://piie.com/publications/working-papers/world-trade-organization-and-climate-change-challenges-and-options.
4. Allen, Linda. 2018. *The Greening of US Free Trade Agreements: From NAFTA to the Present Day*. Routledge.
5. Breakwell, L., A. R. Gerber, A. L. Greiner, D. L. Hastings, K. Mirkovic, M. M. Paczkowski, S. Sidibe, et al. 2016. "Early Identification and Prevention of the Spread of Ebola in High-Risk African Countries." *MMWR Supplements* 65 (3): 21–27.
6. The "Rio Protocol" refers to the *Protocol of Peace, Friendship, and Boundaries Between Peru and Ecuador*, an international agreement signed in Rio de Janeiro, Brazil, on January 29, 1942, by the foreign ministers of Peru and Ecuador, with the participation of the United States, Brazil, Chile, and Argentina as guarantors. The Protocol was intended resolve the

long-running territorial dispute between the two countries, and it brought about the official end of the Ecuadorian–Peruvian War of 1941–1942.

7. Refer to the ITTO website for further details: www.itto.int.
8. USAID has funded numerous efforts for collaborative development along the Peru-Ecuador border. See, for example, Allen, A. et al. 2004. *Assessment of the USAID Peru-Ecuador Border Region Development Program*. https://pdf.usaid.gov/pdf_docs/PDACD980.pdf.
9. Agier, M. 2013. *Borderlands*. Polity Press, 23.
10. Rushdie, S. 2013. *Step Across the Line*. Vintage, 412.
11. Newman, David. 2006. "The Lines That Continue to Separate Us: Borders in Our 'Borderless' World." *Progress in Human Geography* 30 (2): 143–161.
12. University of California Irvine, School of Social ecology website, https://socialecology.uci.edu/pages/conceptual-social-ecology. Accessed, February 15, 2019.
13. Wapner, J. 2020. *Wall Disease: The Psychological Toll of Living Up Against a Border*. The Experiment.
14. Martin, G. T., and E. H. Brown. 2021. *The Earth Constitution Solution: Design for a Living Planet*, edited by L. M. George. Peace Pentagon Press.

10
From International to Global Order in the "Anthropocene"

> If the misery of the poor be caused not by the laws of nature, but by our institutions, great is our sin.
>
> —**Charles Darwin**

On a warm Barcelona evening in the autumn of 2008, I was seated at an al fresco dinner table to honor awardees of the Equator Prize at the World Conservation Congress. This is an award given by the United Nations Development Programme to recognize community efforts to reduce poverty through environmentally sustainable practices. Just before the award ceremony was to start, a silver-haired well-dressed gentleman sat down next to me and pleasantly introduced himself: "I'm Ted Turner." My first inclination was to say, "the CNN guy?" But then I paused: Mr. Turner was at this event not for his global media reach but because a decade earlier he had established the United Nations Foundation with an unprecedented gift of $1 billion—a third of his total net worth at the time. The ensuing conversation with Mr. Turner was memorable in many ways, but what stood out most was his commitment to "global" rather than "international" order. He chose his largest philanthropic gift not to establish a foundation in his own name or for a particular cause, but to celebrate and support an institution which stood for multilateralism in support the full range of planetary problems. The United Nations operated within the primacy of a "nation-state" model (hence international) but with aspirations for a more global order. It was not a global government, but it recognized the importance of global issues in an international world.

A decade before he established the United Nations Foundation, Turner established another novel philanthropy called the Captain Planet Foundation (along with television producer Barbara Pyle), themed after the environmental superhero series *Captain Planet and the Planeteers.* The focus of this charity has been grassroots ecological literacy to inculcate knowledge and values about planetary order and the individual actions needed by citizens for sustainability. Turner himself appeared in an episode of Captain Planet in the 1990s called "Who's Running the Show?" as an environment-friendly media mogul inconspicuously named "Fred Lerner." In 2019, CNN did a documentary on the life of Turner and termed him "the real Captain

Earthly Order. Saleem H. Ali, Oxford University Press. © Oxford University Press 2022.
DOI: 10.1093/oso/9780197640272.003.0011

Planet." One of the defining aspects of the cartoon series is that it encourages children to realize that we are living in a particular time where humans have more agency than ever before. Although it started airing before the term "Anthropocene" was popularized by Nobel Laureate Paul Crutzen, this is very much a saga of redemption, a new epoch for Earth with the progenitors of a new order. The opening narration of each episode reads,

> Our world is in peril. Gaia, the spirit of the Earth, can no longer stand the terrible destruction plaguing our planet. She sends five magic rings to five special young people: Kwame, from Africa, with the power of Earth. From North America, Wheeler, with the power of Fire. From the Soviet Union/Eastern Europe, Linka, with the power of Wind. From Asia, Gi, with the power of Water. And from South America, Ma-Ti, with the power of Heart. When the five powers combine, they summon Earth's greatest champion, Captain Planet![1]

While any form of hero worship is problematic, the role of sensible leadership in facilitating functional order is important and has thus created an epic industry of books on its own. The young planeteers saw Captain Planet as a parental leadership figure. Environmental youth activist Greta Thunberg would perhaps offer a contemporary critique of the planeteers in that they used their power rings with good intentions, but, at the end of every episode, they ended up summoning Captain Planet, with the message being for children under parental supervision that an adult or hero has to be there to fix the problem. While youth and children's empowerment has received a great boost in recent years, including a Nobel Peace Prize for my Pakistani compatriot Malala Yousafzai at age 17, a case can be made for experience as a measure of leadership as well.

Turner told me that he sees a convergence of grassroots and global initiatives moving toward a planetary order. Similarly, the convergence of youth empowerment and adult experience and life-long learning needed convergence. He was also mildly dismissive of the adage "think globally, act locally." In his characteristic Southern style: "Hell we have to think and act locally and globally!" I was reminded of what the former Prime Minister of Norway and Chair of the World Commission on Sustainable Development, Gro Harlem Brundtland, said about Turner in a promotional video about the UN Foundation: "Ted has positive aggression!" The Equator Prize was particularly appealing to Turner because it recognized the connections between global environmental efforts and local poverty alleviation. I reflected on the use of the term "equator" in its caption: a demarcation of order created by us humans through a system of longitudes and latitudes that now delineate ecosystems. The prime meridian, the International Date Line, the tropics of Cancer and Capricorn: all these were ways of marking earthly order on the physiology of the planet. Yet the order they impose is inherently contrived for functionality. I realized this most acutely in a 45-minute flight from Pago Pago (American Samoa) to Apiah (independent country of Samoa) in 2014, when attending a United Nations summit on small island states. This rare flightpath traverses the International Date Line; in 35 minutes of real time, you are transported almost 24 hours forward in functional time. The Date Line itself has two functional orders, *de jure* and *de facto.* The de jure representation of time is based on nautical necessity, and the location was chosen to cause minimal disruption for terrestrial inhabitants through the 1917 Anglo-French Conference on Time-Keeping at Sea, which was later adopted more broadly. However, the de facto date line depends on individual national decisions, as is the choice of time zone. For example, China has kept one time zone for the entire country for national identity reasons even though this is not aligned with any particular planetary process. Similarly, island states could choose their time zones, too. I learned from my pilot on that short flight in time travel that only a few years earlier Pago Pago and Apiah had been in the same time zone. The Samoan president made a decision to change his country's time zone allegiance closer to Asia than to America as a show of economic solidarity. This alone was enough to cause a functional change in how we perceive time in global affairs.

Twenty-five community projects were awarded the Equator Prize that year—a refreshing change from the usual small-order elite prizes where there are only one or two winners. Among the organizations which received the prize were two marine management organizations, one from the Solomon Islands and another from Indonesia. As we discussed global order on the table, maritime law became a particularly interesting point of conversation. Island states, in particular, might have small terrestrial borders but large exclusive economic zones in the sea, as delineated by the Law of the Sea Convention. The open oceans were beyond the usual demarcation of national boundaries and a space where global governance mechanisms could be prototyped. Indeed, the open oceans were also being compared with outer space as a "planetary commons," similarly to how nations considered resource extraction from asteroids or the moon. An international lawyer at our dinner table also interjected with the example of Antarctica, where science had ostensibly been the guiding force for a global treaty at the height of the Cold War. Unlike the open oceans, however, Antarctica had contested territorial claims which were simply put in abeyance. Cynics might say it was less for science and more out of a lack of economic viability for any particular resource enterprise on that continent.

At our table was also a young Palestinian activist who noted that his challenge was first to be part of an institutional mechanism that would place him at the negotiating table. The inability of the United Nations system to resolve the Arab–Israeli conflict remained one of the clearest examples of the limitations of international order and one where the term "global order" may be more helpful. Ted Turner had been a vocal supporter of Palestinian autonomy and had been criticized a few years earlier for harsh words toward Israel. He kept quiet on this part of the conversation but noted his support for the Peace Parks Foundation, a group based in South Africa which had also worked in supporting transboundary conservation across political borders in the Middle East. A year earlier, I had edited a book on *Peace Parks*, which had received endorsements from both scientists like E. O. Wilson and policymakers like Achim Steiner (former environment minister of Germany and past head of the

United Nations Environment Programme). Could the Middle East be a region where hybridity in political order may be better aligned with ecological factors? Were there new solutions to this intractable problem which could be opened up if we began to have a more nested view of earthly order even in this acrimonious part of the world?

Confederations of Peaceful Ecological Order

Around the time that the anthology on peace parks was published in 2007, I received an unusual call from a lawyer in Los Angeles, Josef Avesar, who wanted me to come to the West Coast for a symposium to brainstorm a new set of solutions to the Arab–Israeli conflict. As a Pakistani American who had visited Israel and Palestine, I was intrigued by this prospect and yet also cautious of trying to inject myself into a fraught area of academic inquiry. However, I had just received tenure at the University of Vermont at the time and felt a bit more comfortable venturing forth into the disorder of Middle Eastern political science. I arrived at the symposium, where Mr. Avesar opened the event with a video message from Harvard Law School Professor Alan Dershowitz praising efforts at hybrid solutions to the Arab–Israeli conflict. A few years earlier, I had attended a presentation by Dershowitz where he had suggested redrawing the two-state solution incorporating settler enclaves within the West Bank. He had artfully compared such a state to Denmark, which has numerous islands a short distance from the mainland; perhaps Palestinians could consider the gaps between their territorial spaces in a similar vein. An interesting ecological metaphor, though I wondered if by that measure Israel could be compared to inlets of the Baltic and North Seas!

After the opening video, Mr. Avesar presented his core suggestion: instead of a one-state or a two-state solution, what was needed was a hybrid Confederation of the two states. The term has ominous rings for contemporary Americans, but there are precedents for such forms of hybrid political order. The Iroquois Tribes of present-day New York and Southern Quebec had a confederacy; Switzerland is officially a confederation of cantons. However, unlike earlier efforts, this idea was to be implemented from the grassroots using the internet as a platform to recruit candidates for a "virtual parliament" while the policymakers remained deadlocked. In the following 6 years, Mr. Avesar, to his great credit, managed to get more than 700 Israelis and Palestinians (including in Gaza) to run in a virtual election which was be held on December 12, 2012. Those who may dismiss this as a gimmick should note that even a willingness to run in an election of this kind posed peril to the candidates, but they were willing to do so because they saw this as the most tangible effort to "think outside the box" and move beyond the stagnation of one-state/two-state fixes.

The *New York Times* published a full-page advertisement regarding the Israeli Palestinian Confederation and its election plan. Yet the level of suspicion, cynicism, and contempt on all sides have remained intense. Despite their effort at novel thinking, many across the political spectrum dismiss such initiatives merely as a means of social-climbing or prize-fishing. Often, visionaries like Avesar attempting to disrupt existing order through novel peacebuilding are labelled as "sell-outs" or "conspirators." For those who like to offer a patronizing pat on the back, they are simply dismissed as "well-intentioned idealists." Fast forward to the most recent Peace Plan proposed by the Trump administration, in which efforts at a two-state solution continue, but where a viable map for such an outcome is vague (Figure 10.1). Within this current plan, one that has thus far not been entirely rejected by the Biden administration, natural resources and ecological factors do not appear to be salient.

There is no direct access to the Jordan River for the Palestinians, and the plan would require a vast network of tunnels and bridges to be functional. From the standpoint of economic and natural resources, the new Palestinian state would be considerably dependent on Israel. Conversely, from the standpoint of labor, Israel would still be dependent on Palestinians to provide human capital. Might such a situation create opportunities for cooperative order and comparative advantage as envisaged by David Ricardo (Chapter 4)?

In my visit to Israel in 2010, on an invitation from Tel Aviv University and the US Embassy, I was alarmed to find how much of the narrative of peacebuilding had eroded—to use an environmental metaphor. Because of an existing atmosphere of uneasy calm and a surprisingly sanguine sense of security because of the general absence of violent conflict, efforts such as the Israeli Palestinian Confederation were seen as an attempt to disrupt this calm. Some even suggested that if there were such a confederation, it should exist between Palestine and Jordan instead, given that the ethnic demographic of those two nations is more homogenous.

This, my last visit to Israel a decade before the Trump/Kushner plan, was at a time when the United States had not yet recognized the Golan Heights as a part of Israel and before the start of the Syrian civil war. There were thus discussions also of Israeli peace-building efforts with Syria regarding the Golan region, including Mount Hermon. Even though Syrian participation at the meeting was not possible due to a prohibition of professional contact between the two sides, it was perhaps constructive to have Israelis discussing the issue independently since they are the occupying force in the region and would have to first resolve internal political differences on the issue. I was asked to attend as the keynote speaker, given my previous research on such efforts worldwide and my background as a Pakistani American who has explored such issues in the context of regional peacebuilding in South Asia. Some "Realists" might roll their eyes at such a prospect, but the concept of "peace parks" is more than an idealist's ramblings and has shown promise in resolving territorial disputes.

Warring parties can be made to realize quite pragmatically that joint conservation is economically beneficial and a politically viable exit strategy from a conflict. As noted in Chapter 9, the United States used such a strategy in the mid-1990s to resolve a decades-old armed conflict between Ecuador and Peru in the Cordillera del Condor region. However, a key difference in the Middle East, and one that makes the conflict more intractable, is that it is not merely a matter of borders but a much larger claim to contested place-based identity. Nevertheless, the Obama administration's deputy

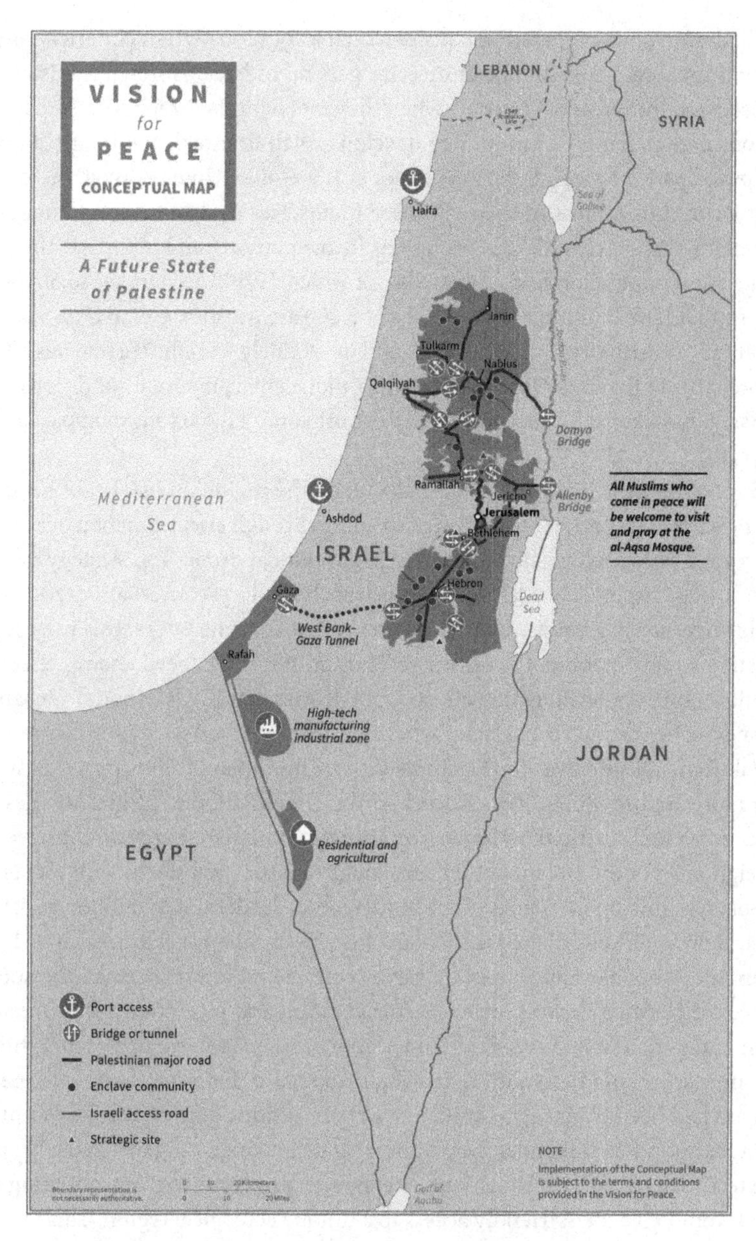

Figure 10.1 Map of Trump/Kushner Peace Plan for Israel/Palestine, 2020.

envoy to the Middle East at the time, Fred Hof, proposed the Golan Peace Park effort as a means of peacebuilding with Syria as well in a formal paper written for the US Institute of Peace in 2008. In Hof's plan, water guarantees to Israel (which currently gets 30% of its water from the region) could be exchanged for return of sovereignty to Israel. It was an idea that policymakers seriously considered, and detailed maps and plans had even been prepared to consider such a solution. Syrian American negotiator Ibrahim Suleiman and former director general of Israel's foreign ministry Alon Liel discussed this prospect in 2007, when they met with the Israeli Knesset's Foreign Relations and Defense Committee to develop a plan to establish a jointly administered peace park between Syria and Israel in the Golan. (Interestingly, the original ethnic Druze inhabitants of the region see themselves as distinct from Israelis and Palestinians since their religious group has its own culture and identity.) The Golan Heights has a population of about 38,900, of which 19,300 are Druze, 16,500 are recently settled Jewish immigrants, and about 2,100 are Muslim. Golan is also an environmentally sensitive region with a cool and moderately wet climate that has allowed fruit orchards to flourish. Underscoring the unique environmental conditions of this area, Israel has allowed Druze farmers to export some 11,000 tons of apples to Syria each year since 2005.

This confluence of interests could have made the region an ideal case for implementing a novel dispute-resolution strategy and, through environmental peacebuilding, create a new ecologically aligned hybrid political order. The strategy involved transforming disputed border areas into transboundary conservation zones with flexible governance arrangements. To some realist commentators, this may suggest idealistic or naïve notions of conflict resolution, but it has been championed even by military officers, such as the late Indian Air Marshal K. C. "Nanda" Cariappa, a former prisoner of war in Pakistan, who called for such a strategy to resolve India's and Pakistan's dispute over the Siachen glacier. The Golan Heights proposal was initially motivated by Robin Twite's work at the Israel-Palestine Center for Research and Information during the 1990s. According to one plan, Syria would have been sovereign in all of the Golan, but Israelis could visit the park freely, without visas. In addition, territory on both sides of the border would be demilitarized along a 4:1 ratio in Israel's favor. When I visited the Golan after the conference, it also occurred to me that another possible solution might have been to find a way to make the spectacular Mount Hermon area a particular conservation and recreation zone, one where Israelis and Syrians could enter to visit without visas but when exiting from this special zone visas would be required. Israel already has a major ski resort on one side, and Syria had been planning to build a resort on its side of the divide. The summit of Mount Hermon was still under Syrian sovereignty and including this in the proposed peace territory would give Israelis an incentive to also come to the negotiating table, since it would give them friendly access to a unique ecological region. This would be similar to the status of the eastern Sinai under the Israeli-Egyptian peace treaty or that of Hong Kong and Macau in China, whereby there are separate entrance concessions for these areas as compared to mainland China.

Similar proposals have also been initiated by the pioneering environmental organization EcoPeace Middleast along the Jordan River, at a "peace island" where Israelis and Jordanians can visit without visas and where the original peace treaty between the two countries was signed (and which is currently under deliberations for expansion). This case is particularly intriguing since, under the treaty, an Israeli kibbutz is allowed to grow crops on Jordanian sovereign territory. A Yale University architecture class worked on the design of the expanded park in collaboration with neighboring Jordanian and Israeli communities. There is also a marine peace park agreement among Jordan, Israel, and Egypt in the Gulf of Aqaba (which was established as part of the first round of Oslo negotiations).

Yet communities are not always aligned with ecological order and can be highly suspicious of conservation efforts due to a tragic history of exclusion rather than inclusion in such demarcations. For example, Native Americans have been suspicious of the US National Park system whose establishment often excluded them from their land. Thus, any peace park must be one where access and economic development are concurrent with conservation. Ultimately, ecology defies political borders, and the governments of the Middle East will need to become aware of this natural reality. Many countries in the region are signatories to international environmental agreements, such as the Convention on Biodiversity and the Convention on Desertification. Perhaps these agreements will provide another avenue through which to pursue ecological cooperation as well. At the end of the day, as erstwhile adversaries realize that they are inherently confined by their ecologies, the chances of cooperation are likely to rise. Yet fortifying trust between essentially primal orders of tribalism are most likely when there is a convergence of top-down efforts at global environmental awareness and grassroots organizing.

Networks and the Realignment of Global Order in the Anthropocene

Fragrant jasmine flowers are ubiquitous among street vendors of Tunis, and they grace most ceremonial events in Tunisia, reminding us of their namesake 2011 Jasmine Revolution, which is considered the progenitor of the "Arab Spring." The self-immolation of a Tunisian vegetable vendor due to police harassment is considered the proverbial spark that lit the flame of the revolution across numerous Arab countries from 2011 onward. The past decade or so has seen the metaphoric spring turn to a very dark winter in Syria and Yemen and the autumnal rise and fall of "Islamic democracy" in Egypt. Tunisia is the only country that has survived this stormy spring with a nascent democratic system in place. Yet the country's predicament remains paradoxical and polarized on many accounts. Soon after the revolution, an estimated 6,000 Tunisian youth went to Syria to fight for radical Islamist movements—the highest number per capita of any country. At the same time, the vast majority of the country, as much as 86% according to 2016 polling (a rise from 72% in 2011), feel that democracy is the best form of government, even with economic woes having increased since the revolution.

Tunisians take great pride in their Phoenician heritage and invoke the pioneering spirit of the legendary Queen Dido, who is believed to have founded the city of Carthage. Subsequent great military leaders such as Hannibal are also venerated, alongside a history of religious pluralism which gained particular fame when Jews and Muslims from Iberia fled to North Africa. Even now, the island of Djerba is a sterling example of Jewish-Muslim coexistence. Tunisia's political culture has also been able to keep the army's meddling in day-to-day politics at bay even during the despotic period of its post-colonial history. The Tunisian military supported the Jasmine Revolution, and analysts credit this to the exceptional role of the armed forces in keeping the country on a democratic path. Given this history of pluralism, Tunisian society is willing to embrace civic dissent—constructive confrontation—internally more so than other parts of the Arab world.

A culture of civic entrepreneurship, coupled with a strong tradition of organizing around fundamental rights, has manifest itself in strong unions of workers and managers alike, as well as professional organizations of socially conscious lawyers and human rights activists. It was these organizations that were able to rescue the country from its darkest days in the post-revolutionary period when two politicians were assassinated and fears of a civil war were on the horizon. For their work, the National Dialogue Quartet was awarded the Nobel Peace Prize in 2015. No other organization of its kind has existed in the Arab world thus far. Tunisia clearly has certain specific structural attributes in its society that make it "an Arab anomaly," as noted by Safwan Masri in his eponymous book published after the Arab Spring. However, despite these exceptional elements, pan-Islamism, which many considered a form of tribalism that transcends territorial borders due to its unity of ideas and not geography, has touched Tunisia as well. Neighboring Libya's unraveling has led to an estimated 2,000,000 immigrants coming into this country of only 11 million citizens. Though these migrants, many of whom come loaded with cash, have created a consumer-driven positive impact on the economy, there is also concern that some have more radical Islamist leanings. Therefore, Tunisian exceptionalism in a pan-Islamist Arab world will certainly continue to be tested.

The Arab Spring emerged out of the ability of grassroots groups to organize through the new "global" order of social media networks that were able to help in organizing agents of social change. The causes per se of the movements are still being debated by complex systems analysts, with variables as far afield as climate change and religion all potentially playing some role. What is more intriguing from the perspective of this final chapter is how the disaffected individuals were able to organize as a movement so quickly. The structural power of grassroots networks in the age of social media, which is credited for this process, is by no means an assurance of a more qualitatively functional order that will deliver on desired sustainability outcomes. Far from Tunisia, a glut of misinformation can also spread noise and confusion within political order. The rise of populism and divisive memes from Brazil to India to Europe and the United States can be credited to the power and speed of memetic transfer across physical borders of communication. Just as energy density at particular nodes in a physical system has the power to mobilize change more efficiently, the same is true of political systems. Nodes of power in this context are particular media companies, private special interest groups with access to information, or celebrities with charisma. The internet developed as a series of nodes to devolve computational power, and it has also allowed for further applications of this resilience through which data can also be devolved to localized devices. Interestingly, the original internet was meant to follow a communal order without the motivation for profit—until 1991, the government banned internet-based business development. There was an anarchic quality to earlier internet proponents such as John Perry Barlow, the lyricist of the band Grateful Dead, who penned the "Cyberspace Declaration of Independence." Barlow became one of the founders of the Electronic Frontier Foundation to support the independence and public service mission of the internet. Yet the allure of advertising revenue through a targeted means of ordering demand and supply was too captivating, and commercial applications took root rapidly and are now prompting calls for increased regulatory action. Wikipedia and the Mozilla Foundation (that provides the nonmonetized browser Firefox) are notable exceptions of large-scale internet platforms that have maintained that original gratis community spirit.

In his book *Network Power*, briefly discussed in Chapter 6, David Singh Grewal lays out the magnificent orders of globalization that have appeared in the twentieth century. Yet he notes an important cautionary note: "everything is being globalized except politics." He is referring to our tendency to move toward common norms in language, dress, and other harmonizing influences of globalization while political systems remain diffuse. This has major implications for how economic power can transcend borders in a "race to the bottom," basically channeling capital toward paths of least resistance. Thus, billionaires can find tax havens, and corporations can find pollution or labor havens. The misalignment of network power between economic order and political order is estimated to have a staggering toll. In 2019, the International Monetary Fund estimated that between $8.7 trillion and $36 trillion in capital were held outside tax jurisdictions by the elite, a phenomenon known as "capital flight."[2] Hence opportunities for long-term compatibility between national nodes of power and economic nodes of power remain elusive, particularly in the context of environmental problem-solving. Environmental governance necessitates making connections across intrinsic ecological networks that are endowed by nature and often influenced negatively by anarchic human behavior. This is where making as many connections as possible between individuals and societies in a systems-oriented approach to politics is so vitally consequential. No longer can we operate under the paradigm of "world order" in a political vacuum, but we must instead strive toward an *earthly order* that embraces the inevitability of natural constraints for sustainable human societies.

Around the time of the Arab Spring, I was invited to attend the annual meeting of the World Economic Forum in Davos, Switzerland, in 2012. This is by no means the most comfortable place to have a conference, and fatigue does set in after a couple of days of walking on icy pavements and being bussed around small traffic-clogged mountain roads. Yet this is among the few major events where public and private sectors have a chance for frank conversations about a nascent global order. The year I attended was particularly opportune as this was the year of the "Occupy Movement" and a shadow event, or "Open Forum," was organized at a nearby high school for the protesters, who had very little engagement with the main forum. The challenges of multistakeholder engagement were best exemplified to me during a rather acerbic conversation on the last day of the summit with an academic professional from a developing country who works on issues of trade and investment in emerging economies. He expressed sympathy with several prominent economists who were hesitant to attend the Forum because there were too many "non-experts" allowed on panels. Furthermore, this distinguished professional expressed disdain for peace activists and others with no direct connection to "economics" being invited to discuss development issues. It was quite alarming for me to learn that while the Forum has tried to respond positively to a criticism of exclusion, it has incurred the wrath of many such professionals with a highly limited view of "expertise!" Networks and the nodes which constitute them may be inherently devolved in their full scope, but they still have a propensity for agglomeration at key hubs. Just like natural monopolies in economics, there is a tendency in networks to also form key hubs, as we saw in the form of the rise of the major tech giants Facebook, Google, and Amazon. This clustering into hubs or even superhubs is a manifestation of what is sometimes termed "the Matthew Effect," after the parable of talents in the gospel of Matthew—a "winner takes all" warning perhaps.

> For to every one who has, will more be given, and he will have abundance; but from him who has not, even what he has will be taken away. (Matthew 25:29)

The World Economic Forum has in essence become a superhub for public–private partnerships. Powerful networking platforms that provide the opportunity to facilitate channeling of motivational energy must not be intimidated by scorn for alternative views. Efforts to bring multiple perspectives and diverse epistemologies to the table will be needed to realign global order in congruence with ecological order. If organizations such as the World Economic Forum lose a few self-absorbed academics in the process, there will happily be many other useful professionals to take their place. It is sheer folly to assume that data and analysis are somehow static and that the last word on economic theory has been delivered. Challenging expertise orthodoxy in one discipline to understand the complexity of human behavior is precisely the role organizations such as the Forum are best fashioned to tackle.

The role of conflict in political order has roots in statistical analysis even before it became embraced as a fertile area of inquiry by complex systems scientists. The British pacifist and polymath Lewis Fly Richardson analyzed the *Statistics of Deadly Quarrels* from 1815 to 1945 in a monograph by that name. He arrived at a logarithmic scale of conflict mortality and scale of the quarrel. This led to the development of "scale-free theories" of conflict that followed "power laws" which show changes between two variables independently of their scale. However, interactions between parties and the flows of information which can change perceptions can greatly impact the speed and severity of how such conflicts evolve. Thus, developing from international to global order requires us to consider the forms of networks—comprising nodes and links—and the level of influence that exist between them. From a political order perspective, researchers are applying network analysis for "earth systems governance," a field which is now blossoming with its own professional organization, journal, and panoply of funding sources. Among the most promising researchers in this arena is Rakhyun Kim, a product of a globalized network himself: ethnically from Korea, Kim grew up in New Zealand and is now based at the Copernicus Institute of Sustainable Development at Utrecht University in the Netherlands. Figure 10.2 shows examples of network forms that he has used to exemplify a range of international regimes and the transitions that can occur between them.

Trained as a lawyer with interests in complex systems theory, Rakhyun Kim is always interested in how such structural order can translate into legal mechanisms. His research is particularly timely as several new institutions are being deliberated

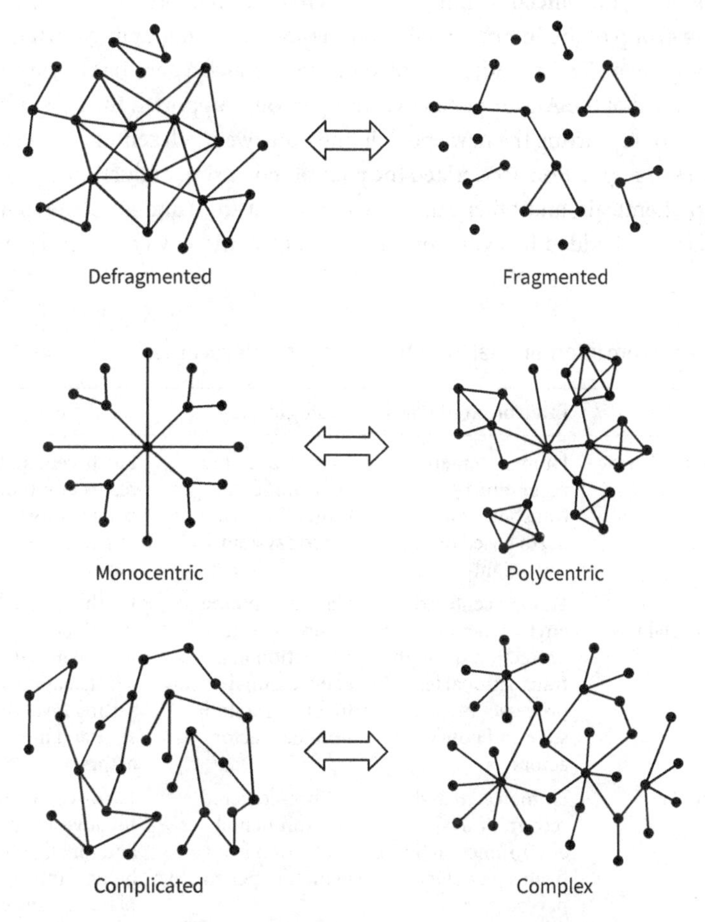

Figure 10.2 Three models of Global Governance Architecture evolution and devolution.
From Kim, R. E. (2020). Is Global Governance Fragmented, Polycentric, or Complex? The State of the Art of the Network Approach. International Studies Review, 22(4), 903–931. https://doi.org/10.1093/isr/viz052.

around the "science–policy interface" for improving congruence between natural order, socioeconomic order, and political order. For example, France has initiated a process at the United Nations General Assembly since 2018 for a Global Pact for the Environment, which would be a legally binding and enforceable agreement between countries with an aim to move environmental law from "soft" to "hard." On the scientific side, the establishment of an Earth Commission, in 2019, aims to reach consensus on "science-based targets" for key ecological indicators, which could then be fed into a global environmental governance mechanism such as the United Nations Environment Assembly. Considering such overtures, Rakhyun Kim partnered with South African researcher Louis Kotze to propose three tiers of legal intersectionality between science and policy in Earth Systems Law (Table 10.1).

Implicit in such a move toward planetary "earth systems" law is a recognition that human impacts are at a scale which has disrupted natural order to the point where we are deliberating the official naming of an epoch for human presence. In March 2019, a working group of the International Commission on Stratigraphy, which delineates such geological order for academia, voted in favor of such a naming. Twenty-nine out of 34 members of the Anthropocene Working Group supported the designation and voted in favor of starting the new epoch in the mid-twentieth century, when a rapidly rising human population accelerated the pace of industrial production, the use of agricultural chemicals, and other human activities. At the same time, the first atomic bomb blasts embedded in sediments and glacial ice, thereby creating new minerals

Table 10.1 From international to global to planetary legal order

	Environmental law	Ecological law	Earth law
International law	Human-centered regulation of transboundary harm based on state sovereignty	Nature-centered environmental protection in a state-centric system	Earth-centered respect for the community of life in a state-centric system
Global or transnational Law	Human centered environmental protection through transnational legal processes involving state and non-state actors	Nature-centered environmental protection in a transnational setting involving state and non-state actors	Earth-centered sustainability governance in a transnational setting involving state and non-state actors
Planetary law	Human centered recognition of environmental limits from a planetary perspective	Nature-centered environmental protection from a planetary perspective	Earth-centered law for governance by and for all living beings from a planetary systems perspective

From Kotzé, L. J., & Kim, R. E. (2019). Earth System Law: The Juridical Dimensions of Earth System Governance. *Earth System Governance, 1*, 100003. https://doi.org/10.1016/j.esg.2019.100003.

which would become part of the geologic record. Moving further into what James Lovelock has called the "Age of Hyper-Intelligence" through machine learning and bioinformatics, we may also move to a "Novacene" epoch. Such tools can be used to assist in vastly amplifying the speed and scale of scientific discovery of new enzymes for biopharmaceuticals, electro-materials for infrastructure, and a range of other approaches urgent human uses. There are highly divergent views about what such a Novacene would mean for humanity. Lovelock is optimistic that humans would be able to coexist with hyper-intelligence and would essentially become like cultivated plants at the behest of cyborgs and other artificial life forms. The global political system would clearly need additional updating in that regard!

Proximately, the power of networks and legal mechanisms is needed to first start a regenerative process for many planetary systems that have been disrupted in the Anthropocene. John Elkington, who is known for his work on incentivizing the private sector to consider sustainability indicators, suggests that we should consciously be seeking "Green Swan" events for intervention. Unlike the Black Swans discussed in Chapter 6, Green Swans are defined as being

> catalyzed by some combination of Black or Gray Swan challenges and changing paradigms, values, mind-sets, politics, policies, technologies, business models and other key factors. A Green Swan delivers exponential progress in the form of economic, social and environmental wealth creation. At worst, it achieves this outcome in two dimensions while holding the third steady. There may be a period of adjustment where one or more dimensions underperform, but the aim is an integrated breakthrough in all three dimensions.[3]

Such events may arise through the advent of new technologies, including artificial intelligence or CRISPR genetic methods. They may also emanate from a reframing of what might have previously been framed as a "wicked problem" (discussed in Chapter 5 from a sociological perspectice). This term was further popularized by policy planning scholars Horst Rittel and Melvin Webber in a presentation to the American Association for the Advancement of Science in 1973. Although they listed ten attributes of these problems, for our purposes of policy insights what matters most is that they have multiple causality and diffuse and uncertain solutions: environmental planning problems are clearly such. Some of the reasons why a problem may seem so intractable, however, is because of its framing in parochial terms of self-interest and without adequate linkages with broader goals being realized. Seeing those nodes of connection in scale and scope in a complex system can provide avenues toward realizing solutions and instantiating Green Swan opportunities. Yet such opportunities are not always self-evident and require an iterative process of understanding scientific premises, social norms, and barriers to trust. Often such iterative processes are circular and loop back, incrementally moving us toward a realization of global order.

Closing the Loop on Global Order

When Justin Trudeau became Prime Minister of Canada in 2015, one of his moves, which received scant attention abroad but raised some eyebrows at home, was to change the name of the Department of Foreign Affairs, Trade and Development to Global Affairs. This name change was more than symbolic for it signaled the view that nation states are ultimately part of a global system. Managing affairs beyond borders as if they were inherently "foreign" or "alien" has set us up for confined solutions to intrinsically globalized challenges. What goes around eventually comes around on a planet with myriad feedback loops. Canada did not always have such a systems approach to considering feedback loops in environmental governance. Trudeau's predecessor, Stephen Harper, had notably withdrawn the country from a previously ratified United Nations treaty to combat desertification. Some of his associates had infamously noted that the rationale for this was that "Canada had no deserts." The potential for land degradation impacting global food supply, changes in commodity prices, migration flows, and numerous other impacts that occur in such complex systems contexts did not appear on Mr. Harper's radar. Trudeau's name change also summarily changed the mission of the department by recognizing a world of feedback loops by stating upfront on their website that "We define, shape and advance Canada's interests and values in a complex global environment." Canada rejoined the convention on desertification in 2017, and it continues to push such a systems approach to global affairs in its interactions with other nation-states in groupings such as the G20, G7, and the United Nations. Another example of this approach has been Canadian funding for and hosting of the Intergovernmental Forum on Mining, Metals and Sustainable Development. The underlying premise is that mineral supply requires a complex network of exploration, extraction, recycling, and refining spread out across the planet, and, without a global approach to mineral supply governance, we would have a far less efficient outcome.

The University of Waterloo in Canada hosts the Environmental Change and Governance Group, where Professor Derek Armitage and his associates have considered "adaptive management" to global environmental change as part of a global project on Earth Systems Governance. Feedback loops do not exist only in terms of material flows and environmental information but also in terms of collective human learning and pathways to ecological literacy. Within educational psychology, there is a common paradox that is presented with reference to creating novel knowledge: in order to truly learn something new, we must know what it is we need to learn. In the context of ecological systems, this can be particularly challenging, and the approach Armitage and colleagues have followed is a "triple loop" learning process which starts with an analysis of outcomes, considers errors, and injects new knowledge through each cycle of the loop (Figure 10.3).[4]

Harmonizing such learning processes across multiple cases of a range of complex socioecological problems across the planet can create a web of learning outcomes

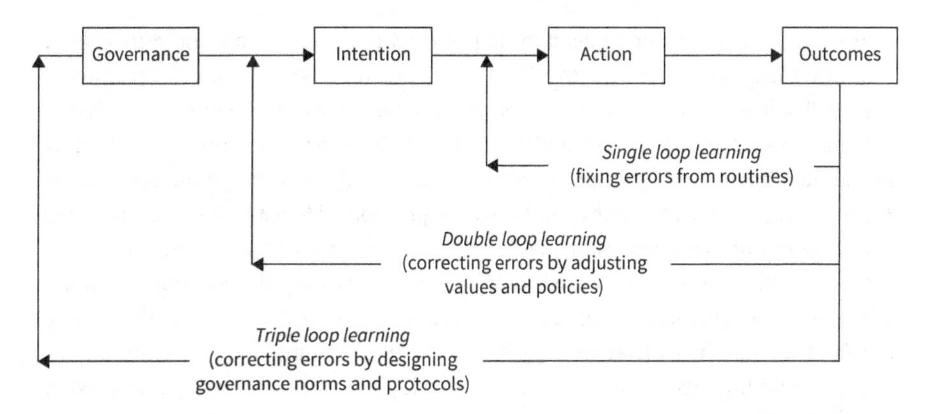

Figure 10.3 Derek Armitage's triple loop learning process.

that can in turn also make a more effective case for effective earth systems governance. Such an approach may lead to a more harmonious skill-based order emerging from the bottom up, as argued by the philosopher Ivan Illich in his book *Tools of Conviviality* (1973). The invention of the personal computer and the development of Youtube teachers and learners have created such "convivial communities" in ways that were previously only deemed possible through a centralized curriculum. A key feature of this approach is the iterative process of error correction, which harkens back to our discussion of the importance of error identification and correction in the structural attributes of scientific inquiry. Some of the lessons of this approach were elaborated by Ben Ramalingam in his award-winning book *Aid at the Edge of Chaos*, in which he highlights the "network of mutuality" that is core to considering an effective use of natural and financial resources for development. His book draws inspiration from Chief Scientist of the Rockefeller Foundation Warren Weaver in the late 1940s. Being a scientist at a philanthropic organization had given him an appreciation for global challenges as well as with working across disciplines to deliver measurable outcomes from grant recipients. The collective learning approach of feedback loops was particularly important in charting the foundation's commitment to resilience research because systems that learn from such feedback increase their resilience to shocks. The creation of the Resilience Alliance in 1998, under the leadership of the late systems ecologist C. S. Holling, as well as recent efforts such as the establishment of the Arsht-Rockefeller Resilience Center at the Atlantic Council in 2019, are examples of the Foundation's continuing commitment in this arena.

The metaphor of loops in providing a mechanism for self-correction and incrementally productive order is powerful. Yet, in a complex world, we should also be cautious about the order or loops themselves and the structures they provide. As we conclude our journey of considering different tiers of order, it is opportune that this last section of Chapter 10 loops back to the personal reflection that prompted the writing of this book. Each one of us yearns for some sense of order in our lives, even if we do not like to be ordered to do so. The sense of "I" versus "Us" in human agency is at the very core of our understanding of earthly order. We are more inclined to make that leap from "me" to "we" if there are greater feedback loops of reciprocity. Yet feedback loops can also create paradoxes for decision-making, where hierarchies and preferences become tangled in what Pulitzer Prize-winning writer Douglas Hofstadter has called "strange loopiness." In his 750-page magnum opus, *Gödel, Escher, Bach*, he considered the fundamental question: "How it is that animate beings can come out of inanimate matter. What is a self, and how can a self come out of stuff that is as selfless as a stone or a puddle?"[5] Using the fantasy story-telling techniques of Lewis Carrol (the mathematician who authored *Alice in Wonderland* and *Through the Looking Glass*), Hofstadter suggests that these "strange loops" of hierarchical decision-making give us agency. The strange loops form structural patterns which could also be conceived as "attractors," a term used in the analysis of chaos going back to the seminal work of Edward Lorenz on weather systems. These strange loops are a set of conditions in which systems may evolve dependent on the underlying fabric of the natural realm. However, to differentiate the scientific from the pseudo-scientific, it is important to note that the notion of such attractors and loops is qualitatively different from the self-gratifying "law of attraction" brandished in bestselling books like Rhonda Byrne's *The Secret*.

Toward the end of *Gödel, Escher, Bach*, which traverses mind-numbing episodes in art, music, and mathematics, there is a small section on "strange loops in government." Eerily, the example given in that section to illustrate entanglement in government hierarchies, in which there may be structural incoherence in the feedback loops, was reminiscent of the aftermath of the 2020 United States election. Hofstadter wrote his book in the mid-1970s, and he chose the example of the Watergate scandal. President Nixon had threatened to obey only the "definitive ruling" of the Supreme Court but also claimed that he had the right to decide what is "definitive." Thankfully, he did not act on the threat, but if he had we would have entered a "strange loop" of confrontation between two branches of government. With contingency of decision authority hierarchies, we can enter strange loops for governance and end up in what may be ultimately termed a "constitutional crisis." If anarchy were to happen in such a context—the outcome in dystopian films about disruptions of earthly order like *Madmax* or *Waterworld*—there would still be new nested hierarchies of rules, only these would be from the bottom up rather than the top down.

To make the full connection with natural analogies in this example, Hofstadter also writes in the next paragraph that the same situation could happen in the world of physics, where gases in equilibrium obey simple laws connecting their temperature, pressure, and volume. However, gases can violate those laws (just as a president might) if they are not in a state of equilibrium. In such contexts, the physicist must resort to statistical mechanics: "to level a description which is not macroscopic, for the ultimate behavior of a gas always lies on the molecular level, just as the ultimate political behavior of a society always lies at the grassroots level."

Loops are thus gratifying loci of informational exchange when we know how to navigate their paths and have guard rails for their exigencies. Global political order, in aligning with earthly order, requires us to have a clearer sense of the natural rails, noting that the loop does not lead us into a dead-end and that we are conscious of path dependency. The prospect of path dependency has its origins in the physics of magnetic order but is most commonly used as a term in political science. Many of the strange loops, when represented as physical pathways, seem like impossible objects, such as Max Escher's mind-bending paintings such as *Order and Chaos* or *Relativity*. However, if we consider them statically or through the lens of statistical probability of location with a particular point in space and time, they are plausible in two dimensions. Political order tends to operate in such a context and can distort reality most potently. As we continue to spin through our lives on a planet looping through a solar system, which in turn is part of series of celestial circuitry, let us commit to learn deeply before we act politically.

Notes

1. Opening narration of Captain Planet cited in Terrace, V. 2013. *Television Introductions: Narrated TV Program Openings since 1949*. Scarecrow Press.
2. Shaxson, N. 2019. "Tackling Tax Havens." *Finance and Development* 56: 3. https://www.imf.org/external/pubs/ft/fandd/2019/09/pdf/tackling-global-tax-havens-shaxon.pdf.
3. Elkington, J. 2020. *Green Swans: The Coming Boom In Regenerative Capitalism*. Fast Company Press, 5.
4. Ramalingam, B. 2015. *Aid on the Edge of Chaos: Rethinking International Cooperation in a Complex World* (reprint edition). Oxford University Press.
5. Hofstadter, D. R. 1999. *Gödel, Escher, Bach: An Eternal Golden Braid* (20th anniversary edition). Basic Books.

Conclusion

Reconciling Orders

> The willingness to redraw boundaries, to notice that a system has shifted into a new mode, to see how to redesign structure—is a necessity when you live in a world of flexible systems.
>
> —**Donella Meadows**

This book has journeyed through the human quest for finding order in nature and constructing order for social, economic, and political systems, which might not always be in congruence with natural order. Throughout these pages are stories of order, which is perceived and functionally useful even when it may not exist in physical terms. Whenever the cost of believing a false pattern is less than the cost of not believing a real pattern, evolution will favor pattern recognition. Thus, the mere perception of order is not enough—we need to be willing to consider the science behind such perception. Most significantly, I have tried to focus on those aspects of this quest for order which can be functionally most useful in helping our species live more sustainably. Ultimately, this is the kind of orderly quest that is most consequential. At the same time, we need to be wary of imposing order in social systems which may thrive on diversity and variation.

The search for new explanatory mechanisms for natural and social processes will continue. Humanity still yearns for a "theory of everything" to explain the origins of order in the universe. The biophysicist Harold Morowitz considered such a task across scales rather than at the level of the archetypes of material existence. Earlier in his career, he considered the essential role of energy in organizing systems, and, toward the end of his career, he wrote a book titled *The Emergence of Everything: How the World Became Complex*. In this book, he laid out 28 steps from a "primordium" (functional origin point for the existing universe) all the way to "spirituality." Steps 1 to 7 dealt with the physical sciences. Step 8 is the biosphere and was noted as "transitional." Steps 9 through 20 were all biological, Step 21 refers to the evolution of "The Great Apes" and is again noted as "transitional." Steps 22 through 26 are cultural within hominid societies. Step 27 is termed "reflective thought" and again deemed "transitional," leading to the final step of "spirituality." More recently, biologist Tyler Volk has elegantly presented a similar Grand Sequencing in his book *Quarks to Culture*

Earthly Order. Saleem H. Ali, Oxford University Press. © Oxford University Press 2022.
DOI: 10.1093/oso/9780197640272.003.0012

but with greater interplay of harmonics between the various sequences in a concept he refers to as "combogenesis," through which the physical, biological, and cultural realms animate the creative process in humanity.

In essence, the connection between a constituent part of any system leads to behavior that would not be possible at the individual level—whether for a particle or a person—and is referred to as "weak emergence" within philosophical thought. This is the kind of emergence with which complex systems science is concerned in situations like swarm behavior or physical phenomena like conductivity, viscosity, or friction—or indeed the development of life itself. However, philosophers consider "strong emergence" to be more structurally robust because it suggests a pattern that has already emerged but is not explained by individual constituents. Such emergence is scientifically hypothetical and not empirically observed thus far but is a key feature of theological and spiritual belief systems.[1] The closest analogy to strong emergence that the physicist Sabine Hossenfleder has provided is that of a digital photo montage, as from the work of Pakistani visual artist Rashid Rana. The macro-image emerges from individual photographs that have no connection to the ultimate design. However, this is a synthetic image rather than one developed through natural laws. If strong emergence were to be observed, new permutations of natural laws and conceptions of order would need to be considered.

Another way to consider the trajectory of human–environmental relations is the transformational trajectories presented in times of crisis, such as the pandemic during which much of this book was written. Thumbing through my library, I found a neglected book titled *Evolutions Edge* by Graeme Tayler, a futurist based in my second-home city of Brisbane, Australia. On its cover was a butterfly with the Earth's map on its wings—an apt metaphor for complex linkages on the planet. Taylor helped to establish the Biosocial Evolution Systems Theory (BEST) Initiative, which presented a compelling schematic of how human systems can oscillate within equilibria but then reach critical bifurcation points in crisis (Figure C.1). The term "schematic" in the preceding sentence is used in its psychological connotation—"schema"—which means a model that can make information useful for human cognition. Murray Gell-Mann, in his classic work *The Quark and the Jaguar: Adventures in the Simple and the Complex* (1994), noted that choosing the correct schemata for action is key to achieving sustainable behavioral change.

Underlying the potential for collapse at Point 3 in the diagram is the tension which social and natural systems must contend with regarding growth trajectories. Complex systems doyen Geoffrey West warns us of this disjuncture in the final chapter of his book *Scale,* where he attempts to craft a "grand unified theory of sustainability." He notes that while biological systems follow "sublinear scaling" in their growth trajectory, social and economic systems can follow a "superlinear scaling" (increasing return to scales in corporations or in cities' growth patterns, for example). This leads to a fundamental disjuncture for earthly order because mathematically superlinear systems have what is termed a *finite time singularity*, meaning that the growth equation can become infinitely large at a finite time. Since this is not physically possible

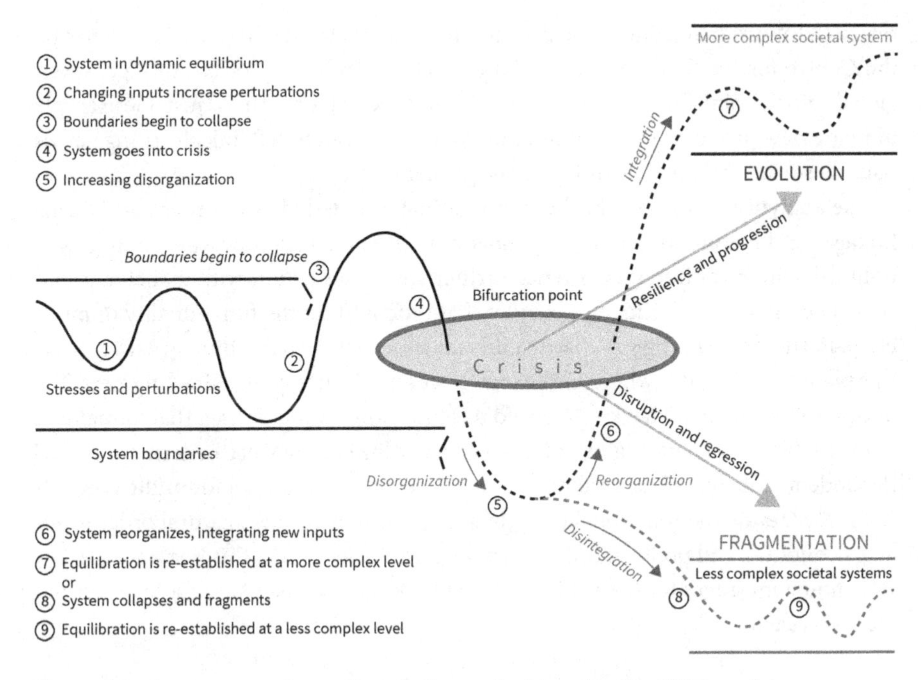

Figure C.1 Systems transformations in times of crisis using the BEST model.
From Taylor, G. (2008). *Evolution's Edge: The Coming Collapse and Transformation of Our World.* New Society Publishers.

to sustain, a sudden collapse becomes inevitable unless there is an intervention. The need for a suitable transition thus needs to be planned and can't just be assumed to emerge from a self-correction mechanism. Scholars such as Ziauddin Sardar have also argued that we are in a "postnormal time" wherein "postnormal science" approaches that dispense with conventional cost-benefit analyses need to be considered.[2] Complexity necessitates a multiscale approach to problem-solving but that is only possible if we understand the underling natural constraints in developing social, economic, and political order.

Ultimately, we should be concerned about applying these lessons to mitigate key risks to sustainability that are within our control in terms of human behavior but also be concerned about protecting against risks that are beyond our control. Fortunately, there has been a remarkable growth in the range of research centers focused on existential risks to humanity, aimed at finding the most effective leverage points for constructive intervention.[3] For example, many of the most dire scenarios for human extinction due to a major calamitous occurrence, such as an asteroid impacts or a pathogen, relate to food depletion. Thus, if artificial means of food synthesis in the absence of sunlight could be developed, we could tackle multiple risks to earthly order. Similarly if we can be aware of the interactions between naturally generated risks like volcanic eruptions and their specific interlocking with social networks, we can divert resource toward managing those threats with greater alacrity. In one recent study by the Centre for the Study of Existential Risk at the University of Cambridge, seven global "pinch points" were identified that meet these criteria.[4] This is just one example of many new promising lines of interdisciplinary research that links learning across natural, social, and political order for the greater good.

The analytical stories in this book reveal that a nested view of order and finding linkages and hierarchies of interdependence between natural, social, economic, and political spheres of human existence further elucidates action within such a grand sequence. As an environmental planner, I am focused on the functionality of models, patterns, and paradigms—particularly for a more sustainable human existence on this planet—and that is where the lessons of this book attempt to shine brightest. The late physicist Stephen Hawking warned us in his book *Grand Design* that humanity's quest for an understanding of universal order is limited by what he termed "model dependent realism." We should not get too hung up on the search for unified models but rather refine the functionality of those models that connect natural order to social, economic, and political order. Summing up the lessons of this book are five key take-home messages for the reader to consider, with some key futurist ideas associated with each.

Lesson 1: Patterns Raise Useful Questions but Not Always Useful Answers

The allure of discerning meaning from ordered patterns in the universe has been a survival imperative for our species. It is through patterns that our brain is able to make the connections and memories that give us positive and negative associations. Evidence from brain injury patients has even suggested that there are parts of the brain which are naturally hardwired to distinguish natural patterns in comparison with synthetic patterns.[5] The tapestry of patterns and relationships in socioecological systems that help us in adapting to environmental changes was termed a "meta-pattern" by Gregory Bateson.[6] However, some of these patterns are incidental, and, while they may fascinate us, their functionality must always be questioned carefully. An intriguing example of such a numerical patterning process, attributed to US Navy researcher Simon Newcomb and physicist Frank Benford, was recently popularized by the Netflix series *Connected*, hosted by Dr. Latif Nasser from RadioLab. The Newcomb-Benford Law observes that in a vast range of unrelated and random numerical listings, such as a table of statistics, the digit 1 tends to occur with a probability of approximately 30%, which is much greater than the expected 11.1% (i.e., one digit out of nine). This observation of recurring patterns can be used to consider if a dataset has been doctored—whether it be reported numbers in tax returns or pixels in a digital image that has been digitally altered. However, misapplication of such pattern identification schematics can also lead to confusion or be put to use to confound facts. An example is the investigation of election fraud allegations after the

2020 US presidential election. The Newcomb-Benford "Law" does not apply to the US electoral process because there is a two-party political system and voter precincts are drawn up to be roughly the same size within a given district. If precincts register 50,000 votes, for example, and were split roughly evenly between Donald Trump and Joseph Biden, you'd expect most tallies to start with 4's, 5's, and 6's, not 1's and 2's, as the Newcomb-Benford Law would have predicted.[7]

Thus, the seductive lure of patterns must always be considered within a particular analytical context. At the same time, we should always be willing to consider that there are still many unknown features of causal and acausal patterns and perceptions which science should continue to investigate. For example, in the early twentieth century, the Swiss psychologist Hermann Rorschach experimented with using more than 300 different kinds of inkblots images on symmetrically folded paper to consider their diagnostic usage for psychiatry. Rorschach was also an artist, and he was interested in how patterns were processed by different neural circuitry. He followed a methodical scientific process for coding his diagnostic results. Sadly, he died a year after his coding was published in 1922, and his test was subsequently misused without proper scientific follow-up of the lessons that could be gleaned from his data. Further research on evolutionary and pathological signals of pattern recognition in neuroscience continues.

A key insight with reference to how our brain makes sense of the world comes from following the anatomy of pattern recognition. The dichotomy between the left side of the brain as being logic-oriented while the right is emotion-oriented is somewhat contrived, but what is functionally accurate is that the left lobe engages in more detail-oriented analytics while the right lobe considers macro-level analytics. The two hemispheres are thus providing complementary views of reality to give us essential functionality. The corpus callosum is a set of neural fibers which connect the two hemispheres: these connections operate like traffic signals to allow for the most meaningful collaboration between the lobes. Hence, when playing music, there is an involuntary macroscopic processing of the sounds that we consider appealing and sonorous, but, to achieve them, a high level of microprocessing is required as well. Effective musical output by a piano player requires this signaling of order between the two hemispheres. Neurologist and psychiatrist Ian McGhilchrist has noted in his book, *The Master and His Emissary*, that one of the challenges of modern society is that we have given more significance to the left lobe's function (the emissary) as compared with the right lobe's panoramic function (the master). Through a remarkable array of clinical examples of peculiar symptoms of people who suffer strokes and injuries in either lobe, he lays out the importance of this complementarity. For example, a patient who loses function in the right lobe may be unable to move their left arm. Astoundingly, they process this lack of function by believing that their left arm simply does not belong to them! Conversely, a left hemisphere stroke may lead to loss of function in the right arm, but if the right lobe is functioning, the processing of this loss is more cogent. The macro-level view provides that broader context and meaning which the micro-level technicality simply cannot provide. In this essential balancing of the macroscopic and the microscopic, the brain is not like a computer.

Computational tools of modern technologies and artificial intelligence (AI) can be an important way for us to sift through large datasets and consider functional pathways for action on global environmental change problems. Microsoft Research established the AI for Earth program, in 2018, to use computational power to deliver solutions aimed at planetary sustainability. A key feature of this program since early 2020 is the development of a "Planetary Computer" that will be able to process vast amounts of data, sift through the noise via machine learning, and provide functional outputs. We may be able to get more useful maps of ecosystems and optimize conservation borders accordingly. We could process long-term weather pattern data to develop improved climate models. And, most consequentially, we could perhaps also better connect social and behavioral data from humanity to socioecological change. There is a burgeoning of innovative scholarly research and technological development in this arena. Emblematic of such programs is the MIT Center for Bits and Atoms, which has applied such approaches to develop bioengineering applications. One of their precocious doctoral graduates, Manu Prakash (now leading his own lab at Stanford), is using such an interdisciplinary approach to understanding functional order in complex systems to develop biosensing nanodevices that bridge biology, social behavior, and computational science.

Computational prowess and the desire for evidence have given ascendance to what is broadly termed scientific "positivism" and its various academically debated forms with the "pre" and "post" labels that academics relish deliberating ad nauseum. However, a recognition of evidence-based decision-making should not detract us from imagining and seeing cultural or literary value in artful accounts of what our minds may conjure. So long as we are aware of the limitations of such conjectures and speculations, they can be important stimulants for creativity. The challenge arises when those conjectures are paraded as scientific process and eclipse the evidence, rather than being celebrated as intriguing speculations or valued manifestations of cultural history. We also need to recognize the importance of subjective human judgments in key decision processes, rather than just relying on the elegant order of algorithms from machine learning or AI. The complex subjectivity of human judgment provides context that machine learning is incapable of delivering at this stage in its development. Concepts like justice, inclusion, and reconciliation require nuanced context and learning. Ordered patterns may need to be defied in such cases if we are to achieve a societally preferred outcome.

Lesson 2: Natural Order Is Ultimate but Humanistic Order Is Proximate

The nested Venn diagram in the Introduction of this book was elaborated upon through each of the three parts of this book as a framework for guidance. A key lesson is that elaboration of the ostensible hierarchy of influence in terms of earthly order relates to how humans are able to perceive and act on such learning. Thus, while natural processes and science would ultimately trump social and behavioral choices, the decision span and time horizon for such changes are widely divergent. Natural systems often have much longer time horizons for agency while humans yearn for immediate gratification in their conscious present. Analysts who work on socioecological systems grapple with these proximate issues of material flows between natural and social systems. For example, the concept of *socioeconomic metabolism*, which builds on the work of the Vienna Institute of Social Ecology, has been instrumental in this context.[8] However, we still need to understand the underlying structural constraints imposed by natural systems and seek constructive alignment between ecological and social structures if we are to transition from a "world order" to an "earthy order."

The mineralogist and futurist Vladimir Vernadsky and the New Age French Jesuit priest Pierre Teilhard de Chardin tried to connect natural order to human behavioral responses by suggesting an additional sphere of planetary existence. They proposed the "noosphere" (derived from the Greek root word for "mind" or "reason") as the human consciousness equivalent of the biosphere, whereby human activity in the form of thoughts and sentiments could be connected across the planet. The concept took a strange parapsychology twist through the Princeton Anomalous Engineering Research (PEAR) group, which hypothesized that human neural activity undertaken collectively could have palpable material impacts. The mechanism by which this may occur has thus far no scientific basis, but there are claims of an "electro Gaia gram" (EGG) effect by which random number generators across the planet are impacted by collective human anxieties and emotions. Psychologist Roger Nelson, who shepherded the PEAR project and its current incarnation, the Global Consciousness Project, for 30 years, published a book in 2019 which synthesizes his findings titled *Connected: The Emergence of Global Consciousness.* While it is tempting to trivialize this as nothing more than pseudoscience and a Star Wars-like "Jedi force" , a measure of humility and appreciation for the efforts made by the PEAR team at following fairly detailed data collection procedures is still appropriate.

The concept of *synchronicity*—a term coined by Carl Jung, supposedly during dinner conversations with Einstein—describes an "acausal connecting principle." In essence this is a way to consider coincidences in nature and find meaning in occurrences that appear to have connections without cause. Jung also considered *archetypes* as universal symbols of the unconscious self that often are revealed through "instinct." For functional purposes, if such patterns of connections help individuals to develop meaning at a personal level, that is fine, but trepidation is warranted in trying to assume broader structure through such observations without some explanatory mechanism. Recent experiments with mice have also shown that cooperative behavior can be triggered through an alignment of neural circuitry initiated by some energy source. Whether internal brain impulses or behavioral cues between organisms can trigger such synchronicity remains to be seen.

The occurrence of such harmonization processes in non-human animals is surprising enough, but when such processes occur in unconscious entities like elementary particles or cells they seem almost miraculous or magical. We know now that much of this emergence of order occurs through *coupled oscillators*. Any two entities—living or nonliving—could be coupled if some physical or chemical property allows them to influence one another: this could happen due to electrical charge or physical contact or chemical signaling. The oscillation couplings are simpler in natural systems but become increasingly complex, asymmetrical, multidimensional, and often more chaotic in social systems.[9] The physicist Paul Halpern makes some interesting linkages between the concept of synchronicity and how it can be applied to quantum physics based on an analysis of the conversations that occurred between Jung and quantum physicist Wolfgang Pauli. The essential tension between looking for a determinate cause for some effect versus observing a phenomenon with no apparent cause has been at the core of what we humans call "beliefs." From our functional lens of seeking earthly order, the knowledge of deep natural and social structures provided in this book will hopefully provide a bedrock of ecological literacy. "Belief" may still be personally important and a supplement to knowledge because some patterns, even if they are fictional illusions, still provide meaning and consolation. My favorite quote from Winston Churchill is "whether you believe or disbelieve, it is a wicked thing to take away Man's hope." So long as we do not conflate what is observable, testable, and verifiable with what is speculative but magisterial, we will remain on target toward achieving a more sustainable human existence on the planet.

Lesson 3: Transitions In and Out of Order Are Not Always Reversible in Space or Time

Throughout the pages of this book, I have also highlighted that emergent order does not need to be determined entirely by particular pathways ordained by fixed equations. This insight harkens back to the work of Nobel Laureate Ilya Prigogine who won a rare solo prize in Chemistry in 1977 for his path-breaking research on irreversible, nonequilibrium dissipative reactions—in essence, for noting the limits of predicting trajectories of natural order. In his 1996 book, *La Fin des certitudes* (*The End of Certainty: Time, Chaos, and the New Laws of Nature*), written in collaboration with Isabelle Stengers, Prigogine questioned the determinism of much of established science.

> Mankind is at a turning point. The beginning of a new rationality in which science is no longer identified with certitude and probability no longer with ignorance.... We are observing the birth of a science that is no longer limited to idealized and simplified situations but reflects the complexity of the real world around us; a science

which views us and our creativity as part of a fundamental trend present at all levels of nature.[10]

Embedded in Prigogine's insights is the foundation of complex systems science, but also an essential insight about the functionality of time. The directionality of emergent processes which lead to order in such processes is a testament to why we need to so carefully consider the linear arrow of time's passage across generations in our appreciation of earthly order. In contrast, space has more texture and maneuverability than does time in its functional importance for dissipative processes beyond equilibrium. Prigogine elaborates quite simply: "matter at equilibrium, with no arrow of time, is 'blind,' but with the arrow of time it begins to 'see.'"

In his illuminating book *Your Brain Is a Time Machine*, Dean Buonomano makes an important distinction between "natural time" and "clock time." "Natural time" is embraced by physicists, in which the past, present, and future are all part of a fabric of "eternalism." However, for neuroscientists who are more interested in functional order—and indeed, which has more bearing on earthly order—subjective or "clock time" is more important to consider. There is thus a commodification of time, which is why it is either "spent" or wasted." As I was completing this manuscript, a new magazine called *Noema* published an article by Iranian-British journalist Joe Zadeh titled "The Tyranny of Time," in which he deftly summarized how the imposition of clock time led to a presumption of exactitude that is defied by nature.[11] Nature works more so through probabilities than through temporal precision. When an iceberg will calve and fall into the sea or when an earthquake will occur can never be precisely predicted. Many human societies have resented clock time because of this imposition of order. Zadeh starts his article with a nineteenth-century terrorist attack that was attempted by a French anarchist on the Greenwich Observatory, which had become a hallmark of global time-keeping. He notes the resistance from Australian Aborigines to the colonial clockmaster's drumbeat. In a pithy phrase he notes, "the clock does not measure time, it produces it."

This is where short-term versus long-term thinking in fields like ecological economics, which we considered in this volume, have more relevance for organizing our lives. Our perception of time passing in social and economic terms is very different from *biospheric time*, which determines the rise and fall of species, or *geospheric time*, which determines plate tectonics. Human beings, unlike other animals, are able to think about the long-term future, but we often act instinctively on short-term impulses as a survival strategy. However, the advancement of modern societies and higher degrees of order have allowed us to now consider longer-term thinking. Geologist Martha Bjornerud has termed such an approach *Timefulness*, whereby we order our lives with recognition for our place within geological time (termed "chronotopia" by Bjornerud) rather than the artful but meaningless ways of "timelessness."[12] Environmental writer John McPhee gave us the term "deep time" to consider such scales of human activity in congruence with natural order; such an approach, however, need not just be about analyzing our place in Earth's past, but also how we can engage with Earth's future.

Lesson 4: Tipping Points Can Lead to Accelerated Order or Disorder

In April 2021, the Nobel Foundation held its first ever global online Nobel Prize Summit, which was themed "Our Planet Our Future." Among the key topics that were discussed as game-changers across disciplines was the notion of *tipping points*, which have been addressed at various junctures in this book as well. The term was popularized by the writer Malcom Gladwell in his best-selling compendium of essays. Yet the kind of functional tipping points relevant to earthly order requires us to have a less discursive and more time-sensitive analysis. Among the speakers at the Nobel summit were some of the leading lights of global change science who also authored a collective article on "Our Future in the Anthropocene Biosphere," which braided social and ecological systems around points of change but also points of leverage. Building on earlier work, the authors identified nine "sleeping giants" of planetary tipping points. The good news is that just as tipping points can lead us down a slope of accelerated despair, they can also lead to a virtuous velocity of progress. The braiding of the various forms of order provides us with what has been termed "sensitive intervention points," a term coined by scholars at Oxford's Martin School, which brands itself as a "research organization of urgency and optimism."[13] There are two forms of these points proposed—*kicks* and *shifts*—the former suggests a nudging toward a transition within an existing structure, while the latter suggests a change in the underlying order itself. If we are to gain amplification and nonlinear change after a positive tipping point, it is more likely to emerge with a shift than with a kick. The Global Systems Institute at the University of Exeter, directed by Tim Lenton—one of the doyens of tipping point literature in climate science—has been working on specific policy interventions that could trigger a "tipping cascade."

A key identifying feature of such an effort is the recognition that one needs "response diversity" to safeguard against negative contingencies of action. Having multiple pathways to drive positive change toward a favored tipping point also prevents a negative reversal into a downward trend. Social and ecological systems can also behave in ways whereby a resource can easily convert from an asset to liability in terms of its input–output balance. The Amazon rainforest is a prime example in this regard. Environmental scientists have considered the immense rainforest as a net carbon sink asset, but, in early 2021, research revealed that from 2010 through 2019, the Amazon Basin gave off 16.6 billion tons of carbon dioxide while drawing down only 13.9 billion tons. A tipping point was reached due to excessive deforestation and forest burning. The next phase of research will consider when such a change may be irreversible on human time scales for mitigating catastrophic climate change impacts.

Transformations in socioecological systems—for better and for worse—can face inertia, but through processes of what may be termed "social contagion" and other mechanisms of diffusion, they may overcome such hurdles (as shown in Figure C.2). A key insight of this analysis in Figure C2, which comes from a group of systems

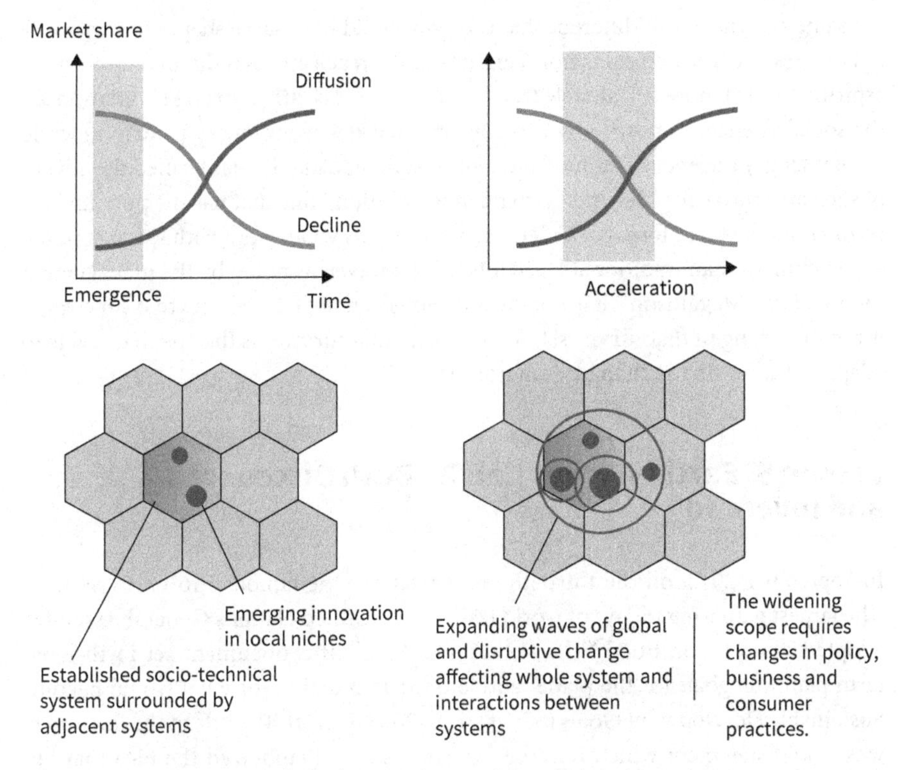

Figure C.2 Emergence and acceleration in transitions in human systems.
From Markard, Jochen, Frank W. Geels, and Rob Raven. "Challenges in the Acceleration of Sustainability Transitions." Environmental Research Letters *15, no.* 8 (August 2020): 081001. https://doi.org/10.1088/1748-9326/ab9468.

scientists based in business schools, also helps us differentiate emergent and acceleration phases in transitions. Such processes reflect what material scientists are trying to engineer through *four-dimensional printing*—developing materials which can harness natural structures and transform themselves in sync with natural processes. Designer Skylar Tibbits has pioneered this through the Self-Assembly Lab. A complex range of materials have been developed, including coastal adaptive infrastructure that can be designed or "programmed" to change and self-order with particular wave patterns.

At the social scale, ensuring that the transitions noted happen with equity concerns in mind has also led to the field of *just transitions.* In particular, the hexagonal cells in the diagram represent a range of ordered subsystems and the resonating impact of accelerated transitions on these adjacent entities. These subsystems can themselves be products of natural emergence or synthetic human intervention. Yet the processes which empower the phenomenon of tipping points are structurally relevant regardless of their points of origin. The subsystems should be evaluated, however, for both their size and their shape. You could have a triangular or hexagonal subsystem with the same size (area) but different shape, or you could have same shapes but with different sizes. Such structural factors can impact the reach and breadth of our efforts in aspiring toward more sustainable transitions. There is clearly a strong inclination for the social sciences to now more directly embrace key evolutionary principles while recognizing the agency of cultural attributes. Paul McLaughlin has termed the advent of such an approach a "Second Darwinian Revolution," one that has the potential for reconciling the three forms of order discussed in this volume but with less positional entrenchment than traditional Darwinism.[14] However, as noted by the management theorist Leon Megginson, "it is not the most intellectual of the species that survives; it is not the strongest that survives; but the species that survives is the one that is able to adapt and adjust to the changing environment."[15]

Lesson 5: Earthly Order Can Be Both Discovered and Invented

In September 2012, on the thirtieth anniversary of the famous Rio Earth Summit (the largest gathering today) of world leaders, the United Nations General Assembly adopted a resolution titled "The Future We Want." This document set forth long-term planning goals for the planet and laid the foundation for what would become Sustainable Development Goals (SDGs) from 2015 to 2030. The title of the document was a bold statement which rejected determinism and embraced the idea that humanity has the ability to craft a desirable trajectory for earthly order. Whether such a trajectory comes about through individual behavioral change or a realignment of natural systems through technology is for us to determine. There will inevitably be a need to prioritize action, and one of the recurring themes of this book is the salience of energy as the ultimate determinant of order or lack thereof. This harkens back to the seminal insights of the late Nobel Laureate Richard Smalley, one of the discoverers of the "buckyball" form of carbon. When asked to prioritize the world's top challenges for the next 50 years around the turn of the millennium, he noted energy at number 1 because all other challenges, such as poverty alleviation or even disease eradication, require energy flows. But the amount of energy we need physically for metabolic processes is vastly less than what we now use for our social existence. Geoffrey West estimated that a human needs around 90 watts of power for metabolic existence while needing around 11,000 watts for social existence in a developed economy. For futurists, the order of human civilizational advance is also categorized by the level of energy harnessed, harkening back to the three-point scale developed by Soviet astrophysicist Nikolai Kardeshev in the 1960s. Humanity is currently a Type 1 civilization, with the ability to harness energy that falls from our parent star. Type II civilizations could use and control energy at the scale of their planetary systems. Energy harnessing would be possible with proposed devices such as the Dyson Sphere or a solar sail that could absorb far greater stellar energy. Type III civilizations could control energy at the scale of their galaxy using sources such as quasars or "white holes" (hypothetical opposites of blackholes, or, as some have proposed, like mini big bang episodes). At present, human sustainability discourse is largely at the level of a Type 1 civilization, and we need to keep our ambitions grounded on harnessing energy sources on Earth—many of which are still beyond our grasp.

The need to prioritize action based on impact likelihood is perhaps a way to recognize these hierarchies that some controversial "environmental modernists" like Bjorn Lomborg or Michael Shellenberger put forth. Yet their tone and bombast can often occlude the pragmatic goals to which they aspire. The details of structural and functional attributes of natural systems and subsequent human constructed systems in this book are presented so that the reader can better grasp the limits and opportunities for what can be "discovered" or "invented." The difference between these two terms is reminescent of Plato's ancient metaphor, in which he suggested that the most efficient and effective approach to human ideas was to "carve nature at its joints." Plato suggested that there were indeed natural laws and forms (the joints in a carcass) which, if found and dissected properly, could lead humanity quite literally toward a "cleaner cut" of resource usage. His carnal metaphor may seem rather disturbing and destructive, but it is perhaps apt in the context of human interactions with other living systems—interactions that have largely been a story of carving out resources for sustenance (or, more accurately, engorgement). William Ophuls has termed our current ecological crisis, which results from a lack of appreciation for the existing demarcation of natural joints, *Plato's Revenge*. The radical premise of his book of the same title is that "sustainability" as usually understood is an oxymoron. The quandary between invention and discovery, or between intuition and formalism, has also been posited by mathematics as the universal language of science. The Golden Ratio or the appearance of the Fibonacci sequence in so many natural systems may suggest that there is indeed an quotient of discovery in mathematics. At the same time, we have also invented a range of mathematical tools and methods from the digit zero to a variety of numbering systems and even an imaginary Cartesian plane to help us decipher that latent order. At times our invented tools of mathematics can also create a self-consistent order without empirical validation—as is the concern with concepts like string theory or even certain theoretical models of economic growth.

As our study of emergent systems has revealed, human beings have the ability to also act as agents of nature in shaping the trajectory of nonequilibrium processes, so long as they are cognizant of underlying structures. Thus, we should be willing to consider the "invention" of new joints or indeed permutations of new biotic systems so long as those underlying structures are well understood. The advances in genetic engineering, whereby there is now growing consensus on the viability of many genetically modified organisms in our lives, is an example of such inventive prowess that has come after a careful appreciation for the underlying biotic order. Each intervention toward order poses ethical questions for which universal laws are not possible to derive, but we can have some guiding principles and methods for evaluation. One such is *life cycle analysis*, a technique learned from industrial ecology that is used to compare choices between materials and products based on environmental and social criteria.[16]

At much larger scales we will be faced with similar quandaries as planetary life support systems are challenged. The same appreciation for underlying structures should chart our deliberations on *geoengineering* as we consider stabilization of Earth's climate or hydrological systems. Milder manifestations of such interventions have been tried through indirect human experiments with alien species introductions worldwide and even directly through terraforming experiments. A notable and often neglected experiment of this kind was carried out by Charles Darwin himself after a visit to the remote Ascension Island in the South Atlantic. Through the introduction of more than 300 species of plants on the relatively barren volcanic island's high slopes, Darwin and Sir Joseph Hooker were able to terraform an ecosystem that is able to capture water from misty wind flows to this day. A small pond eventually formed and made the island more habitable.[17] Whether we go all-out for engineering drastic changes through new physical interventions like sending giant mirrors into orbit or chemically injecting cooling aerosols into the upper atmosphere, or we pursue the more modest "geomimicry" of geochemical calcium carbonate sequestration, we should be guided by our knowledge of those natural parameters over which we are not likely to have control. Discovering natural and societal fault lines (or joints) as well as having the measured courage to invent and develop new functionally useful paths of intervention is what we all must aspire to.

Coda: Chromatic Order

Returning to the imagery with which I started this book, the role played by mirrors in a kaleidoscope to impart shapely order to shapeless forms is only part of the device's appeal. The different colors of the glass fragments add a layer of chromatic complexity which further stimulates our senses to also consider how the images are a reflection of an even more resplendent order. The co-founder of calculus, Gottfried Leibnitz, considered the universe similarly as an infinitely detailed image of itself—a set of nested microcosms with structural connectivity in form but differentiation in function that ultimately creates the Cosmos. A modern version of this correspondent view of existence between the individual "I" and the universal "all" was also put forward by the physicist David Bohm. A protégé of Robert Oppenheimer, Bohm's views of natural order were eclipsed by the mainstream because of his Marxist inclinations of political order. Exiled from American academia, he lived in Brazil and the United Kingdom and formulated a theory of "implicate" and "explicate" order. Bohm's view of a fragmentation of human society has resonance with the shape and color of chromatic order of a kaleidoscope: "The notion that all these fragments are separately existent is evidently an illusion, and this illusion cannot do (much) other than lead to endless conflict and confusion. Indeed, the attempt to live according to the notion that the fragments are really separate is, in essence, what has led to the growing series of extremely urgent crises that is confronting us today."[18]

The human quest for discerning order in natural systems continues unabated even with the relative saturation of our discoveries of the key fundamental organizing principles of the universe. One of the key areas in this regard is the alignment of units of measurement with natural constants. The exactitude of the seven key base units of measurement that were decided by the Paris Metre Convention in 1875 have become increasingly consequential (time, distance, mass, molecular amount, electrical current, temperature, and luminosity). Among the earliest of international treaties, the Convention created order in measurement by linking these units to palpable objects. For example, a meter was for more than a century the length of a very particular bar of inert platinum-iridium alloy kept at a stable temperature in a vault by the International Bureau of Weights and Measures. But now we need far greater precision in calculating landing trajectories for space flight to Mars, other planets, and moons or for measuring molecular distances. We have thus redefined a meter with one of nature's fundamental constants instead: the length of the path traveled by light in a vacuum in 1/299,792,458 of a second. Similarly, the kilogram was redefined with reference to the Planck constant in 2019, and the Celsius and Kelvin degree unit of temperature to the Boltzmann constant. Such anchoring of measurement in the fundamental tenets of natural order highlights the importance of precision in human endeavors. Striving for such precision and refinement in social, economic, and political decisions is much harder, but by linking neurology to social science, it is becoming more plausible. For the long-term sustainability of our species within the celestial constraints of planetary order, we should be willing to at least consider such refinements, even if we then decide not to act on the suggested alignment for ethical reasons of human choice.

How might the disjunctures between natural, economic, social, and political orders be reconciled? Fundamental to such a process is to highlight the prime dependence of all systems on core natural processes, many of which are beyond human agency despite our contention of living in "the Anthropocene." Furthermore, we must recognize that order is not normatively good or bad and must be considered in terms of functionality toward desired goals of the sustainability of natural systems and the development of progressively harmonious social systems. We could thus be moving to what Thomas Barry has called an "Ecozoic" era or perhaps even an epoch that Glenn Albrecht has termed the "Symbiocene."[19] Both these terms imply a functional view of finding order that achieves particular goals of sustainability in the short- to long-term trajectory. There is implicit hybridity in such an approach rather than simplistic allegiance to natural determinism. For example, as we saw, borders as a form of order have an important ecological functionality in preventing the spread of disease, but they can also be a hindrance to economically and ecologically efficient trade. In this regard, maps have been important delineators of order and also a composite reflection of our knowledge and ignorance. They are highly functional in their efficacy, and the field of study which has best captured such synthesis is geography. It is no wonder that eclectic academics like myself, or the far more eminent Jared Diamond, can chart their discursive career paths eventually to departments of geography. It is also no wonder then that geographical societies worldwide have been at the forefront of efforts at ecological literacy. We must embrace this discipline and its potential for furthering the goals of a society which can think laterally across natural, economic, and political order.[20] Geographers revel in complexity and abjure the tempting ease of simple "self-help" narratives. While enumerating categories is tempting and eases our retention of key knowledge, we should know that such categories like "five stages of grief" or "seven steps to success" are inherently contrived.

Finally, we must always be sensitive to the temporal aspects of the emergence of order at geological, biological, and human time scales and approach studies of order with humility. We still do not have answers to many fundamental questions. For example, why is the universe relatively flat (even though the earth is thankfully not)? Triangles drawn within the space of the universe with angles adding up to 180 degrees are a perplexing feature of this flatness when curved universes may seem more likely. The isotropic constant temperature of much of the universe is similarly enigmatic, and we have devised the theory of rapid inflation of the universe to explain this. But each attempt at explaining order at such scales can lead to other imponderables, which continue to lead to other propositions such as the prospect of an "ekpyrotic" universe (from the original Greek Stoic root of the concept of a "conflagration of earlier worlds" in possibly cyclical processes). The cosmological principle of relative homogeneity of matter across large distances in the universe—on which much of the standard model of physics is premised—is also eroding as increasingly larger clumps and clusters of quasars are discovered. While such origins of universal order will continue to be debated for the foreseeable future, we have ample science to guide us on the functional order which has been provided throughout this book.

The growing use of the term "complex adaptive system" for our planet's ecosphere should go beyond being a buzz phrase and must permeate our educational system at multiple levels. We must remain humble, too, in knowing that the fundamental fabric of being and carefully proven mathematical concepts confirm that some true statements might never be proven as such. Nature only reveals to us the answers to questions we ask its elemental emissaries, and, as Gödel and Heisenberg showed us in different ways, there are certain questions we cannot ask simultaneously.

We seek patterns because they make learning more efficient and have been an effective survival strategy for our species. Harkening back to the wisdom of Greek architectural polymath Vitruvius, for human well-being and the drive toward sustainability, we need to keep a tenuous balance between beauty (*venustas*), usability (*utilitas*), and durability (*firmitas*). This has also meant that sometimes even illusory patterns can connote meaning when they may not inherently have such. We may fear a shadow or a superstition because, from a survival perspective, it might have been a useful risk management strategy to signal against broader patterns of predation, even if the specifics were not accurate. Keeping perspective on the divergence in order and equilibrium processes between natural versus human-social systems can avoid cognitive disjunctures and lead to more effective planning. Ultimately, this book will have served its purpose if readers can test their own comfort zones about what is orderly in the natural world while interrogating the structures in our social, economic, and political lives. We should feel empowered to still act as agents of positive change toward sustainability when a system seems out of order to our immediate senses.

Notes

1. For an excellent discussion of Emergence, see Clayton, Philip, and Paul Davies, eds. 2008. *The Re-Emergence of Emergence: The Emergentist Hypothesis from Science to Religion.* Oxford University Press. https://doi.org/10.1093/acprof:oso/9780199544318.001.0001.
2. Sardar, Ziauddin. June 1, 2010. "Welcome to Postnormal Times." *Futures* 42 (5): 435–444. https://doi.org/10.1016/j.futures.2009.11.028.
3. In particular, research centers supported by the Future of Life Institute and the Global Challenges Foundation are prolific. The University of Oxford's Future of Humanity Institute and the University of Cambridge's Center for Existential Risk have applied many of the techniques from complex systems to understand these risks and ways of mitigating them.
4. Mani, L., A. Tzachor, and P. Cole. 2021. "Global Catastrophic Risk from Lower Magnitude Volcanic Eruptions." *Nature Communications* 12 (1): 4756. https://doi.org/10.1038/s41467-021-25021-8.
5. In clinical studies, the case of a British patient from the 1980s called "JBR" suggests that injury to one part of his brain entirely impacted natural pattern recognition but not synthetic pattern recognition. However, this case has been questioned in terms of the extent of its findings; refer to Funnell, Elaine, and Paul De Mornay Davies. November 1, 1996. "JBR: A Reassessment of Concept Familiarity and a Category-Specific Disorder for Living Things." *Neurocase* 2 (6): 461–474. https://doi.org/10.1080/13554799608402422.
6. A good contemporary ecological interpretation of this concept can be found in Volk, T. 1996. *Metapatterns* (1st edition). Columbia University Press.
7. "Benford's Law and the 2020 US Presidential Election: Nothing Out of the Ordinary." November 19, 2020. Physics World. https://physicsworld.com/a/benfords-law-and-the-2020-us-presidential-election-nothing-out-of-the-ordinary/.
8. Pauliuk, S., and E. G. Hertwich. 2015. "Socioeconomic Metabolism as Paradigm for Studying the Biophysical Basis of Human Societies." *Ecological Economics* 119: 83–93. https://doi.org/10.1016/j.ecolecon.2015.08.012.
9. These insights are eloquently studied and presented in Strogatz, S. H. 2012. *Sync: How Order Emerges from Chaos in the Universe, Nature, and Daily Life* (1st edition). Hachette Books. Adding a meditative narrative of faith alongside since is Johnson, G. 1996. *Fire in the Mind: Science, Faith and the Search for Order.* Vintage.
10. Prigogine, I. 1997. *The End of Certainty: Time, Chaos and the New Laws of Nature.* Simon and Schuster.
11. Zadeh, Joe. June 2021. "The Tyranny of Time." *Noema.*
12. Bjornerud, M. 2020. *Timefulness: How Thinking Like a Geologist Can Help Save the World* (illustrated edition). Princeton University Press.
13. Farmer, J. D., C. Hepburn, M. C. Ives, T. Hale, T. Wetzer, P. Mealy, R. Rafaty, S. Srivastav, and R. Way. April 12, 2019. "Sensitive Intervention Points in the Post-Carbon Transition." *Science* 364 (6436): 132–134. https://doi.org/10.1126/science.aaw7287.

14. McLaughlin, Paul. June 1, 2012. "Ecological Modernization in Evolutionary Perspective." *Organization & Environment* 25 (2): 178–196. https://doi.org/10.1177/1086026612450870.
15. The quote is often misattributed to Darwin himself but comes from Megginson, Leon. 1963. "Lessons from Europe for American Business." *Southwestern Social Science Quarterly* 44 (1): 3–13.
16. Blokdyk, Gerardus. 2021. *Life Cycle Analysis: A Complete Guide: 2020 Edition*. 5STARCooks.
17. Fieseler, Clare. May 8, 2017. "Mysterious Island Experiment Could Help Us Colonize Other Planets." National Geographic online. https://www.nationalgeographic.com/news/2017/05/ascension-island-terraformed-biology-evolution-conservation/.
18. Bohm, D. 1989/2005. *Wholeness and the Implicate Order*. Routledge.
19. Albrecht, G. A. 2019. *Earth Emotions: New Words for a New World*. Cornell University Press.
20. Harlan Barrows called geography the "mother of sciences" for this reason in his oft-remember lecture in 1923. Barrows, H. H. 1923. "Geography as Human Ecology." *Annals of the Association of American Geographers* 13 (1): 1–14. https://doi.org/10.2307/2560816.

Index

For the benefit of digital users, indexed terms that span two pages (e.g., 52–53) may, on occasion, appear on only one of those pages.